Electromechanical Energy Conversion Through Active Learning

Electromechanical Energy Conversion Through Active Learning

J R Cardoso and M B C Salles
Escola Politécnica da USP, São Paulo, Brazil

M C Costa
Electromagnetics Technologies Ltd, São Paulo, Brazil

IOP Publishing, Bristol, UK

ISBN 978-0-7503-2084-9 (ebook)
ISBN 978-0-7503-2082-5 (print)
ISBN 978-0-7503-2085-6 (myPrint)
ISBN 978-0-7503-2083-2 (mobi)

DOI 10.1088/978-0-7503-2084-9

Version: 20191201

IOP ebooks

British Library Cataloguing-in-Publication Data: A catalogue record for this book is available from the British Library.

Published by IOP Publishing, wholly owned by The Institute of Physics, London

IOP Publishing, Temple Circus, Temple Way, Bristol, BS1 6HG, UK

US Office: IOP Publishing, Inc., 190 North Independence Mall West, Suite 601, Philadelphia, PA 19106, USA

Contents

Preface

Why should the subject of *electromechanical energy conversion* be studied in the 21st century? What is the appeal of this discipline, based on 19th century concepts, that might attract engineering students who were born in this century?

The relevance of this discipline is justified by strong concerns around environmentally sensitive issues. In the past, the subject of electromechanical energy conversion was driven by finding solutions to the control of motion in industrial activities. Although this task is still important, another concern has been presented to the engineers of this century.

The turnaround caused by the impact of fossil fuels on global warming triggered disruption to human mobility, reversing the idea that the automobile was a final solution to it becoming a problem to be solved.

Ambitious projects that were believed to be economically unviable decades ago became the best alternatives for minimizing the environmental impact caused by human mobility.

In this scenario, electromechanical energy conversion became again the center of a new technological advance. The advent of both new materials and the technological improvements on control theory has led to the development of high-performance electromechanical devices, both electrical motors and generators, as well as electromechanical actuators that are lighter, more efficient and accurate.

Not only the electromechanical energy conversion needs, but also techniques for learning this discipline have been changed in this century. The content has improved due to new devices that have appeared and students have acquired competence in actively searching for knowledge.

The main objective of this book is to guide instructors and students on this new technological path, encouraging the use of active learning and providing the right content to support a modern curricular structure for engineering schools.

The initiation of the design of electromechanical devices and their applications is integrated with the content, with the objective of leading the student to practice 'hands-on', that is, facing the challenges that do not have a closed solution, but that require creativity and determination to find a solution that is economically viable.

The proposed projects should involve some kind of competition to challenge the students, not individual competition but those involving team competition. This kind of activity is driven for practicing *teamwork* where the student is still in the undergraduate program.

The authors have been applying with success a *flipped classroom* for more than five years when teaching electromechanical energy conversion. The authors suggest the use of *project based learning* techniques to be discussed with the students. Competitions, that reward the team who won the race, promote an increase in both the teacher/student relationship and classroom efficiency.

The final goal can be achieved if the instructor guides the students to give them the competences and motivation to play with the basics of the design of electromechanical devices involving electromechanical energy conversion.

Acknowledgments

The authors would like to express their gratitude to their students and families for supporting this project.

Author biographies

Jose Roberto Cardoso

Jose Cardoso (University of São Paulo—USP) was born in Marilia, SP Brazil, in 1949. He received his BS, MS and PhD degree in electrical engineering from Escola Politécnica da USP in 1974, 1979 and 1986, respectively. Since 1999, he has been a Professor with the Electrical Energy and Automation Engineering Department, Escola Politécnica of USP. He is the author of four books, more than 70 articles, and has advised more than 40 theses/dissertations. His research interests include numerical methods for electromagnetics, theoretical electromagnetics, electrical machines, electromechanical biomedical application, electrical railway research and engineering education. He is the president of SBMAG—Brazilian Electromagnetics Society and former Dean of Escola Politécnica da USP. He is the founder of the LMAG—the Electromagnetic Applied Laboratory. Cardoso was a recipient of the 2013 Emeritus Engineer of the Year in São Paulo State. He was awarded Doctor Honoris Causa by the Grenoble-INP Institute of Engineering in 2019.

Mauricio Barbosa de Camargo Salles

Mauricio Salles (University of São Paulo—USP) was born in Guarulhos, SP Brazil, in 1975. He has been Assistant Professor with the Electrical Energy and Automation Engineering Department, Escola Politécnica of the University of Sao Paulo, since 2010. In 2004, he obtained the MSc degree from the State University of Campinas (UNICAMP), Campinas/SP, Brazil. From 2006 to 2008, he was part of the research team of the Institute of Electrical Machines at the RWTH Aachen University. In 2009, he received the Doctorate degree from the University of Sao Paulo (USP). Between 2014 and 2015, he was Visiting Scholar at Harvard John A Paulson School of Engineering and Applied Sciences. In 2018, he was an Invited Professor at Ecole Centrale de Lille in France. His main interests are on distributed generation, power system dynamics, control and stability, renewable energy, energy storage and electricity markets. He is one of the founders of the Laboratory of Advanced Electric Grids—LGrid.

Mauricio Caldora Costa

Mauricio Costa was born in Sao Paulo, SP Brazil, in 1973. He has been Director at Electromagnetics Technology since 2008, where he works on consulting projects involving electromagnetic field simulation. In 1998, he obtained the MSc degree from the Escola Politécnica of the University of Sao Paulo (POLI-USP), Brazil. In 2001, he received his Doctorate degree from the Grenoble Institute of Engineering (Grenoble INP), France. From 2005 to 2008, he worked at IPT (Institute for Technological Research of Sao Paulo) as the leader of Electromagnetic Compatibility team at Electrical and Optical Equipment Laboratory. His main interests are finite element and other numerical methods, electromagnetic devices simulation and optimization, software development and power systems.

IOP Publishing

Electromechanical Energy Conversion Through Active Learning

J R Cardoso, M B C Salles and M C Costa

Chapter 1

The magnetic circuit

1.1 Some magnetic properties of the material

Associated with electron displacement, we can explain several magnetic properties of the material. The circular movement of the electron could be imagined as an elementary current loop that produces an elementary magnetic flux density. In the same way, we can imagine that the spin movement of the electron is a source of an elementary magnetic flux density as well.

For simplicity, it is possible to associate a net elementary flux density produced by the movement of all electric charges of the atom or molecule of the material (figure 1.1).

Normally, the net elementary flux density is null in most materials.

Although some atoms or molecules present a non-null elementary flux density field, the final result of the vectorial summation of all elementary flux density fields of the material is also null when the material was never magnetized.

The composition effects of several substances in the same molecule or the arrangements of several atoms in a crystal could be classified based on its reaction to the effect of a given magnetic field.

Diamagnetic materials: (i) These become weakly magnetized in the opposite direction of the magnetic field intensity in which they are placed; (ii) the flux density in diamagnetic materials varies linearly with the magnetic field intensity.

Examples of diamagnetic materials: Cu, Ag, Zn, Bi, Au, water, etc (figure 1.2).

Paramagnetic materials: (i) These become weakly magnetized in the same direction as the magnetic field intensity in which they are placed; (ii) the flux density in paramagnetic materials varies linearly with the magnetic field intensity (figure 1.3).

Examples of paramagnetic materials: Pt, Mg, Al, Cr, etc.

Ferromagnetic materials: (i) These become strongly magnetized in the same direction as the magnetic field intensity in which they are placed; (ii) the flux density in

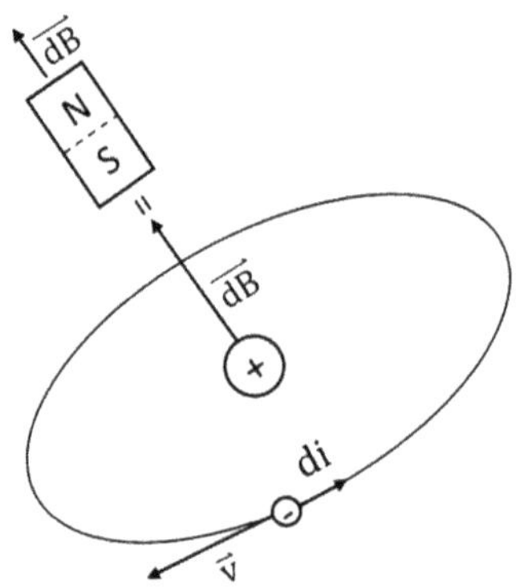

Figure 1.1. Elementary magnetic flux density produced by an atom. No elementary magnetic flux density is produced by most atoms in Nature.

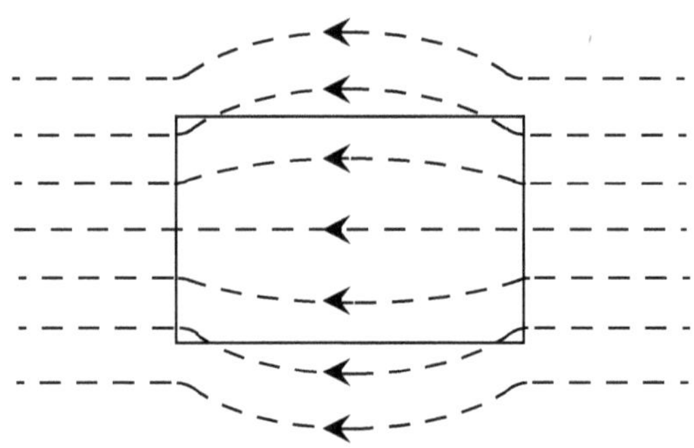

Figure 1.2. Slab of diamagnetic material embedded in a uniform flux density field. Observe that the diamagnetic material spreads the magnetic flux density line from inside to outside of the slab.

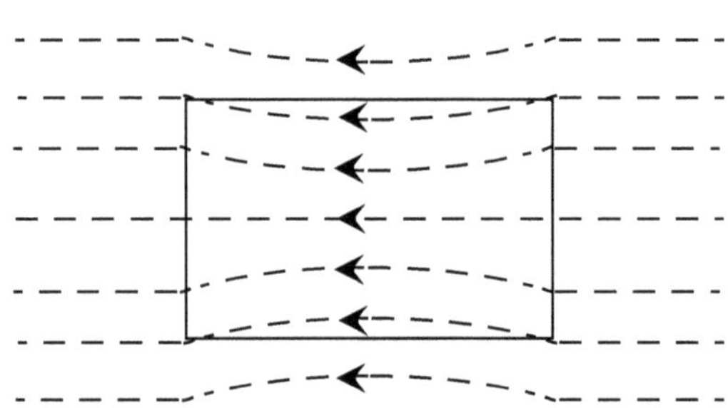

Figure 1.3. Slab of paramagnetic material embedded in a uniform flux density field. Observe that the paramagnetic material attracts the magnetic flux density line from outside to inside of the slab.

ferromagnetic materials varies non-linearly with the magnetic field intensity. For a small range of magnetic field intensity, ferromagnetic materials present linear variation behavior. (iii) Ferromagnetic materials present saturation (non-linearity), hysteresis (behavior dependent of the sense of flux density) and retentivity (property of a permanent magnet). (iv) The magnetic properties of ferromagnetic materials are very sensitive to both thermal treatment and mechanical stress (figure 1.4).

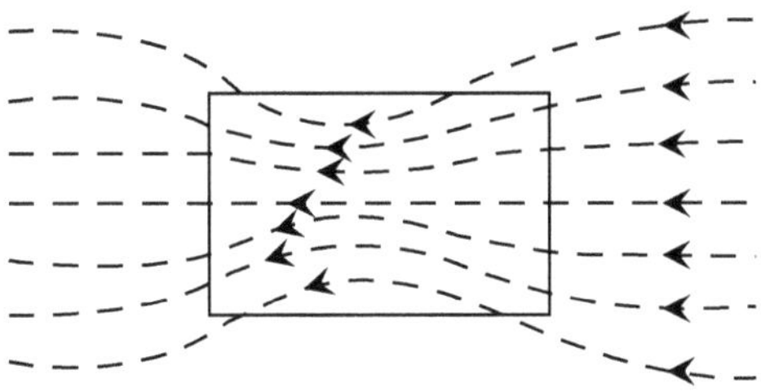

Figure 1.4. Slab of ferromagnetic material embedded in a uniform flux density field. Observe that the ferromagnetic material strongly attracts the magnetic flux density line from the outside to the inside of the slab.

Examples of ferromagnetic materials: Fe, Co, Ni, etc.

The property associated with the magnetic behavior of materials is called magnetic permeability, symbolized by the Greek letter μ (mu).

For engineering, the vacuum has a smaller magnetic permeability the value of which, in the International System (IS), is $\mu_0 = 4\pi \times 10^{-7}$ H m^{-1}.

Indeed, the magnetic permeability of diamagnetic materials is smaller than the vacuum, but the difference between them is negligible. The magnetic permeability of some diamagnetic materials is: $\mu_{\text{Cu}} = 0.999\,99 \times \mu_0$; $\mu_{\text{Ag}} = 0.999\,97 \times \mu_0$; $\mu_{\text{Au}} = 0.999\,96 \times \mu_0$.

The magnetic permeability of paramagnetic materials is a little bit bigger than the vacuum, but also the difference between them is negligible. The magnetic permeability of some paramagnetic materials is: $\mu_{\text{Al}} = 1.000\,008 \times \mu_0$; $\mu_{\text{W}} = 1.000\,002 \times \mu_0$.

Hence, for all engineering effects the permeability of both diamagnetic and paramagnetic materials, including the air, is $\mu_0 = 4\pi \times 10^{-7}$ H m^{-1}.

From electromagnetism, the following relation:

$$\mu_r = \frac{\mu}{\mu_0}$$

is defined as the relative magnetic permeability (non-dimensional). For both diamagnetic and paramagnetic materials, the relative magnetic permeability is (virtually) unitary.

Most materials used in the construction of electrical engineering apparatus are made by ferromagnetic materials due to their high value of magnetic permeability. From electromagnetism, the magnetic permeability of ferromagnetic materials is non-linear, i.e., the magnetic permeability is dependent on the flux density or $\mu = \mu(B)$. As the magnetic permeability of ferromagnetic materials is not constant, the magnetization characteristics curve is not linear either. Both curves of the magnetic permeability and the magnetization characteristics of ferromagnetic materials are, respectively, represented in figures 1.5 and 1.6.

1.2 The basis for the linear magnetic circuit

Let us suppose a rectangular cross-section toroid with N closely spaced turns fed by a DC current source, as shown in figure 1.7.

Due to both the symmetry and the right-hand rule, the magnetic flux density presents a circular path in the clockwise sense.

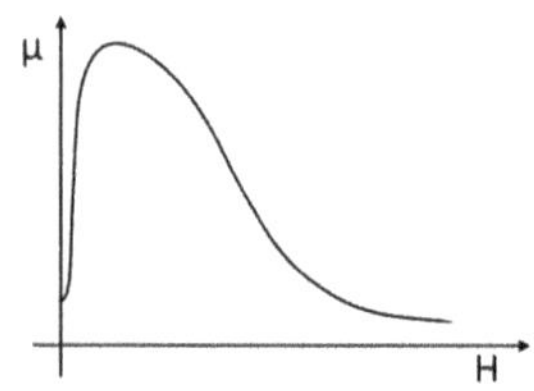

Figure 1.5. Magnetic permeability curve $\mu \times H$.

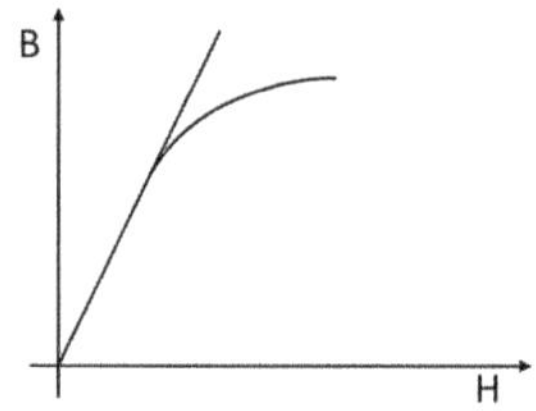

Figure 1.6. Magnetization characteristics curve $B \times H$. For small range of flux density, the $B \times H$ curve is straightforward.

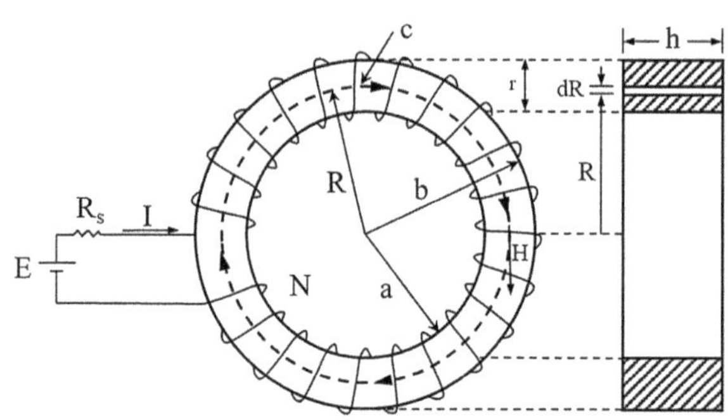

Figure 1.7. A toroid with N closely spaced turns.

Ampere's law establishes that:

$$\oint_C \boldsymbol{H} \cdot \boldsymbol{dl} = I_t \tag{1.1}$$

where:

$\boldsymbol{H}$: magnetic field intensity vector (A m^{-1})

I_t: the total amount of linkage current by the closed path C (A)

Let us suppose that the circle of radius R is oriented in the clockwise sense as shown in figure 1.8.

Once the intensity of $\boldsymbol{H}$ is constant along the path C, Ampere's law can be written as:

$$H \oint_C \boldsymbol{dl} = NI \tag{1.2}$$

where NI is the total amount of linkage current by the circle R.

So, we can write:

$$H \times 2\pi R = NI \tag{1.3}$$

and the magnetic field intensity inside the toroid is:

$$H = \frac{NI}{2\pi R}\ (\text{A m}^{-1}) \tag{1.4}$$

Outside the toroid, the magnetic field intensity $\boldsymbol{H}$ is null because there is no linkage current with the path.

The relationship of $\boldsymbol{B}$ and $\boldsymbol{H}$ is the constitutive relation:

$$\boldsymbol{B} = \mu \boldsymbol{H} \tag{1.5}$$

The magnetic flux density inside the toroid is given by:

$$B = \frac{\mu NI}{2\pi R}\ (\text{Wb m}^{-2}) \tag{1.6}$$

Figure 1.9 shows the curve $B = B(R)$. If we suppose that $a \gg r$, there is no big variation of B inside the toroid. Under this condition, we can admit without any loss of accuracy that B is constant in a cross-section of the toroid and obtain its value at the mean radius R_m as follows:

$$B = \frac{\mu NI}{l_m} (\text{Wb m}^{-2}) \tag{1.7}$$

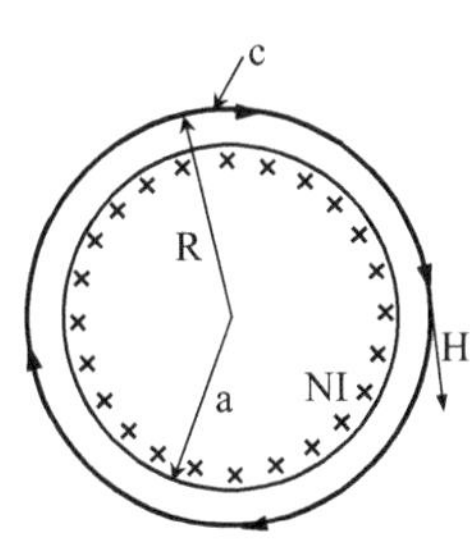

Figure 1.8. The linkage current by the circle of radius R.

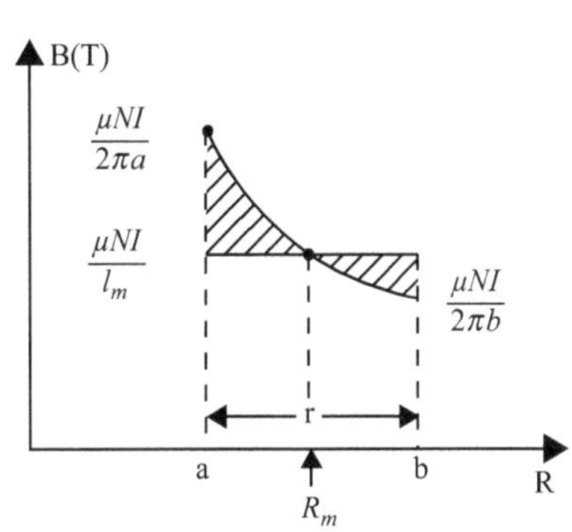

Figure 1.9. B variation inside the toroid. For $a \gg r$ we can admit $B = B(R_m)$ in the cross-section.

with:

$l_m = 2\pi R_m$: The mean length of magnetic path inside the toroid.

In all electromagnetic designs, the condition $a \gg r$ is very common. In a transformer design, the bigger dimension of the core cross-section is smaller than a tenth of the main dimension of the transformer.

It is easy to evaluate the magnetic flux in the toroid cross-section if we suppose B constant inside it.

The magnetic flux is given by:

$$\varnothing = \int_S \boldsymbol{B} \cdot \boldsymbol{dS} (\text{Wb}) \tag{1.8}$$

For $\boldsymbol{B}$ constant inside the toroid and aligned with $\boldsymbol{dS}$, expression (1.8) can be rewritten as:

$$\varnothing = BS \text{ (Wb)} \tag{1.9}$$

Substituting B by its value expressed in (1.7), the magnetic flux $\varnothing$ is given by:

$$\varnothing = \frac{\mu NIS}{l_m} = \frac{NI}{\frac{1}{\mu}\frac{l_m}{S}} \text{ (Wb)} \tag{1.10}$$

At this point, it is convenient to compare this result with one extract from the solution of a simple electric circuit like the one shown in figure 1.10.

The electric current in this simple circuit is:

$$I = \frac{E}{\frac{1}{\sigma}\frac{l}{S}} = \frac{E}{R} (A) \tag{1.11}$$

If we compare both expressions (1.10) and (1.11), we can easily identify the analogy between their values that can be summarized in table 1.1:

Following this analogy, we describe the toroid magnetic circuit by the analog electric circuit of figure 1.11.

Only to justify the approach of considering the flux density B constant in the cross-section, let us evaluate the exact value of ϕ applying (1.8). As both flux density and elemental surface vector are aligned, equation (1.8) can be expressed in its scalar form as:

$$\varnothing_{\text{exact}} = \int_S B \cdot dS (\text{Wb}) \tag{1.12}$$

Substituting B for its value obtained in (1.6) and considering that $dS = h\,dR$ extracted from the figure 1.7, we have:

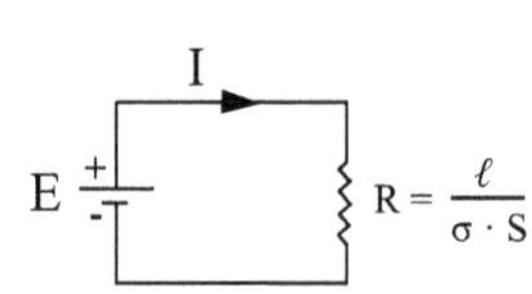

Figure 1.10. Electric circuit.

Table 1.1. Analogy electric circuit × magnetic circuit.

Electric circuit	Magnetic circuit
E or V: Electromotive force (emf) (V)	$\mathcal{F} = NI$: Magnetomotive force (mmf) (A)
σ: Electric conductivity (S m^{-1})	μ: Magnetic permeability (H m^{-1})
I: Electric current (A)	$\varnothing$: Magnetic flux (Wb)
$R = \frac{l}{\sigma S}$: Electrical resistance (Ω)	$\mathcal{R} = \frac{l}{\mu S}$: Magnetic reluctance (A Wb^{-1})
$G = \frac{1}{R}$: Electrical conductance (S)	$\mathcal{P} = \frac{1}{\mathcal{R}}$: Magnetic permeance (Wb A^{-1})
$J = \frac{I}{S}$: Current density (A m^{-2})	$B = \frac{\phi}{S}$: Flux density (Wb m^{-2})
E: Electric field (V m^{-1})	H: Magnetic field intensity (A m^{-1})

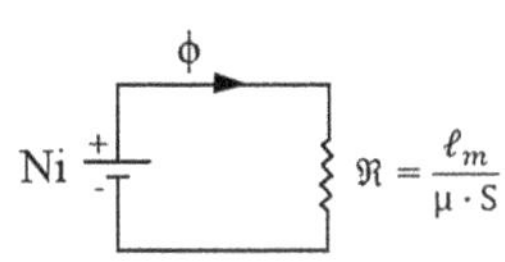

Figure 1.11. Magnetic circuit.

$$\varnothing_{\text{exact}} = \frac{\mu NIh}{2\pi}\int_a^b \frac{dR}{R} = \frac{\mu NIh}{2\pi}\ln\frac{b}{a}\ \text{(Wb)} \tag{1.13}$$

Considering that the toroid has the following dimensions:

$$a = 25\ \text{cm}$$
$$b = 30\ \text{cm}$$
$$h = 10\ \text{cm}$$

we have the requirements established for (1.7). Comparing its value with the one extract from (1.9), we obtain the relation:

$$\frac{\varnothing}{\varnothing_{\text{exact}}} = 0.997$$

whose error is less than 0.4%!

1.3 The real world

The toroid assemblage is not a simple task. There are several difficulties forbidding the use of this kind of assemblage in electromechanical devices.

Even with the complex geometry of all electrical apparatus, their studies become easier since they are made with high permeability ferromagnetic materials. This kind of material has the capacity of concentrate the flux density that makes it easy to apply Ampere's law as we will see later.

The general aspect of a magnetic structure is the one shown in figure 1.12. Despite using a classical toroid assemblage, the magnetic structure is normally assembled as a polygon with a uniform cross-section on each leg.

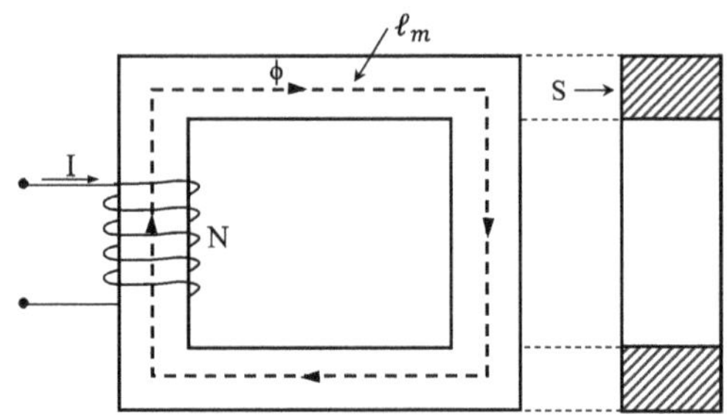

Figure 1.12. Magnetic structure—each leg has a uniform cross-section. The flux density is constant on each leg of the structure.

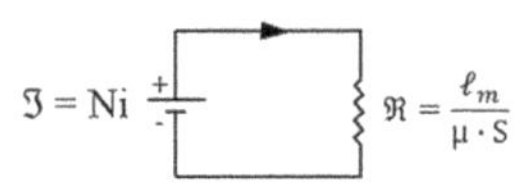

Figure 1.13. Analog electric circuit of magnetic structure: l_m Mean length of magnetic path. S Cross-sectional area: $\mathcal{R}$ Equivalent reluctance of magnetic structure.

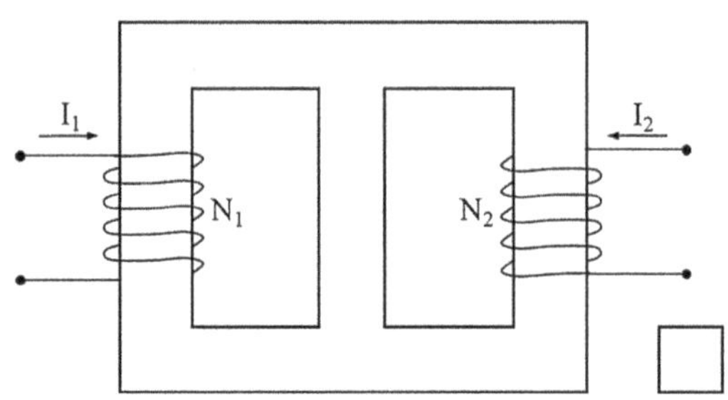

Figure 1.14. Magnetic structure with more than two legs. Two sources of mmf.

For each leg, we evaluate its reluctance by the expression:

$$\mathcal{R}_i = \frac{l_i}{\mu_i S_i} \tag{1.14}$$

Using this procedure for this example, the final analog circuit (figure 1.13) for it is where the equivalent reluctance for this magnetic structure is:

$$\mathcal{R} = \sum_{i=1}^{4} \mathcal{R}_i \tag{1.15}$$

If we have a more complex magnetic structure, a more complex analog circuit should be assembled. Figures 1.14–1.17 show some examples of more complex magnetic structures and their analog circuit.

All types of analysis of an electric circuit are applied to the analysis of the analog electric circuit of a magnetic structure.

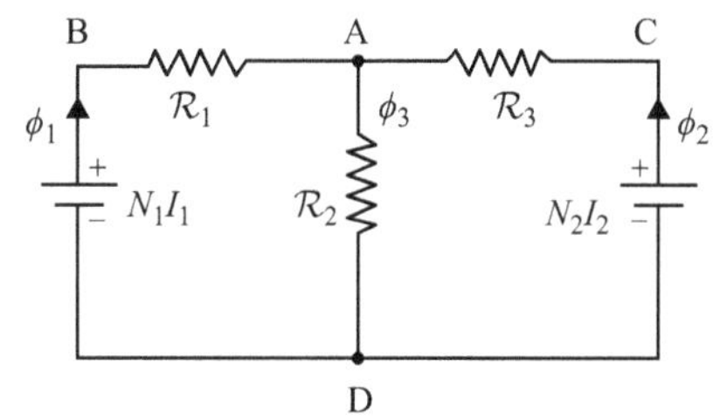

Figure 1.15. Analog electric circuit of figure 1.14. The right-hand rule gives the sense of mmf.

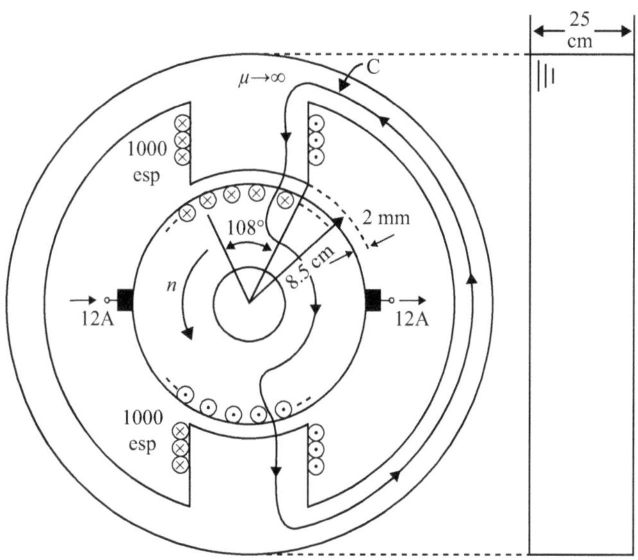

Figure 1.16. Magnetic structure of a DC machine. Two sources of mmf.

From the nodes law, we have:

$$\sum_n \varnothing_i = 0 \text{ at node } i$$

From the voltage law, we can write:

$$\sum_{i=1}^{n} (\mathcal{F}_i - \mathcal{R}_i \phi_i) = 0 \text{ at the mesh } i$$

Applying these laws at the analog circuit of figure 1.15, we have:

$$\phi_1 + \phi_2 - \phi_3 = 0 \text{ at the node A}$$
$$N_1 I_1 - \mathcal{R}_1 \phi_1 - \mathcal{R}_2 \phi_3 = 0 \text{ at mesh BAD}$$
$$N_2 I_2 - \mathcal{R}_3 \phi_2 - \mathcal{R}_2 \phi_3 = 0 \text{ at the mesh ACD}$$
$$N_1 I_1 - \mathcal{R}_1 \phi_1 + \mathcal{R}_3 \phi_2 - N_2 I_2 = 0 \text{ at the mesh ABCD}$$

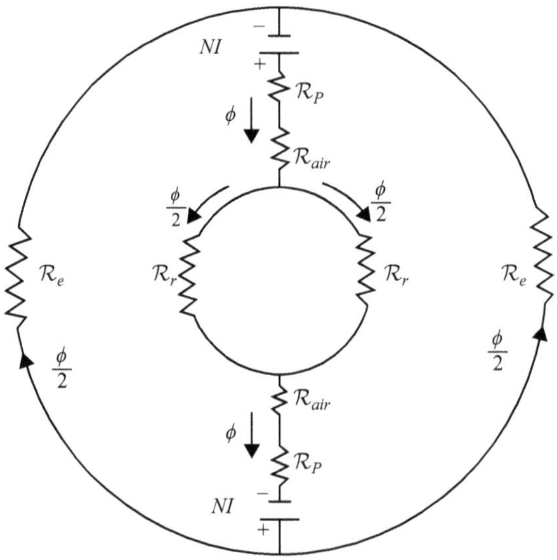

Figure 1.17. Analog electric circuit of a DC machine. The right-hand rule gives the sense of mmf.

For the analog electric circuit of figure 1.17, the mesh law applied on one side of the circuit gives:

$$-NI + (\mathcal{R}_p + \mathcal{R}_{ar})\phi + \mathcal{R}_r\frac{\phi}{2} + (\mathcal{R}_p + \mathcal{R}_{ar})\phi - NI + \mathcal{R}_e\frac{\phi}{2} = 0$$

or else:

$$2NI - 2(\mathcal{R}_p + \mathcal{R}_{ar})\phi - \mathcal{R}_r\frac{\phi}{2} - \mathcal{R}_e\frac{\phi}{2} = 0$$

Due to the symmetry only, this equation is enough to evaluate the magnetic performance of the magnetic circuit.

1.4 Air-gap in a magnetic structure

The presence of an air-gap in a magnetic structure is very common. There are two possibilities: intentional or unintentional. In the first case, the air-gap is necessary for separate moving parts from the static ones, as we can observe in electrical machines. In the DC machine of figure 1.16, the stator (static part) is separated from the rotor (moving part) by a small air-gap.

The second case is the unintentional air-gap. This kind of air-gap appears during the assemblage of closed magnetic structures, like the one used in transformers, when a natural, very small, air-gap arises in the junction of the steel strips, as shown in figure 1.18.

As the unintentional air-gap is normally less than a tenth of millimeter, it is often neglected in the magnetic circuit solution, but the intentional air-gap should be considered because it has a very important role in the magnetic circuit's performance.

Two undesirable effects appear around the air-gap. The first one is the fringing effect on the flux lines that makes the crossing surface of the magnetic flux bigger.

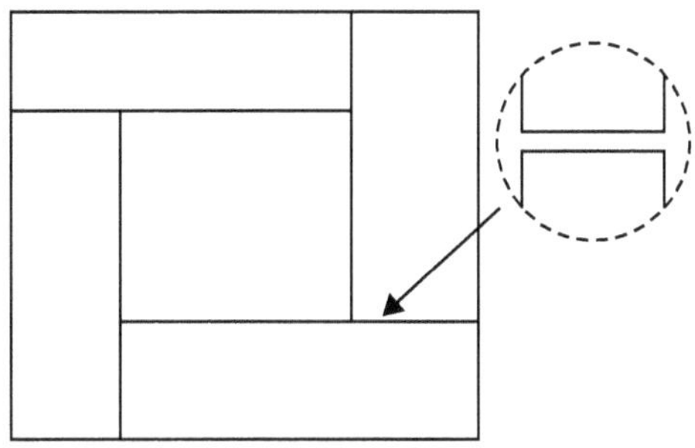

Figure 1.18. Unintentional air-gap. This air-gap is normally neglected in the magnetic circuit solution.

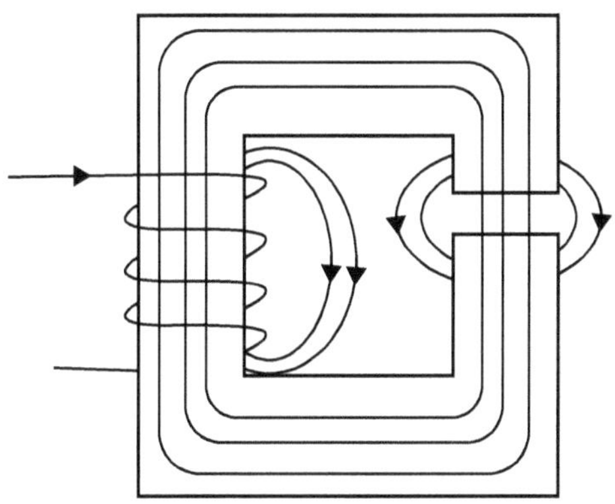

Figure 1.19. Leakage and fringing effects in the air-gap.

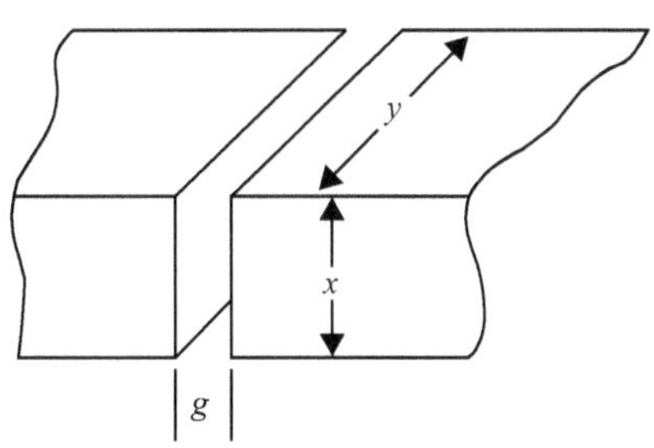

Figure 1.20. The fringing effect evaluation $g \ll x, y$.

The air-gap also increases the leakage flux, which means that some lines of the total flux produced by the coil are not self-closed by the main magnetic circuit.

Figure 1.19 shows both effects. The fringing effect could be considered in the magnetic circuit analysis by using an empirical formula that works well for small air-gaps. On the other hand, the evaluation of the leakage flux is not a simple task and we will not discuss it in this book.

The fringing effects are considered when correcting the crossing surface of the magnetic flux using the empirical formula (figure 1.20):

$$S_g = (x + g)(y + g) \tag{1.16}$$

when the surface of both sides has the same dimension. If the dimensions are different, we must use the following equation:

$$S_g = (x + 2g)(y + 2g) \tag{1.17}$$

where x and y are the dimensions of the smaller face bounding the air-gap.

Example 1.1

The magnetic structure of figure 1.21 was built with cast steel. As the final flux density is low, the relative magnetic permeability may be taken as 1000.

Determine: (a) The reluctance of all legs of the structure; (b) the electric current necessary for establishing a magnetic flux of 3.2×10^{-4} Wb in the air-gap.

Note: All dimensions are in millimeters. The fringing effect should be considered and the leakage flux neglected. $\mu_0 = 4\pi \times 10^{-7}$ H m^{-1}.

Solution

The analog electric circuit of the magnetic structure (figure 1.22) is:

Reluctance of lateral legs

$$\mathcal{R}_1 = \frac{l_1}{\mu S_1}$$

where:

$$\mu = 1000\mu_0 = 4\pi \times 10^{-4} \text{ H m}^{-1}$$
$$l_1 = 2(10 + 60 + 10) + (10 + 80 + 10) = 260 \text{ mm}$$
$$l_1 = 0.26 \text{ m}$$
$$S_1 = 20 \times 20 = 400 \text{ mm}^2$$
$$S_1 = 400 \times 10^{-6} \text{ m}^2$$
$$\mathcal{R}_1 = \frac{0.26}{4\pi \times 10^{-4} \times 400 \times 10^{-6}} = 5 \times 10^5 \text{ A Wb}^{-1}$$

Reluctance of central leg

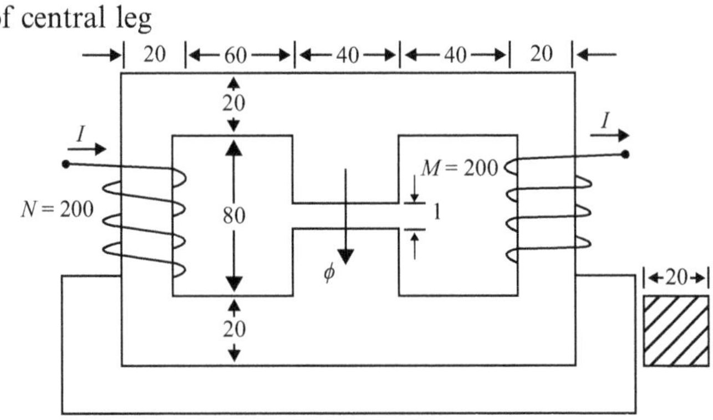

Figure 1.21. Magnetic structure of example 1.4.1.

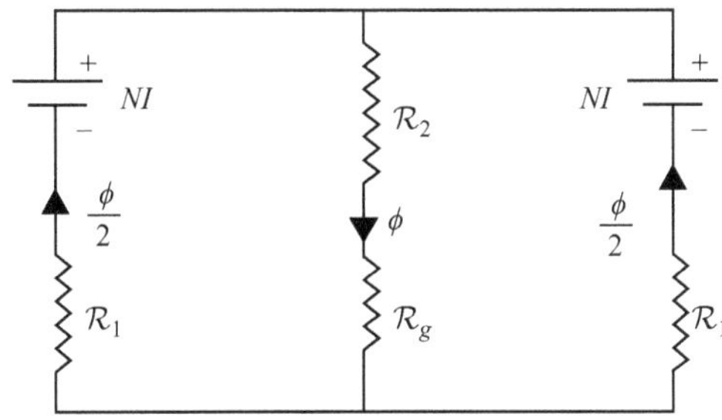

Figure 1.22. Analog circuit of magnetic structure. The magnetic circuit is symmetric.

$$\mathcal{R}_2 = \frac{l_2}{\mu S_2}$$

$$\mu = 1000\mu_0 = 4\pi \times 10^{-4}\ \text{H m}^{-1}$$

$$l_2 = 10 + 80 + 10 = 100\ \text{mm}$$

$$l_2 = 0.1\ \text{m}$$

$$S_2 = 40 \times 20 = 800\ \text{mm}^2$$

$$S_2 = 800 \times 10^{-6}\ \text{m}^2$$

$$\mathcal{R}_2 = \frac{0.1}{4\pi \times 10^{-4} \times 800 \times 10^{-6}} = 10^5\ \text{A Wb}^{-1}$$

Reluctance of air-gap

$$\mathcal{R}_g = \frac{g}{\mu_0 S_g}$$

$$\mu_0 = 4\pi \times 10^{-7}\ \text{H m}^{-1}$$

$$g = 10^{-3}\ \text{m}$$

$$S_g = (40 + 1) \times (20 + 1) \times 10^{-6} = 861 \times 10^{-6}\ \text{m}^2$$

$$\mathcal{R}_g = \frac{10^{-3}}{4\pi \times 10^{-7} \times 861 \times 10^{-6}} = 9.2 \times 10^5\ \text{A Wb}^{-1}$$

As the magnetic structure is symmetric, just one mesh is enough to solve the magnetic circuit. For that,

$$-NI + (\mathcal{R}_2 + \mathcal{R}_g)\phi + \mathcal{R}_1\frac{\phi}{2} = 0$$

or else,

$$-200I + (1 + 9,2) \times 10^5 \times 3,2 \times 10^{-4} + 5 \times 10^5 \times 1,6 \times 10^{-4} = 0$$

that results in

$$I \cong 2\text{A}$$

1.5 Axisymmetric geometry

Until now, we have learned how to evaluate the performance of a magnetic circuit with a planar symmetry. In this case, the flux density distribution inside the structure can be represented by a bi-dimensional (2D) symmetry, i.e., the flux density distribution is the same at any section parallel to the plane of the paper (x, y plane). No field variation is observed in the z-direction.

Although this kind of symmetry is the most common, we find several kinds of electromagnetic devices with axisymmetric geometry (or cylindrical geometry). In this kind of symmetry, the flux density distribution remains the same in all planes that contain the axis of rotation.

Example 1.2

Figure 1.23 shows a cylindrical reactor built with an ideal magnetic material ($\mu \to \infty$). The magnetic flux in the air-gap is 25×10^{-4} Wb. Determine the current in the 100 turns coil for establishing the magnetic flux required. Both the fringing effects and leakage magnetic flux are neglected.

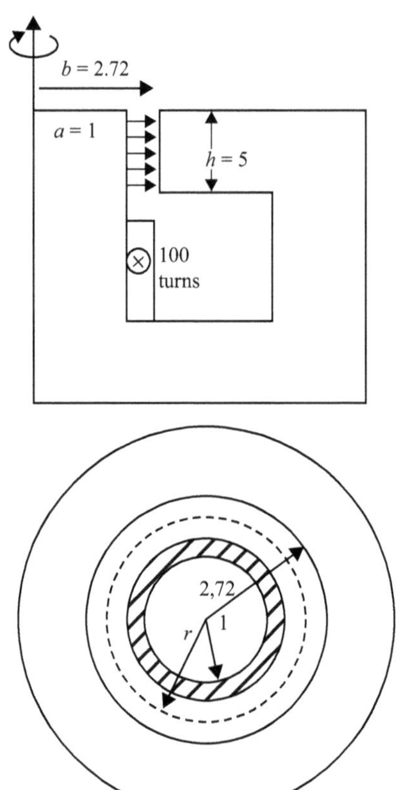

Figure 1.23. Figure of example 1.5.1. Dimensions in centimeters.

Solution
The magnetic circuit of this electromagnetic device is composed of two reluctances. One of them is due to the air-gap ($\mathcal{R}_g$) and the other one is due to the magnetic material ($\mathcal{R}_m$), as shown in figure 1.24.

As the magnetic material has a very high magnetic permeability, the magnetic reluctance of the structure ($\mathcal{R}_m$) is considered null in this example.

To evaluate the air-gap reluctance ($\mathcal{R}_g$), we first need to evaluate the elemental reluctance ($d\mathcal{R}_g$) based on the small ring shown in figure 1.25 located at a generic position $a < r < b$.

The expression for evaluating the elemental reluctance $d\mathcal{R}_g$ is such that:

$$d\mathcal{R}_g = \frac{dr}{\mu_0 \times 2\pi r h}$$

By integration, we get:

$$\mathcal{R}_g = \frac{1}{2\pi\mu_0 h}\int_a^b \frac{dr}{r}$$

That results in:

$$\mathcal{R}_g = \frac{\ln\frac{b}{a}}{2\pi\mu_0 h}$$

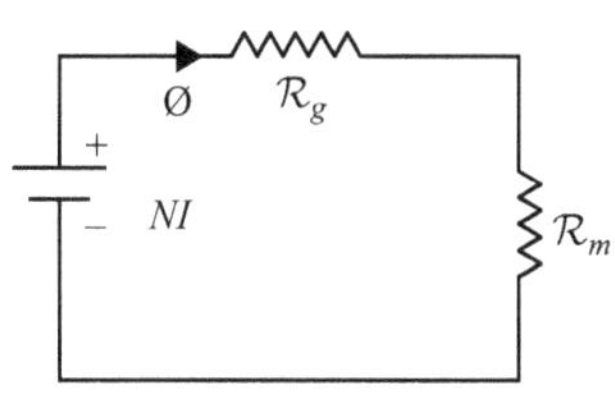

Figure 1.24. Analog electric circuit. $\mathcal{R}_g$: air-gap reluctance; $\mathcal{R}_m$: the magnetic material reluctance. Ø: magnetic radial flux.

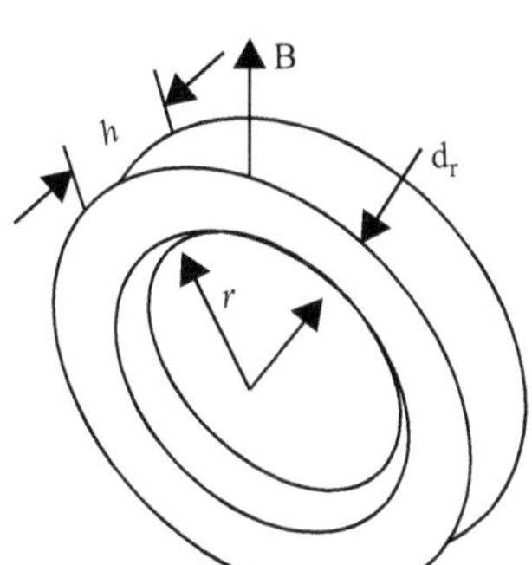

Figure 1.25. Elemental reluctance. This ring is located at a generic position $a < r < b$.

Once $a = 1$ cm and $b = 2.72$ cm, substituting for its value, the final result is:

$$\mathcal{R}_g = 0.253 \times 10^5 \text{ A Wb}^{-1}$$

Based on the analog circuit of figure 1.24, we have:

$$NI = \mathcal{R}_g \varnothing$$

so,

$$100I = 0.253 \times 10^5 \times 25 \times 10^{-4}$$

Resulting in:

$$I = 0.63 \text{ A}$$

1.6 The permanent magnet

Some materials have the property of retaining a residual magnetic flux density even without any electrical current excitation.

When submitted to a high excitation, some of the atoms of this kind of material remain oriented in the direction of impressed mmf even after the mmf is kept down. This is due to the properties of the crystalline structure of the material that gives enough cohesion between groups of atoms to keep these crystals oriented to give a resultant magnetic flux density that is non-null.

The technology of permanent magnets (PM) grew considerably in the late 20th century with the development of the rare-earth permanent magnet. This kind of PM has a very high energy product (the $B \times H$ product) that makes viable the development of several kinds of devices, not only for special applications like high efficiency electrical motors and others, but also for common applications in electrical appliances.

The PM in the magnetic circuit is analogous to a 'battery' in the electric circuit. The battery releases an emf between its plates and a PM releases a mmf inside the material. Figure 1.26 shows the analogy between the battery and PM.

The magnetization characteristics of a PM extracted from its hysteresis curve has the aspect shown in figure 1.27.

B_r is the residual magnetic flux density obtained from the orientation of atoms of the crystals after both a strong and sudden excitation is applied, and H_c is the coercive magnetic field that represents the intensity of the magnetic field intensity necessary for the PM demagnetization.

In the magnetic circuit of figure 1.28, the shaded area is a PM. A linear magnetic material of magnetic permeability μ of mean length l and uniform cross-section S is excited by this PM.

Let us evaluate the magnetic flux created by the PM inside both the PM and the magnetic material.

The analog electric circuit of this magnetic circuit is shown in figure 1.29.

Solving the electric circuit of figure 1.29, we get:

$$\mathcal{F} = \mathcal{R} \varnothing$$

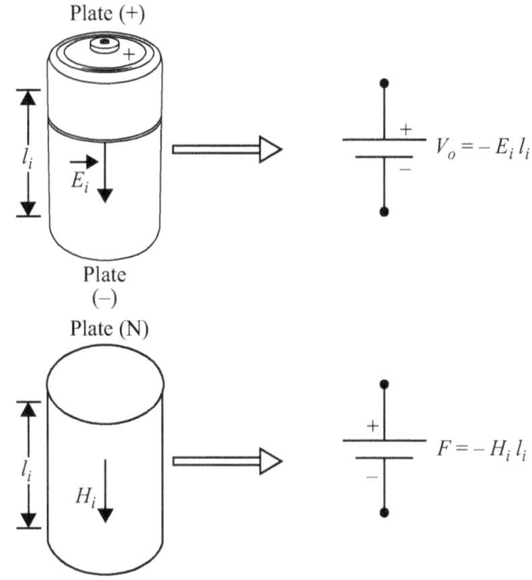

Figure 1.26. Analogy between a battery and a PM. The electric field inside the battery is in the opposite sense to the emf. The magnetic field intensity inside the material is in the opposite sense to the mmf.

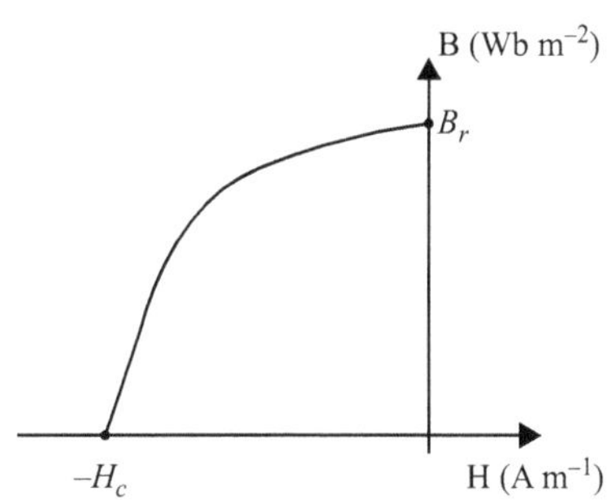

Figure 1.27. Magnetization characteristics of PM. B_r: residual flux density Wb m^{-2}. H_c: coercive magnetic field A m^{-1}.

or,

$$-H_i l_i = \mathcal{R} B_i S$$

that results in:

$$B_i = -\frac{\mathcal{P} l_i}{S} H_i \tag{1.18}$$

where:

$$\mathcal{P} = \frac{1}{\mathcal{R}} [\text{Wb A}^{-1}] \tag{1.19}$$

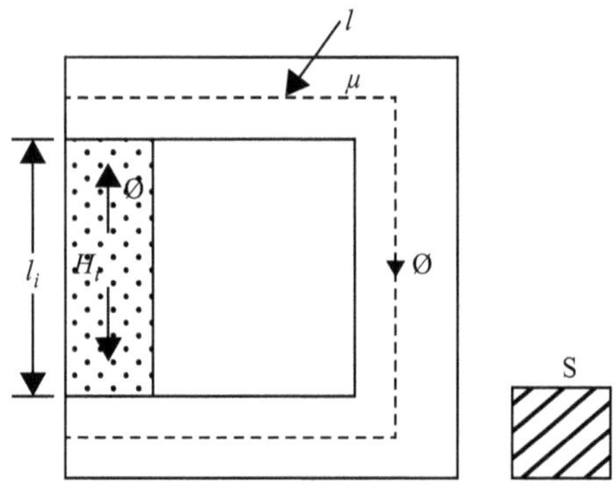

Figure 1.28. Magnetic circuit + permanent magnet. The shaded area is a PM with length l_i.

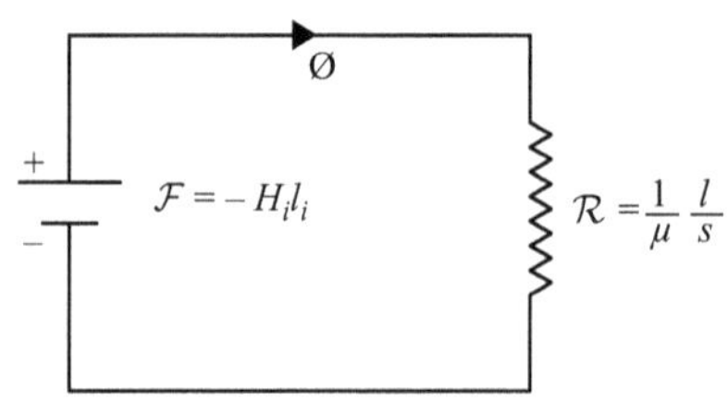

Figure 1.29. The analog electric circuit of magnetic circuit. $\mathcal{F} = -H_i l_i$: The mmf created by a PM. The magnetic flux is in the opposite sense of the magnetic field intensity at the PM.

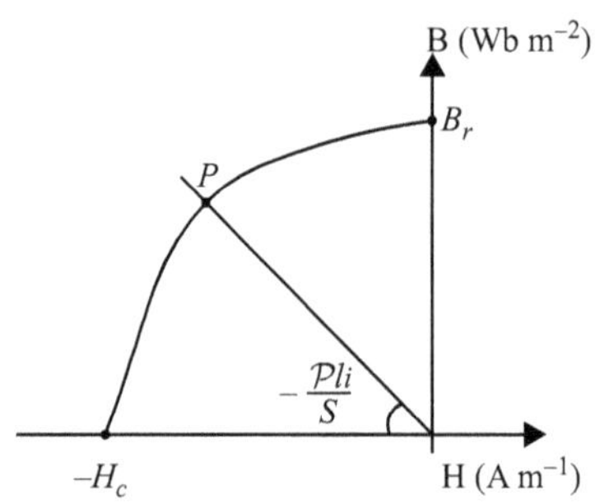

Figure 1.30. Operation point of a PM.

is the permeance of the magnetic circuit.

The relation (B_i, H_i) for the magnetic circuit is a straight line of angular coefficient $-\frac{\mathcal{P}l_i}{S}$. The operation point of the PM is the intersection of the straight line of the magnetic circuit with the magnetization characteristics of PM represented by the point P in figure 1.30.

1.7 The inductance

We can associate with each coil a parameter called 'self-inductance' or simply 'inductance' that characterizes its capacity for producing a magnetic flux.

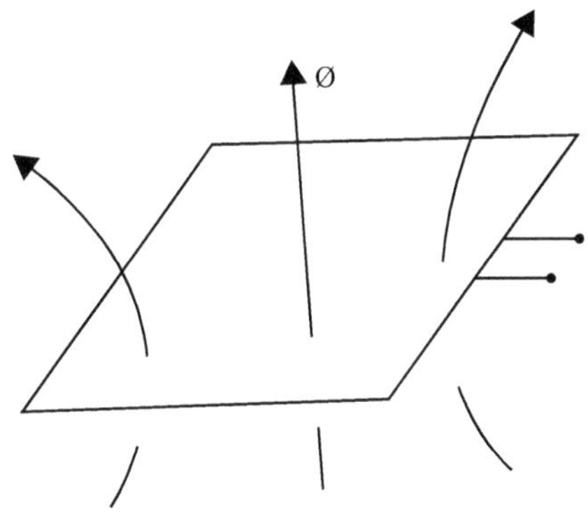

Figure 1.31. Magnetic flux crossing one single turn coil.

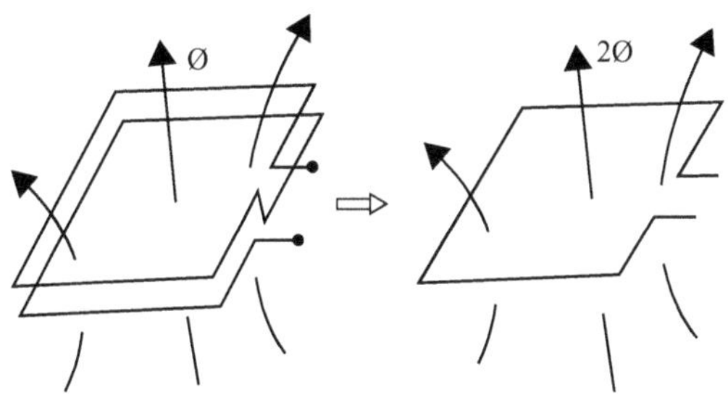

Figure 1.32. Two turns coil crossing by magnetic flux $\varnothing$. Equivalent to a single coil crossing by magnetic flux $2\varnothing$.

Let us consider $\varnothing$ the magnetic flux that crosses one turn coil as shown in figure 1.31.

It is convenient to consider this single turn as a planned electric circuit.

Let us consider now the same magnetic flux crossing a two turns coil as shown in figure 1.32. As the electric circuit has two plans (one for each turn), the total amount of magnetic flux that crosses the circuit is $2\varnothing$, i.e., the magnetic flux crossing a two turns coil is equivalent to double the magnetic flux crossing one single turn coil.

So, if we have N turns coil crossed by the same magnetic flux, as shown in figure 1.33, the total amount of magnetic flux crossing the coil is $N\varnothing$.

The product $N\varnothing$ is called 'concatenate magnetic flux' with the coil, and it is represented by the Greek letter λ:

$$\lambda = N\varnothing \tag{1.20}$$

The magnetic flux crossing the coil can be created by an external source or created by the coil itself when it is carrying an electric current. A linear magnetic circuit under this condition shows a direct proportionality between λ and the electric current i. Therefore, we can express the relation of both values as:

$$\lambda = Li \tag{1.21}$$

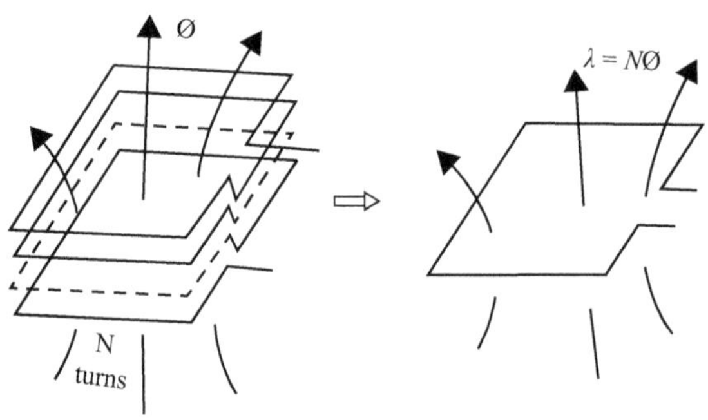

Figure 1.33. N turns coil crossing by magnetic flux $\varnothing$. Equivalent to single coil crossing by magnetic flux $N\varnothing$.

The constant L is called 'self-inductance' of the coil. It is possible to evaluate such a parameter in simple magnetic circuits, remembering that,

$$Ni = \mathcal{R}\varnothing \tag{1.22}$$

From equations (1.20)–(1.22), we can extract the 'inductance' making:

$$Ni = \mathcal{R}\frac{Li}{N}$$

That results in:

$$L = \frac{N^2}{\mathcal{R}} \tag{1.23}$$

1.8 Summary

The concept of magnetic circuits introduced in this chapter is the first step for a complete understanding of the theory of electromechanical energy conversion. Almost all electromagnetic devices are built with magnetic material that can be split by a prism with both a uniform cross-section and uniform magnetic permeability.

For each prism, it is required to evaluate its reluctance given by:

$$\mathcal{R}_i = \frac{l_i}{\mu_i S_i} \tag{1.24}$$

In the next step we compose the analog electric circuit with all reluctance at the same topology as the magnetic circuit.

In the analog electric circuit, all currents are associated with the magnetic flux in each branch and all drop voltages are equivalent to the drop magnetomotive force in the component (figure 1.34).

When the magnetic circuit has a permanent magnet, the operation point is given by the intersection of the straight line of the magnetic circuit with angular coefficient $-\frac{\mathcal{P}l_i}{S}$ with the magnetization characteristics of a PM.

The inductance of an N turns coil embedded in a linear magnetic circuit of equivalent reluctance $\mathcal{R}_{eq}$ is given by:

$$L = \frac{N^2}{\mathcal{R}_{eq}} \tag{1.25}$$

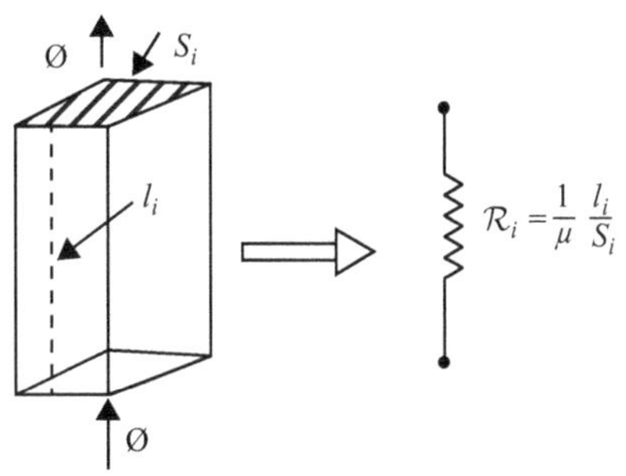

Figure 1.34. Equivalence magnetic circuit × analog electric circuit. The approximations should be done at the corners and at the circle arcs. l_i: the mean length of magnetic path of prism i. S_i: cross-sectional area.

Problems

1.1 A circular cross-section toroid (figure 1.35) is made with linear magnetic material of relative permeability = 5000. The toroid dimensions are: $a = 50$ cm; $b = 60$ cm and $N = 500$ turns. The coil carries a 2 A direct current. Determine:
   (a) The flux density distribution for $a \leqslant r \leqslant b$.
   (b) Which value of B can be considered constant in the toroid cross-section?
   (c) The magnetic flux in the cross-section using the value of flux density obtained in (b).
   (d) The reluctance of the magnetic circuit.
   (e) The self-inductance of the coil.

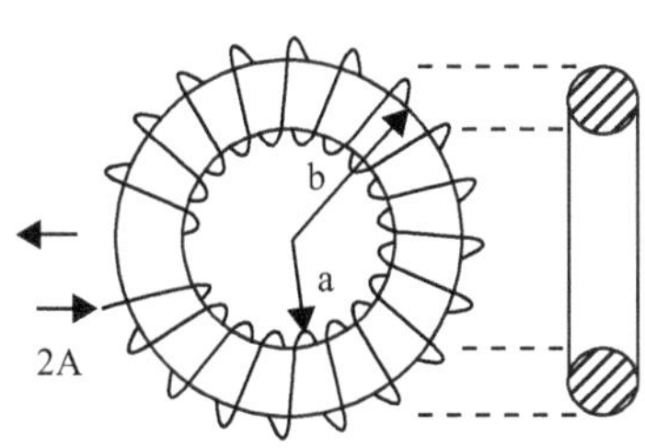

Figure 1.35. Problem 1.1. Circular cross section toroid.

Answers: (b) $B_{cte} = 1.82$ Wb m^{-2}; (c) $\phi = 14.3 \cdot 10^{-3}$ Wb ; $R = 70\,028$ A Wb^{-1}; $L = 3.57$ H.

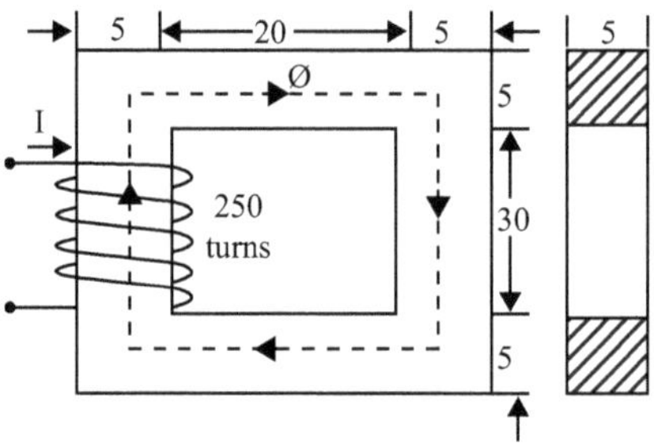

Figure 1.36. Problem 1.2. Sheet steel reactor.

1.2 A reactor shown in figure 1.36 was built of sheet steel. The maximum value of flux density inside the reactor is less than 0.4 Wb m^{-2}, to make this possible consider it as a linear magnetic material of relative magnetic permeability of 5500 ($\mu_0 = 4\pi \times 10^{-7}$ H). The coil has 250 turns and all dimensions are in cm.

Determine:

(a) The DC electric current for establishing a magnetic flux ∅ of 7.5×10^{-4} Wb in the core.

(b) The self-inductance of the coil.

Answers: (a) I = 208 mA ; (b) 900 mH.

1.3 In the reactor of figure 1.36 an air-gap of 1 mm is inserted in the right leg of the core. Neglecting the leakage flux but considering the fringing effect, determine:

(a) The DC electric current for establishing a magnetic flux ∅ of 7.5×10^{-4} Wb in the core.

(b) The self-inductance of the coil;

Answers: (a) I = 1.16 A ; (b) L = 161 mH.

1.4 The symmetric magnetic structure of figure 1.37 is built with two different magnetic materials. The first one has the relative magnetic permeability of 4000 (left side) and the other one 6000 (right side). All dimensions are in cm.

Determine:

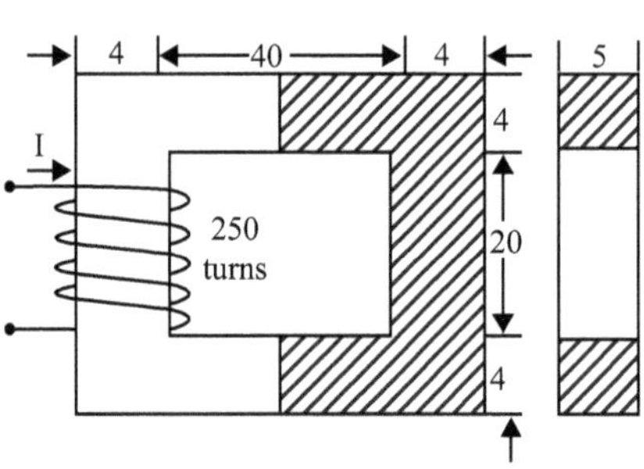

Figure 1.37. Problem 1.4.

(a) The DC electric current in the 250 turns coil for establishing 10^{-3} Wb in the core.
(b) The percentage distribution of mmf in each part of the structure.
(c) The self-inductance of the coil.

Answers: (a) $I = 451$ mA ; (b) 60% (left side) and 40% (right side) ; (c) $L = 554$ mH.

1.5 The magnetic structure of figure 1.38 is made by transformer steel and is operating at a low range of flux density. In this condition, the material can be considered linear with a relative magnetic permeability of 15 000. All dimensions are in centimeters and all leakage flux is neglected.

The flux density inside the coil is 0.4 Wb m^{-2}.

Determine:

(a) The electric current in the coil for establishing this flux density.
(b) The magnetic flux and flux density in all legs of the structure.
(c) The inductance of the coil

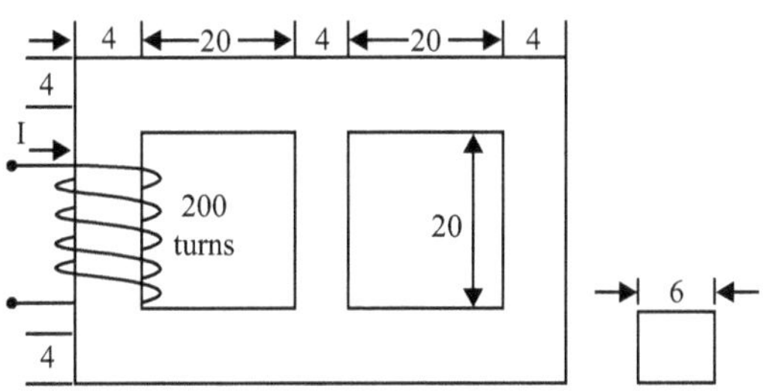

Figure 1.38. Magnetic structure for problem 1.5.

Answers: (a) $I = 95$ mA; (b) $\phi_1 = 9.6 \cdot 10^{-4}$ Wb; $B_1 = 0.4$ T; $\phi_1 = 7.2 \cdot 10^{-4}$ Wb; $B_1 = 0.3$ Wb m^{-2}. $\Phi_3 = 2.4 \cdot 10^{-4}$ Wb; $B_3 = 0.1$ Wb m^{-2}; (c) $L = 2$H.

1.6 The magnetic structure of figure 1.39 is built with a linear magnetic material of relative magnetic permeability 10 000. The magnetic flux in the central leg of the core is 1.0 Wb m^{-2}. The right leg has a 100 turns coil and

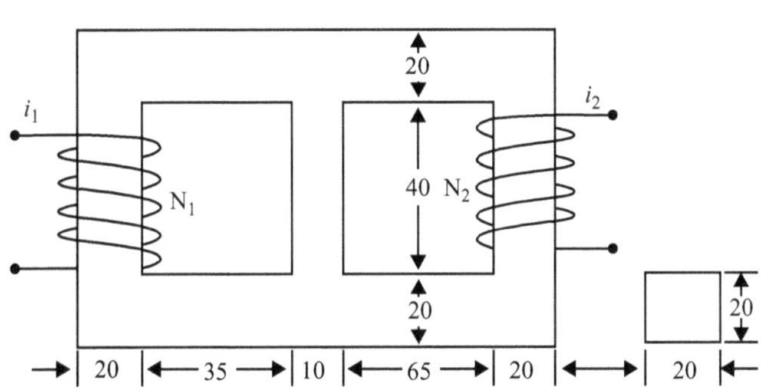

Figure 1.39. Magnetic structure for problem 1.6. The mmf in the left leg ($\mathcal{F}_1 = N_1 i_1$) is 100 A.

the *mmf* in the left leg is ($\mathcal{F}_1 = N_1 i_1$) 100 A. All dimensions are in centimeter and all leakage magnetic flux should be neglected.

Determine:

(a) The DC electric current in the 100 turns coil.

(b) The flux density in all legs of the structure.

Answers: (a) $i_2 = 635$ mA; (b) $B_1 = 0.41$ T; $B_2 = 1$ T; $B_3 = 0.09$ T.

1.7 The electromagnet of figure 1.40 is made of sheet steel magnetized at a low range of flux density. In this condition the sheet steel could be considered a linear magnetic structure of 5000 relative magnetic permeability. The mmf developed by the coil is 1500 A. All dimensions are in centimeters, the leakage magnetic flux is neglected, and the fringing effect should be considered.

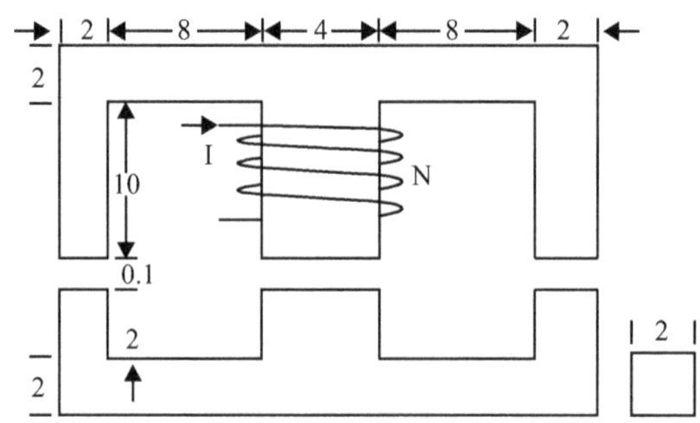

Figure 1.40. Electromagnet of sheet steel. All dimensions in centimeters.

It will be demonstrated later that in each air-gap a magnetic force per unit area of boundary is:

$$\frac{F}{S} = \frac{1}{2}\frac{B^2}{\mu_0} \text{N m}^{-2}$$

Determine:

(a) The magnetic flux in each air-gap.

(b) The magnetic force in each air-gap and the total amount of magnetic force in the electromagnet.

Answers: (a) $\phi_1 = \phi_3 = 3.9 \cdot 10^{-4}$ Wb; $\phi_2 = 7.8 \cdot 10^{-4}$ Wb.

(b) $F_1 = F_3 = 151.3$ N; $F_2 = 302.6$ N.

1.8 The DC electric machine shown in figure 1.41 is built with an ideal magnetic material ($\mu \to \infty$). Determine the electrical current in the 1000 turns/pole field coil for establishing a flux density of 0.6 Wb m^{-2} in the air-gap. No electric current is carried by the armature coil.

Answer: $I = 9.5$ A.

1.9 A kind of covered spool is made of an ideal magnetic material ($\mu \to \infty$), as shown in figure 1.42. This magnetic structure is excited by a 200 turns coil that produces a magnetic flux 25×10^{-4} Wb in the air-gap. Determine the respective current value to produce the required magnetic flux. No fringing effect or leakage magnetic flux should be considered in the solution. All dimensions are in centimeters.

Answer: 1.0 A.

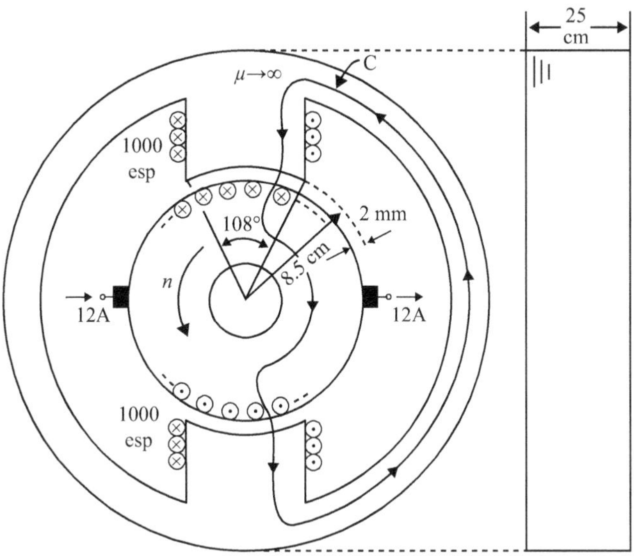

Figure 1.41. DC electrical machine magnetic circuit.

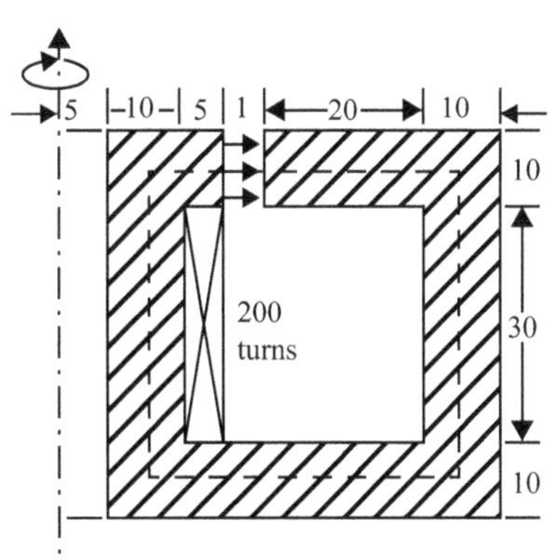

Figure 1.42. Covered spool—cylindrical symmetry.

1.10 The magnetic device shown in figure 1.43 is made of a soft magnetic material that can be considered ideal ($\mu \to \infty$) and a permanent magnet rigid enough that we can consider its magnetic characteristics linear such that $B_r = 0.5$ T and $H_c = 500\,000$ A m^{-1} (figure 1.44).

The length of the PM is 10 mm in the direction of magnetization and a cross-section area is 10 cm^2. The cross-section of the air-gap area is also 10 cm^2. Fringing of flux around the air-gaps may be ignored. Determine for both $l_g = 2$ and $l_g = 5$ mm (figure 1.43).

Compute the magnetic flux density in both the PM and the air-gap.

Answer: 0.27 T.

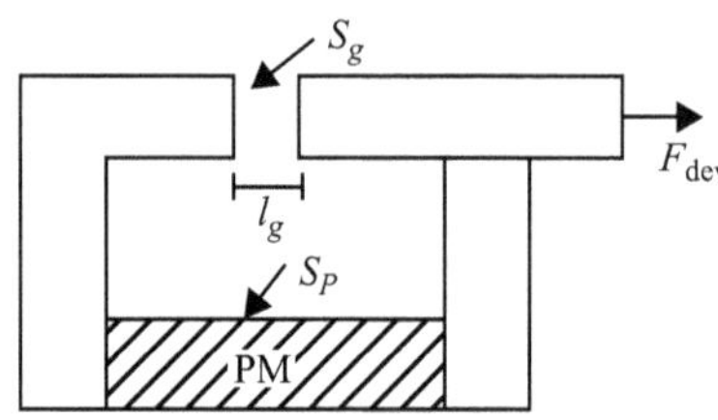

Figure 1.43. Magnetic device of example 5.4. $S_g = 10\ \text{cm}^2$ - $S_p = 10\ \text{cm}^2$. $l_g = 2-5$ mm; $l_p = 102$ mm. $B_r = 0.5$ T; $H_c = 500\ 000\ \text{A m}^{-1}$

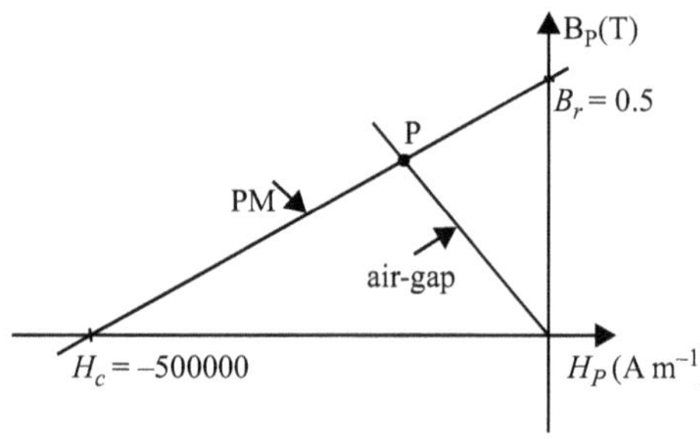

Figure 1.44. PM and air-gap characteristics.

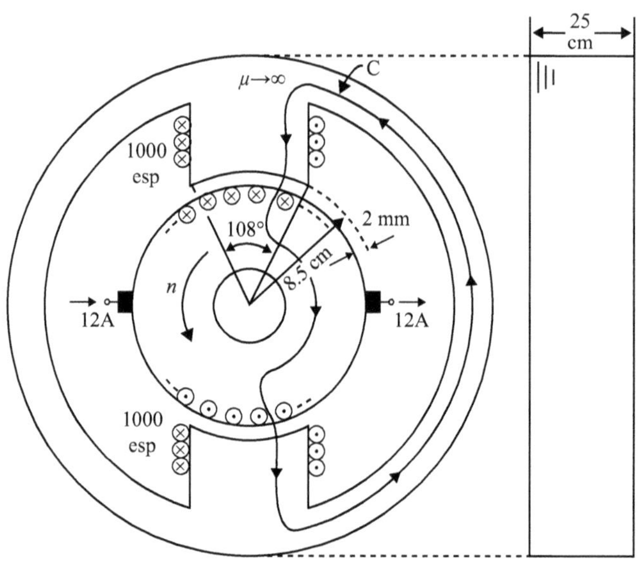

Figure 1.45. Draft for project 1.

Project 1

Design the field coil of the DC electrical machine shown in figure 1.45 considering the stator made with magnetic material of 5000 relative magnetic permeability, the rotor made with magnetic material of 12 000 relative magnetic permeability.

The leakage magnetic flux is neglected, and the fringing effect should be considered.

The field coil is fed by a 220 VDC source. As the duty work of this machine is very high, the current density in the field coil should not exceed 3 A mm^{-2}.

Further reading

[1] Gourishankar V 1966 *Electromechanical Energy Conversion* (Scranton: International Textbook Company)

[2] Skilling H H 1962 *Electromechanics: A First Course in Electromechanical Energy Conversion* (New York: Wiley)

[3] Bansal R (ed) 2004 *Handbook of Engineering Electromagnetics* (New York: Marcel Decker)

IOP Publishing

Electromechanical Energy Conversion Through Active Learning

J R Cardoso, M B C Salles and M C Costa

Chapter 2

The non-linear magnetic circuit

2.1 Some magnetic properties of ferromagnetic materials

Each crystal of ferromagnetic material (Fe–Ni–Co) has a natural non-null elementary magnetic flux density. A group of several crystals with the same magnetic orientation are aligned constituting what is commonly named *the magnetic domain*. The net magnetic flux density of each magnetic domain is a non-null quantity. In Nature, all magnetic domains of ferromagnetic materials is randomly distributed, so the resultant magnetic flux density (B) is null. Figure 2.1 shows a very small sample of ferromagnetic material that was never magnetized, like the one we find in Nature.

If we roll up this slab with a coil as shown in figure 2.2, we can orient each magnetic domain in the same direction of the impressed magnetic field (H) generated by the electric current carried by the coil.

With this procedure, we experimentally observe for a small value of magneto-motive force (mmf = Ni) that a great number of magnetic domains are aligned in the same direction as the impressed magnetic field intensity (H), producing an expressive growth of the magnetic flux density (B). This effect is observed in the straight section of the curve shown in figure 2.3.

For a high value of mmf, the alignment of each magnetic domain becomes more difficult due to the repulsion of the others, which results in the loss of linearity. This effect is shown in the second stage of the curve shown in figure 2.3. Finally, for a very high value of mmf, a non-expressive growth of magnetic flux density (B) is observed even with a big increment of the impressed magnetic field intensity (H), mainly because all magnetic domains are 'practically' aligned with (H). At this stage, the ferromagnetic material becomes 'saturated', as shown in the third stage of figure 2.3.

2.2 Solving a non-linear magnetic circuit

As the magnetic permeability of ferromagnetic materials is dependent on the magnetic flux density, it is not possible to evaluate the reluctance of the magnetic circuit. So, to know the reluctance we need to know the answer to the problem!

doi:10.1088/978-0-7503-2084-9ch2

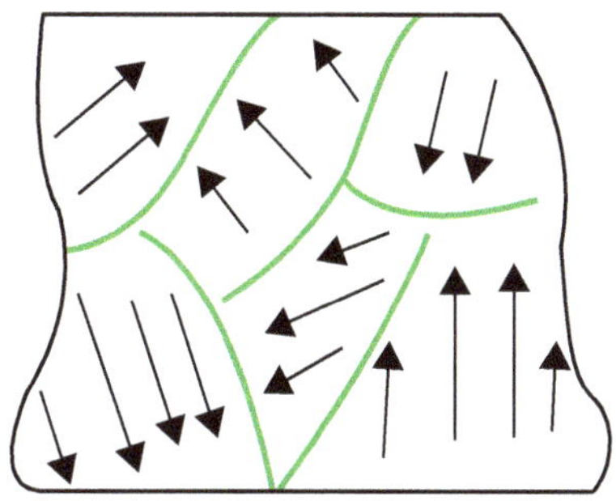

Figure 2.1. Slab of ferromagnetic material. The resultant of magnetic flux density is null.

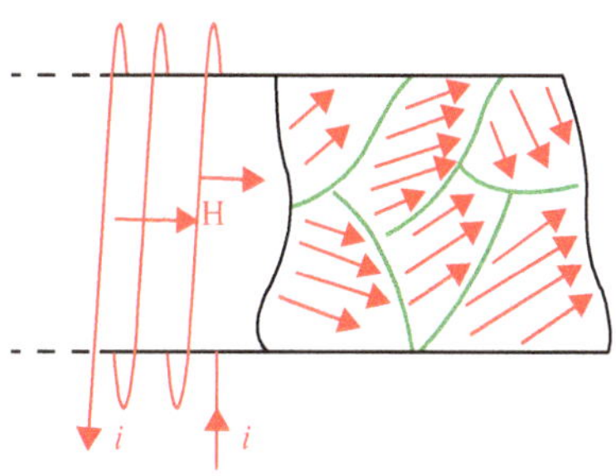

Figure 2.2. Slab of ferromagnetic material excited by an electric current. The resultant of magnetic flux density is non-null. The bigger the electric current is, the greater the number of oriented magnetic domains.

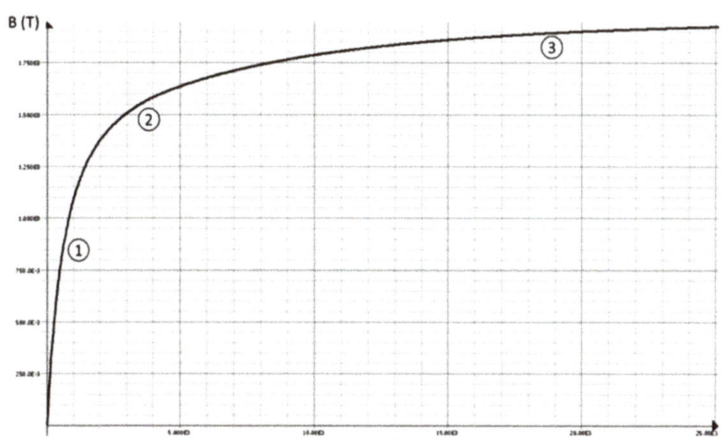

Figure 2.3. The magnetization characteristic of a ferromagnetic material. Stage 1: the linear behavior at low mmf. Stage 2: the repulsion effect among magnetic domains. Stage 3: the saturation effect.

The easiest way to solve a non-linear magnetic circuit is using Ampere's law that establishes that the line integral of magnetic field intensity (H) in a closed path is the magnetomotive force (mmf) involved in this closed path.

$$\oint_C \boldsymbol{H} \cdot \boldsymbol{dl} = i_t \tag{2.1}$$

where

H: magnetic field intensity (A m^{-1});
C: closed path;
i_t: total amount of current that crosses C or the magnetomotive force over C (A).

As we saw in chapter 1, the magnetic circuit of any magnetic structure is comprised of a set of prismatic parts where both the magnetic flux density (B) and the magnetic field intensity (H) are constant, figure 2.4.

With this approach, the left member of (2.1) could be written as:

$$\oint_C \boldsymbol{H} \cdot \boldsymbol{dl} = \sum_{i=1}^{n} H_i l_i \tag{2.2}$$

where:

H_i: the magnetic field intensity at prism i;
l_i: the mean length of prism i.

The right member of (2.1) is the net magnetomotive force (mmf) involved by the closed path C shown in figure 2.5, so:

$$i_t = \sum_{j=1}^{m} N_m i_m \tag{2.3}$$

where:

N_m: the number of turns of the coil located at prism m;
i_m: the current in coil located at prism m.

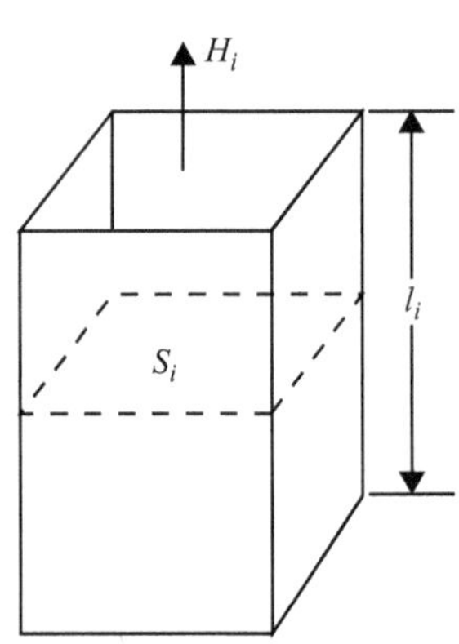

Figure 2.4. Elemental prismatic part of magnetic structure. The generic cross-section is a polygon.

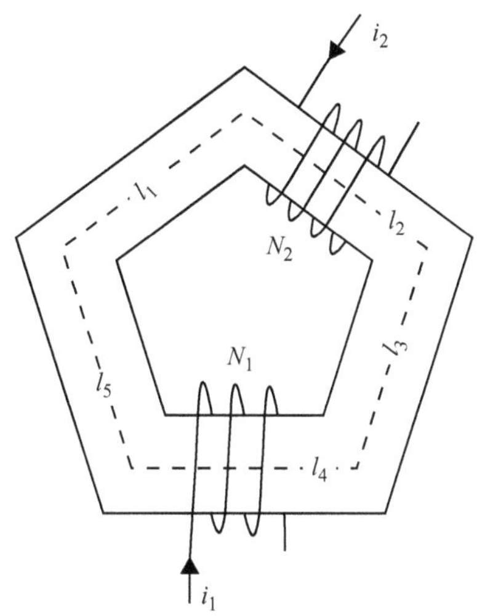

Figure 2.5. Magnetic device as a set of prismatic branch and coils.

$$\sum_{i=1}^{n} H_i l_i = H_1 l_1 + H_2 l_2 + H_3 l_3 + H_4 l_4 + H_5 l_5.$$

$$\sum_{j=1}^{m} N_m i_m = N_1 i_1 - N_2 i_2.$$

Remark: $N_1 i_1$ is in the opposite direction to $N_2 i_2$.

The right member of (2.3) is an algebraic summation because it is common to have mmf in the opposite direction.

As an example, for figure 2.6 we obtain this set of equations extracted from (2.1):

The first step for obtaining the set of equations for a magnetic structure is to choose the positive direction for the quantities involved in the problem.

As the direction of the electric current is defined, we choose the positive direction of magnetic flux according to the right-hand rule applied to the electric current direction in each coil.

The directions of $\varnothing_1$ and $\varnothing_2$ are given using it and the direction of $\varnothing_3$ is chosen by logical thinking.

The directions of magnetic field intensity (H) are the same directions of magnetic flux.

The next step is to establish the direction of the circulation in the closed path defined by the meshes of the magnetic circuit as shown in figure 2.7.

Applying (2.2) for the mesh α:

$$\oint_C \boldsymbol{H} \cdot \boldsymbol{dl} = \sum_{i=1}^{n} H_i l_i = +H_1 l_1 - H_2 l_2 \tag{2.4}$$

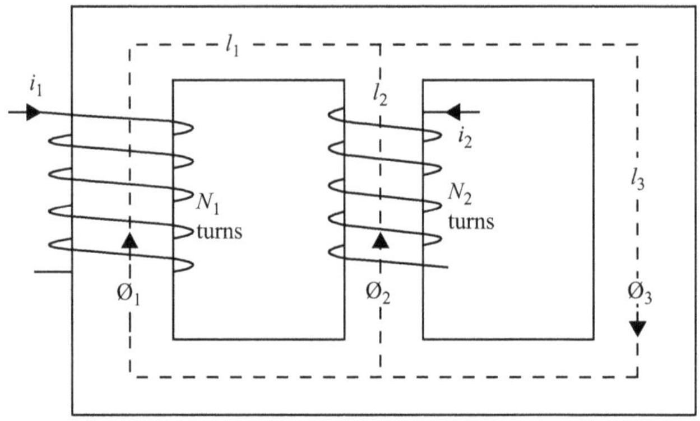

Figure 2.6. Magnetic structure and analog magnetic circuit. H_il_i is the mmf drop in the branch i. The mmf drop is opposite to the magnetic flux sense.

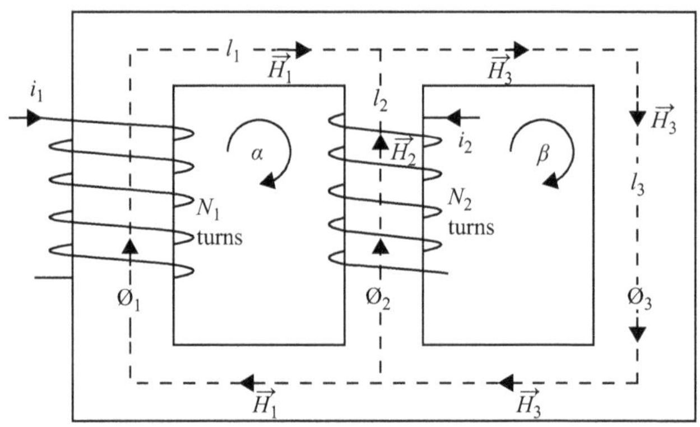

Figure 2.7. Positive directions of $\varnothing$ and H. The directions of $\varnothing$ and H are the same. The directions of the circulations of the mesh are arbitrary.

The first term of (2.4) is positive because the direction of H_1 is the same direction chosen by the circulation and the second one is negative because the direction of H_2 is opposite it.

Applying (2.2) for mesh β results in:

$$\oint_C \boldsymbol{H} \cdot \boldsymbol{dl} = \sum_{i=1}^{n} H_il_i = H_2l_2 + H_3l_3 \tag{2.5}$$

In this case the directions of both H_2 and H_3 are the same as the one chosen by the mesh circulation.

The mmf is easily obtained considering the total amount of current crossing the mesh. Figure 2.8 shows the current crossing the mesh α.

So, applying (2.3) in the mesh α delimited by the closed path defined by the mean length of the magnetic structure results in:

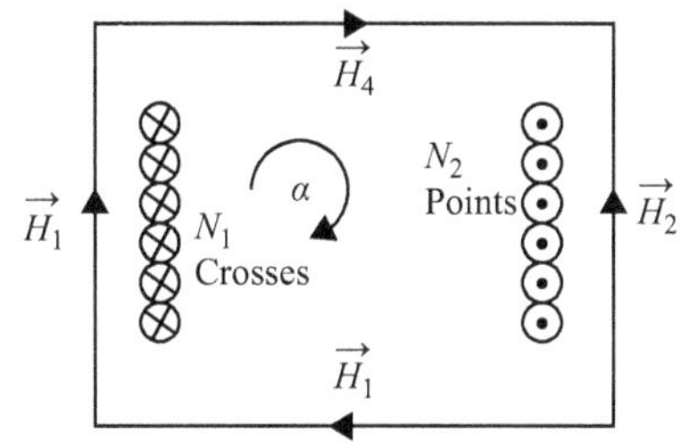

Figure 2.8. The total amount of electric current crossing the mesh α. The signal of the mmf is dependent on the sense of the circulation.

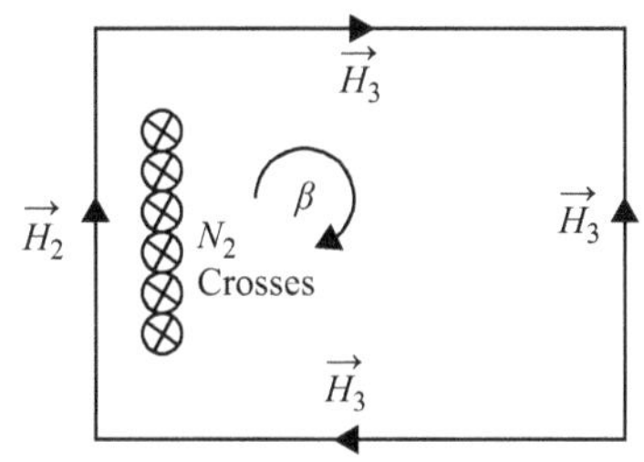

Figure 2.9. The total amount of electric current crossing the mesh β. The signal of the mmf is dependent on the sense of the circulation.

$$i_t = \sum_{j=1}^{m} N_m i_m = N_1 i_1 - N_2 i_2 \tag{2.6}$$

The first term is positive because acting alone this direction of electric current produces a magnetic field in agreement with the chosen circulation direction based on the right-hand rule.

Based on the right-hand rule again, the second one is negative because acting alone this direction of electric current produces a magnetic field not in agreement with the chosen circulation direction.

Figure 2.9 shows the mmf crossing the mesh β. Based on the right-hand rule it is easy to verify that the mmf is positive because acting alone this direction of current is in agreement with the one chosen by the circulation.

$$i_t = \sum_{j=1}^{m} N_m i_m = N_2 i_2 \tag{2.7}$$

Applying Ampere's law for both meshes, we get:

$$N_1 i_1 - N_2 i_2 = H_1 l_1 - H_2 l_2 \tag{2.8}$$

$$N_2 i_2 = H_2 l_2 + H_3 l_3 \tag{2.9}$$

As the magnetic flux has no point source, we complete the set of equations extracted from the magnetic structure including the magnetic flux relation:

$$\varnothing_1 + \varnothing_2 = \varnothing_3 \tag{2.10}$$

2.3 The Kirchhoff law approach

The results of (2.8)–(2.10) give the same results obtained from both voltage and current Kirchhoff Law to both meshes handling the mmf drop ($H{\cdot}l$) as a voltage drop ($E{\cdot}l$) and the emf (V) as a mmf (Ni).

Applying Kirchhoff laws to the circuit of figure 2.10, we get:

$$\begin{cases} -N_1i_1 + H_1l_1 - H_2l_2 + N_2i_2 = 0 \\ -N_2i_2 + H_2l_2 + H_3l_3 = 0 \\ \varnothing_1 + \varnothing_2 - \varnothing_3 = 0 \end{cases} \tag{2.11}$$

These equations, with both the relation $\varnothing = BS$ and the support of the magnetization curve that gives the relation ($B \times H$), are enough to solve any non-linear magnetic problem of the device when the magnetic flux in any branch is known.

2.4 The magnetization curves

Figure 2.11 shows the magnetization curves of several ferromagnetic materials. These curves depend on several factors like, temperature, mechanical stress and manufacturing process.

They are very important for solving non-linear magnetic circuits because they give the correspondence of magnetic flux density (B) and magnetic field intensity (H).

Example 2.1

The magnetic structure of figure 2.12 was built to set up a magnetic flux density of 1.0 Wb m^{-2} in the air-gap. It used an M36 sheet steel for closing the magnetic circuit.

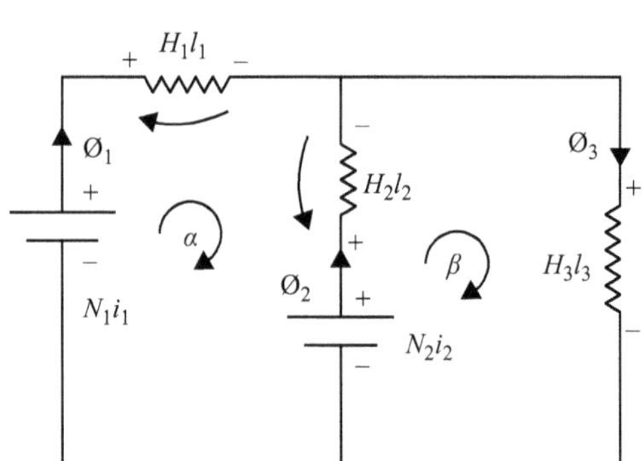

Figure 2.10. Analog electric circuit. Voltage drop is opposed to electric current sense; mmf drop is opposed to magnetic flux.

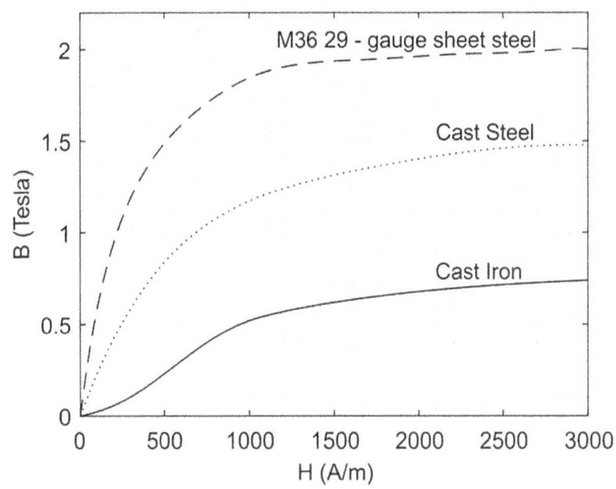

Figure 2.11. Magnetization curves.

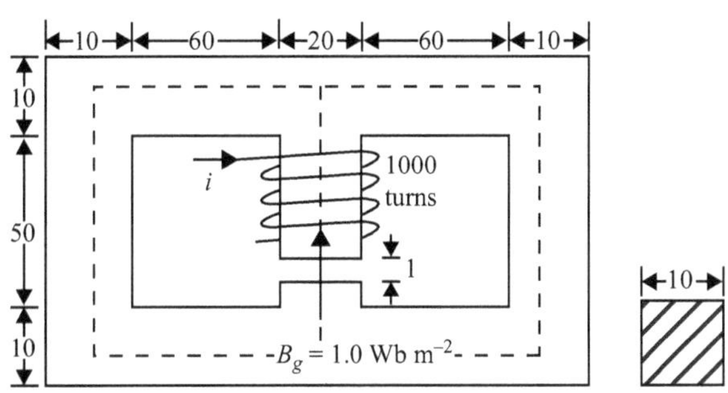

Figure 2.12. Magnetic structure.

The 'stacking factor' for the core is 0.95. Neglecting the leakage magnetic flux, but considering the fringing effect in the air-gap, determine the current that must flow in the exciting winding, which has 1000 turns. All dimensions are in millimeters and $\mu_0 = 4\pi \times 10^{-7}$ H m^{-1}.

Stacking factor: For the best performance of the magnetic circuit the surface of the sheet steel is treated by an insulating process or by coating it with a thin layer of insulating varnish. This makes the effective area of cross-section of a laminated core less than the area obtained from its dimensions. The ratio of the effective area to the gross area of cross-section is called the stacking factor. A typical range of stacking factor is from 0.95 to 0.98.

Solution

Figure 2.13 shows the analog electric circuit of the problem.

The equation system extracted from the analog electric circuit is given by:

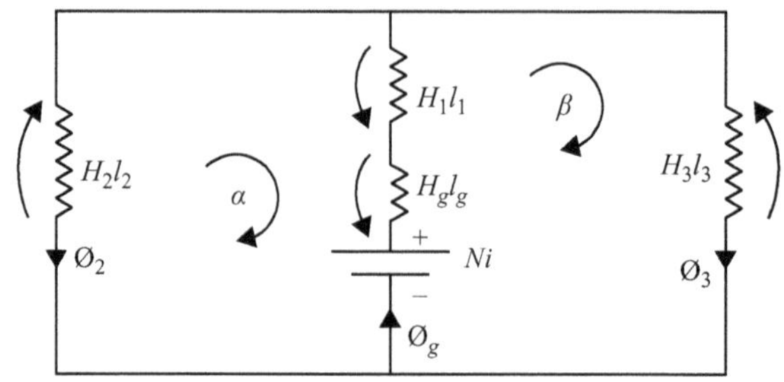

Figure 2.13. Analog electric circuit.

$$\begin{aligned} -H_2l_2 - H_1l_1 - H_gl_g + Ni &= 0 \text{ for mesh } \alpha \\ -Ni + H_gl_g + H_2l_2 + H_3l_3 &= 0 \text{ for mesh } \beta \\ \varnothing_g - \varnothing_2 - \varnothing_3 &= 0 \end{aligned} \quad (2.12)$$

As $B_g = 1.0 \text{ Wb m}^{-2}$ in the air-gap results in:

$$H_g = \frac{B_g}{\mu_0} = \frac{1}{4\pi \times 10^{-7}} = 0.796 \times 10^6 \text{ A m}^{-1}$$

and

$$\varnothing_g = B_gS_g = 1.0 \times (20 + 1)(10 + 1) \times 10^{-6} = 2.31 \times 10^{-4} \text{ Wb}$$

The magnetic flux in the central leg sheet is the same as the air-gap, so:

$$\varnothing_1 = 2.31 \times 10^{-4} \text{ Wb}$$

The magnetic flux density in this same leg is given by:

$$B_1 = \frac{\varnothing_1}{S_{1\text{eff}}}$$

where

$$S_{1\text{eff}} = k_sS_1 = 0.95 \times 20 \times 10 \times 10^{-6} = 1.9 \times 10^{-4} \text{ m}^2$$

So,

$$B_1 = \frac{\varnothing_1}{S_{1\text{eff}}} = \frac{2.31 \times 10^{-4}}{1.9 \times 10^{-4}} = 1.2 \text{ Wb m}^{-2}$$

With the support of the magnetization curve of figure 2.11, we obtain for $B_1 = 1.2 \text{ Wb m}^{-2}$, $H_1 \approx 300 \text{ A m}^{-1}$.

As the magnetic structure is symmetric, it results in:

$$\varnothing_2 = \varnothing_3 = \frac{\varnothing_1}{2} = 1.155 \times 10^{-4} \text{ Wb}$$

The magnetic flux density in both legs is:

$$B_2 = B_3 = \frac{\varnothing_2}{S_{2eff}} = \frac{1.155 \times 10^{-4}}{0.95 \times 10 \times 10 \times 10^{-6}} = 1215 \text{ Wb m}^{-2}$$

With the support of the magnetization curve, we obtain for $B_2 = B_3 = 1.215 \text{ Wb m}^{-2}$, $H_2 = H_3 \approx 350 \text{ A m}^{-1}$.

The mean length of each leg extracted from the figure 2.12, is:

$$l_1 = (50 + 2 \times 5 - 1) \times 10^{-3} = 59 \times 10^{-3} \text{ m}$$

$$l_2 = l_3 = [(50 + 2 \times 5) + 2 \times (5 + 60 + 10)] \times 10^{-3} = 210 \times 10^{-3} \text{ m}$$

Using the first equation of (2.12):

$$-H_2 l_2 - H_1 l_1 - H_g l_g + Ni = 0$$

So,

$$1000i = 350 \times 210 \times 10^{-3} + 300 \times 59 \times 10^{-3} + 0.796 \times 10^6 \times 10^{-3}$$

or,

$$i = 0.887 \text{ A}$$

Remark*:* Even with a very small air-gap (1 mm compared with the main dimension), its mmf drop is big.

2.5 Mapping the magnetic circuit

For a simple excited magnetic circuit, it is common to determine the representative curve $\varnothing = f(\mathcal{F})$ called the 'saturation characteristic'. This characteristic is very useful for solving the converse problem that is the determination of the magnetic flux due to a given electric magnetomotive force.

Figure 2.14 shows a simple magnetic circuit built with a single ferromagnetic material whose magnetization characteristic is shown in figure 2.15.

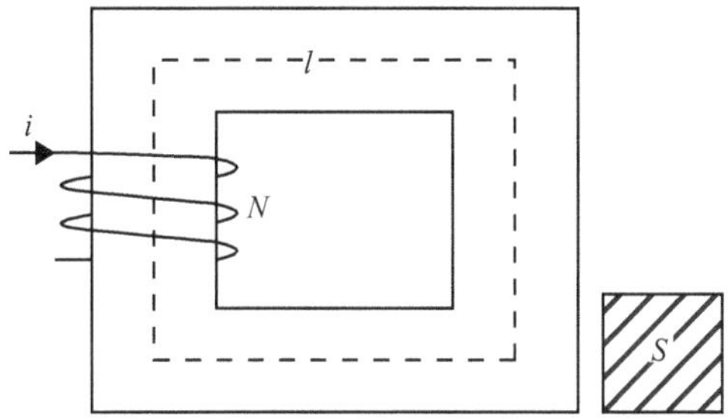

Figure 2.14. Magnetic structure.

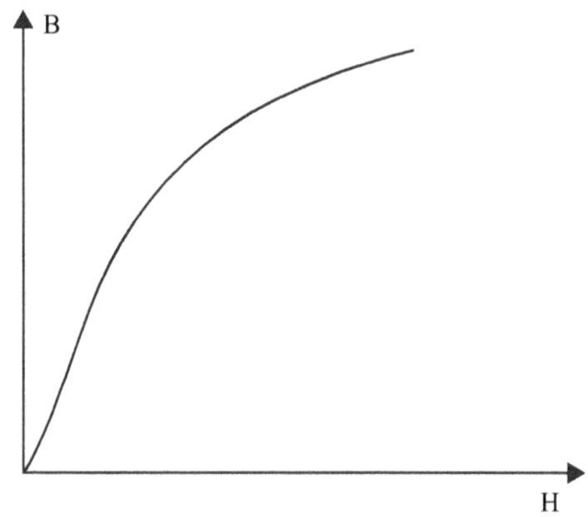

Figure 2.15. Magnetization characteristic.

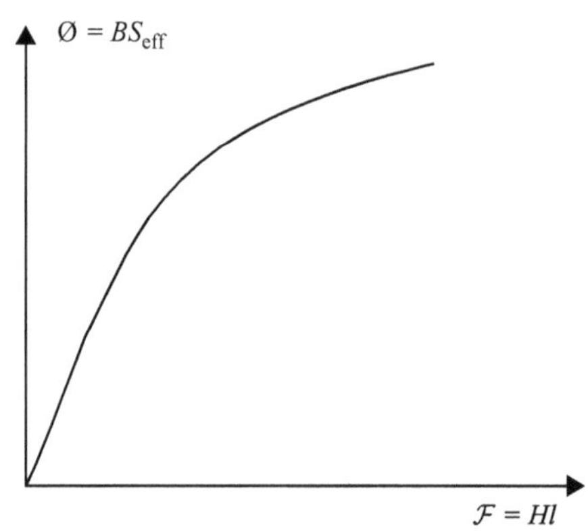

Figure 2.16. Saturation characteristics of the magnetic structure.

The *saturation characteristic* for the magnetic structure is easily obtained changing the axis scale of the magnetization characteristics as follows:

a. The B-axis scale should be multiplied by S_{eff}. As $\varnothing = BS_{eff}$ this axis represents the coordinate of $\varnothing$.
b. The H-axis scale should be multiplied by l (the mean length of the magnetic circuit). As the mmf drop is $\mathcal{F} = Hl$ this new axis represents the coordinate of $\mathcal{F}$.

The saturation characteristics (figure 2.16) of the magnetic structure is the same curve as figure 2.15 but with new scales on the axis.

If the magnetic structure is built with several magnetic materials, we must manipulate the saturation curve of each one to obtain the saturation curve of the magnetic structure.

Figure 2.17 shows a case where the magnetic structure contains two different ferromagnetic materials. As the magnetic flux in both materials is the same, the reluctances of both parts are connected in series.

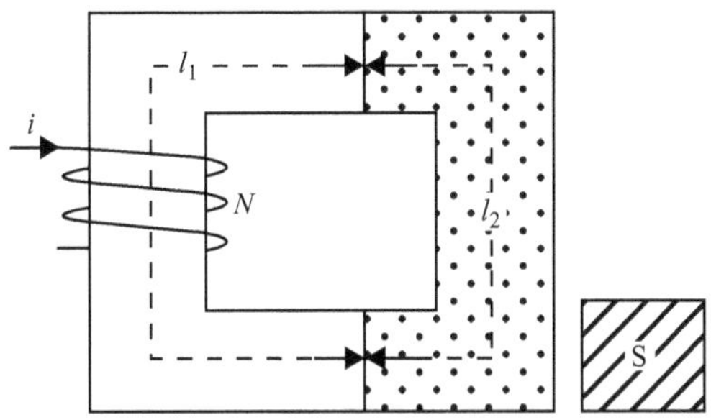

Figure 2.17. Magnetic structure with two different ferromagnetic materials.

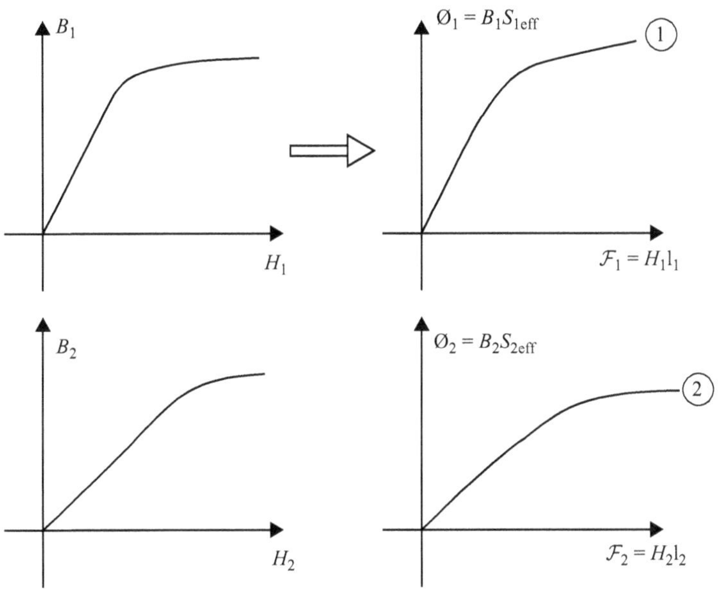

Figure 2.18. The saturation curve for each part of magnetic structure. The dimensions are different for each part.

The saturation curves of both ferromagnetic materials are obtained following the approach described using a suitable dimension as we can see in figure 2.18.

Figure 2.19 shows the final saturation curve for the magnetic structure extracted from both curves obtained with the procedure shown in the figure 2.17.

The saturation curves (1) and (2) are those for each magnetic part of the structure. We will describe how to obtain one point of the final saturation curve.

For a given magnetic flux $\varnothing_a$ the mmf drop in part (1) is $\mathcal{F}_1$ defined by the point (A) in the curve (1). For the same magnetic flux in part (2) the mmf drop is $\mathcal{F}_2$ defined by the point (B) in the curve (2). As the magnetic parts are connected in series, the total mmf necessary for establishing the same magnetic flux $\varnothing_a$ is $\mathcal{F}_a = \mathcal{F}_1 + \mathcal{F}_2$ defining a point (C) belonging to the saturation curve of the magnetic structure.

Reproducing this procedure for several points the saturation curve is assembled.
Remark: When an air-gap is part of the magnetic structure, its saturation curve is a straight line defined by the equation:

$$\varnothing = \mathcal{P}\mathcal{F} \tag{2.13}$$

where:

$$\mathcal{P} = \frac{\mu_0 S_g}{l_g} \tag{2.14}$$

is the permeance of the air-gap. The fringing effect should be considered in the evaluation of this parameter.

Example 2.2

The magnetic structure of figure 2.20 was built with a hypothetical ferromagnetic material with magnetization characteristics shown in figure 2.21. The mean length of

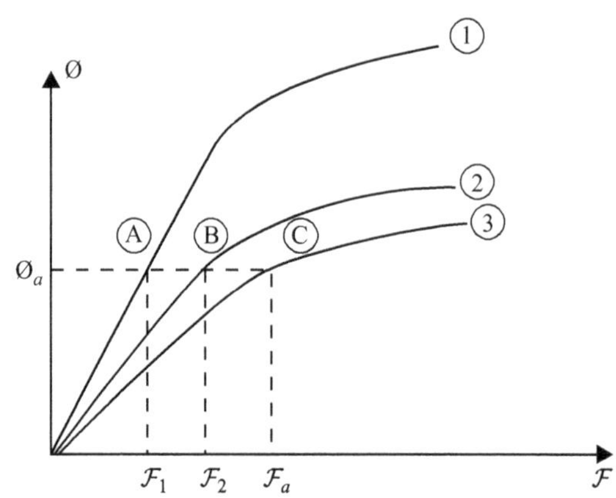

Figure 2.19. The saturation curve for the magnetic structure.

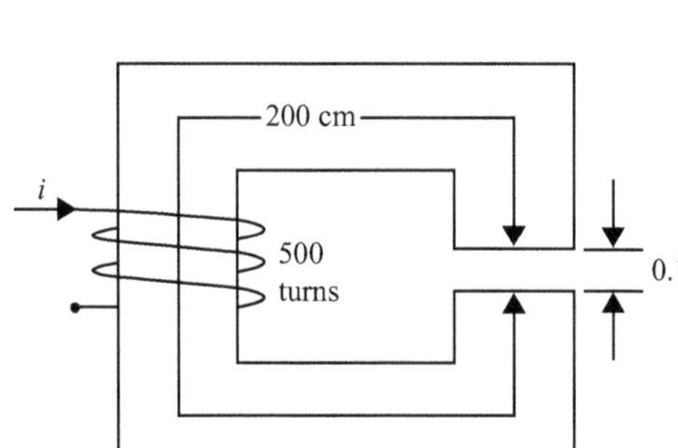

Figure 2.20. Magnetic structure. Dimensions in centimeters.

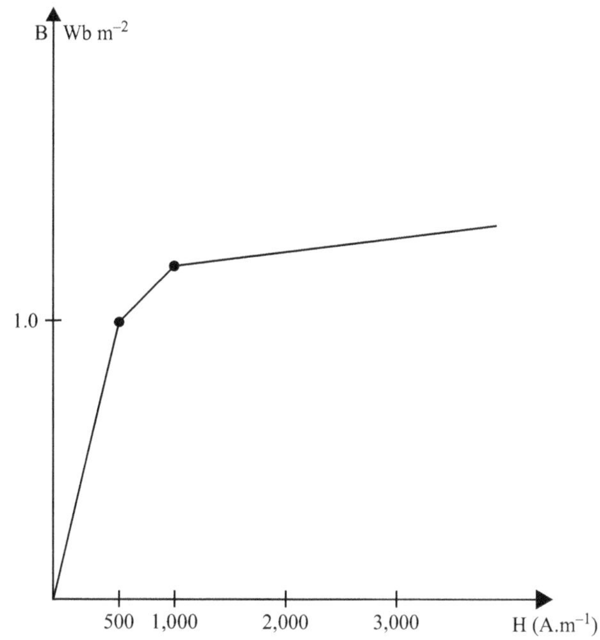

Figure 2.21. Magnetization curve. Hypothetical magnetic material.

the magnetic material is 200 cm and the air-gap 1 mm. The effective cross-section of both magnetic material and air-gap is 40 cm^2. Both leakage flux and fringing effect should be neglected. Determine:

(a) the *saturation characteristics* of this structure;
(b) the magnetic flux density in the core for 15.0 A in the coil.

Solution

(a) The saturation characteristics of the structure.

The first step for solving this kind of problem is to determine the saturation characteristics of both materials.

For the hypothetical magnetic material this characteristic is obtained by changing the scales of the magnetization characteristics as follows:

(i) The scale of H is $[H] = 250$ A/div, so the scale of $[\mathcal{F}] = [H] \times l_m = 250 \times 200 \times 10^{-2} = 500$ A/div.

(ii) The scale of B is $[B] = \frac{0.1 \text{ Wb}}{\text{m}^2}$/div, so the scale of $[\emptyset] = [B] \times S = 0.1 \times 40 \times 10^{-4} = 4 \times 10^{-4}$ Wb/div.

Considering these new scales, the saturation characteristics of the magnetic material is curve (1) of figure 2.22.

The second step is to determine the saturation characteristics of the air-gap. This characteristic is a straight line defined by:

$$\emptyset = \mathcal{P}\mathcal{F}$$

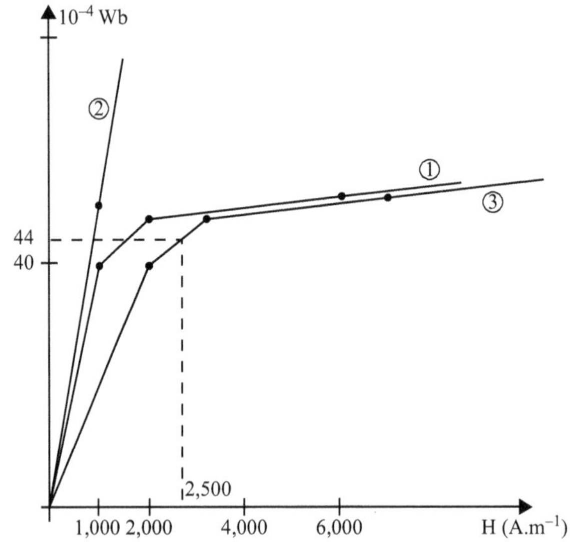

Figure 2.22. Saturation characteristics. (1) Saturation characteristics of magnetic material. (2) Saturation characteristics of air-gap. (3) Saturation characteristics of the magnetic structure: (1) + (2).

where:

$$\mathcal{P} = \frac{\mu_0 S_g}{l_g} = \frac{4\pi \times 10^{-7} \times 40 \times 10^{-4}}{1 \times 10^{-3}} = 5 \times 10^{-6} \text{ Wb A}^{-1}$$

Choosing $\mathcal{F} = 2000$ *Aesp* results in:

$$\varnothing = 5 \times 10^{-6} \times 2000 = 10^{-2} \text{ Wb}$$

So, the straight line (2) represents the saturation characteristics of the air-gap.

The final saturation characteristic of the magnetic structure is the summation of curves (1) and (2) represented by curve (3).

(b) The magnetic flux density in the core for 5.0 A in the coil

With 5.0 A in the coil the mmf in the magnetic circuit will be $\mathcal{F} = 500 \times 5 = 2500$ Aturns, that results from curve (3):

$$\varnothing = 44 \times 10^{-4} \text{ Wb}$$

The magnetic flux density for establishing this magnetic flux results:

$$B = \frac{\varnothing}{S} = \frac{44 \times 10^{-4}}{40 \times 10^{-4}} = 1.1 \text{ Wb m}^{-2}$$

2.6 Numerical method: the finite element method

Some magnetic structures have a very complex geometry that makes it impossible to evaluate the magnetic field distribution applying the approach discussed earlier. For this kind of problem, a numerical solution is the only alternative.

In recent years, numerical methodologies for solving all kinds of engineering problems were developed. With the support of powerful graphical computation, the visualization, accuracy and flexibility allow us to solve, in a few minutes, complex electromagnetics not only in a steady state but also in dynamic and time-dependent situations.

The finite element method (FEM) is considered the most powerful tool suitable for solving the static electromagnetic problem and its application involves six steps.

Step 1: Geometry definition

The geometry definition needs a modeler to draw the cross-section (in the 2D approach) of the device. Two types of geometry should be considered:

(a) The closed domain: This is the situation applied when the magnetic field is totally confined inside the device like the one found in electrical machines (figure 2.23).

(b) The unbounded domain: This is the situation applied when the magnetic field is spreading indefinitely in the space. As FEM requires a closed domain, we must truncate the domain with a closed boundary far enough from the device (figure 2.24).

Figure 2.23. Closed domain. Magnetic field confined inside the device. Source: Modeled on FEM Software FLUX-2D—ALTAIR Engineering, Inc.

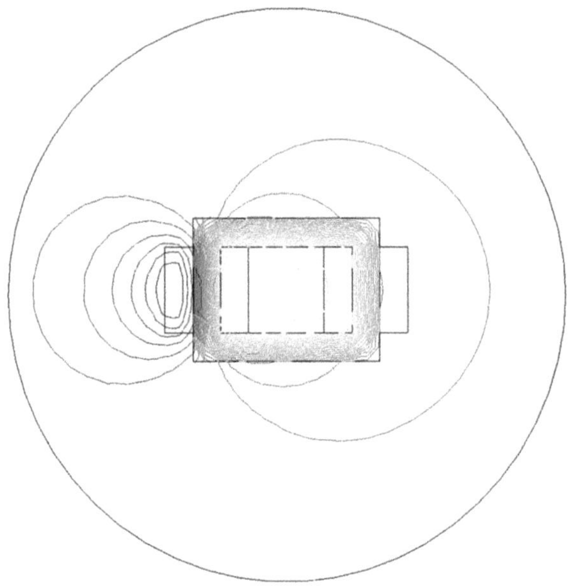

Figure 2.24. Unbounded domain. The truncated line must be far enough from the device. Source: Modeled on FEM Software FLUX-2D—ALTAIR Engineering, Inc.

Step 2: Discretization process
In this step the domain is divided into triangles (the elements) suitably distributed. In the regions where the magnetic field variation is high, the elements' density must be bigger, as shown in figure 2.25.

Step 3: Sources and physical properties specification
This step assigns to each element both the current density and the physical property (permeability or magnetization characteristics)

Step 4: Boundary condition specifications
The boundary condition in a magnetic problem is the assignment of the magnetic potential (A) at part of the domain's borderline (figure 2.26). If ($A = 0$) is assigned it means that the magnetic field outside the borderline is null, if not the magnetic field at this borderline is normal.

Step 5: Solver
After completing all requirements from step 1 to step 4 FEM assemblies, solution of an equation system with NN order (NN = nodes number) gives the magnetic potential of all nodes (vertices of the elements).

Step 6: Results exploitation
With the magnetic potential of all nodes extracted from the solver, it is possible to evaluate not only the magnetic flux density inside all elements but also a multitude of different quantities like forces, parameters, losses and others (figure 2.27).

Figure 2.25. FEM mesh. High variation of magnetic field $\Longrightarrow$ high density of elements. Source: Modeled on FEM Software FLUX-2D—ALTAIR Engineering, Inc.

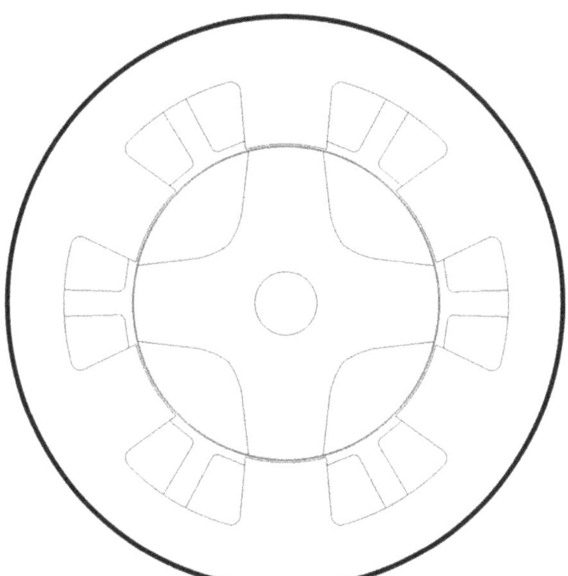

Figure 2.26. Boundary conditions. Source: Modeled on FEM Software FLUX-2D—ALTAIR Engineering, Inc.

Figure 2.27. Results exploitation. Source: Modeled on FEM Software FLUX-2D—ALTAIR Engineering, Inc.

2.7 Summary

In this chapter we introduced the methodology for solving a non-linear magnetic circuit like the one built with ferromagnetic material. The values of both the magnetic flux density (B) and magnetic field intensity (H) are evaluated by using the magnetization curves of the material. These kinds of curve are supplied by the manufacturer or can be obtained by a suitable test.

Different to the magnetic linear circuit we do not use the reluctance for solving the non-linear magnetic circuit because the magnetic permeability is only known when the problem is solved.

As the ferromagnetic structure is made of sheet steel it is convenient to consider the stacking factor for evaluating the real magnetic cross-section of the device.

Problems

2.1 The toroid in figure 2.28 (dimensions is built with M36 sheet steel with stacking factor of 0.98. The electric current in the 300 turns coil is enough to establish a magnetic flux of 5.0×10^{-3} Wb.

The dimensions are: $a = 20$ cm; $b = 25$ cm and $h = 10$cm.

Determine:

(a) the electric current in the coil;
(b) the electric current in the coil after the introduction of an 1 mm air-gap;
(c) the percentage variation of the electric current due to the air-gap introduction.

Answers: (a) $i = 0.94$ A; (b) $i = 3.6$ A; (c) 280%.

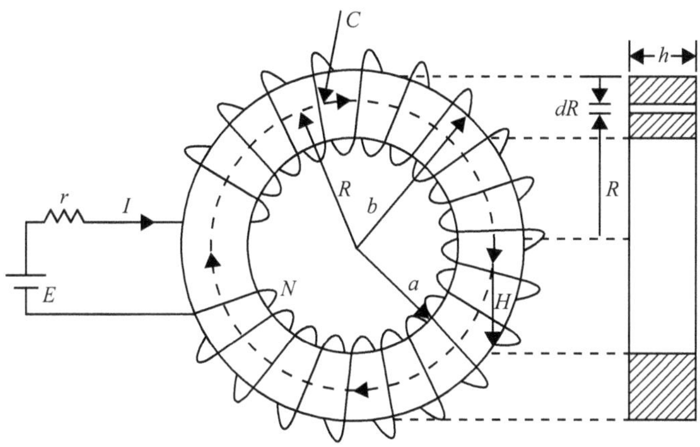

Figure 2.28. Toroid for problem 2.1.

2.2 A sheet steel reactor of figure 2.29 is excited by two coils. The mean length of magnetic path is 200 cm and the effective cross-section 40 cm^2. The leakage flux is negligible.

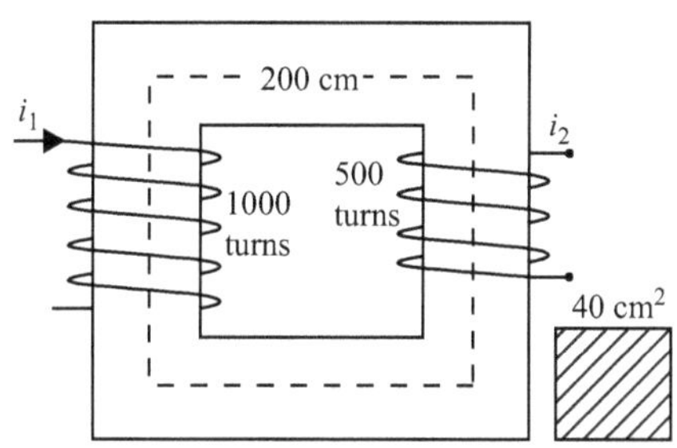

Figure 2.29. Reactor for problem 2.2.

The magnetic flux in the core 50×10^{-4} Wb. Consider the magnetization curve of M36 steel shown in figure 2.11.

Determine:

(a) The electric current in 1000 turns coil for establishing the magnetic flux required with no current in the 500 turns coil.

(b) The electric current in the 1000 turns coil is doubled. Find both the sense and the intensity of the electric current in the 500 turns coil for establishing the same magnetic flux in the core.

(c) The electric currents ratio for (b).

Answers: (a) 0.8 A; (b) 1.6 A; (c) 0.5.

2.3 The magnetic structure of figure 2.30 is symmetric and made with cast iron and cast steel. Determine the electric current in the coil for establishing a magnetic flux of 40×10^{-4} Wb in the air-gap. The leakage magnetic flux

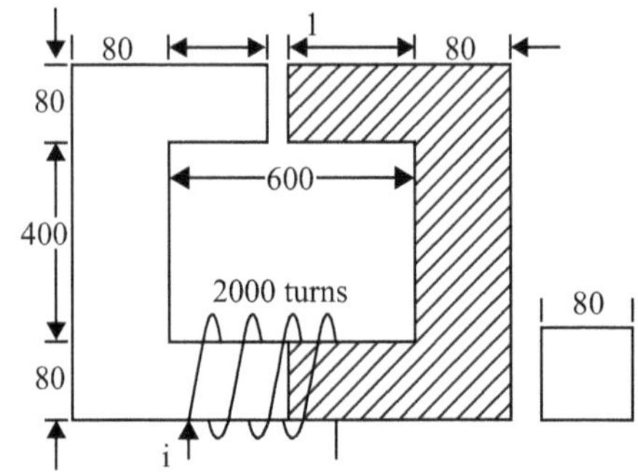

Figure 2.30. Magnetic structure for problem 2.3.

should be neglected but the fringing effects should be considered. All dimensions are in millimeters.

Answer: 1.25 A.

2.4 In the magnetic structure of figure 2.31 the magnetic flux density in the air-gap is 0.75 Wb m^{-2}. A sheet steel lamination with 0.95 stacking factor was used in its construction. Find both the mmf and the electrical current in the 200 turns coil. The leakage flux should be neglected but the fringing effect should be considered. All dimensions are in centimeters.

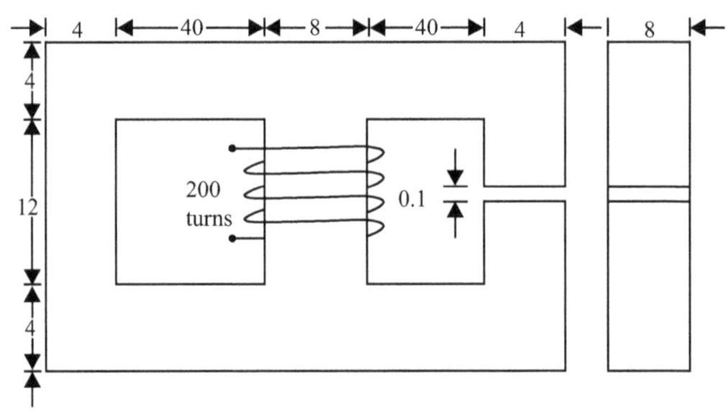

Figure 2.31. Magnetic structure of problem 2.4.

Answer: mmf = 738 A; i = 3.7A.

2.5 The magnetic structure of figure 2.32 is made of sheet steel laminations with 0.95 stacking factor. The flux in three legs are $\varnothing_a = 4 \times 10^{-4}$ Wb, $\varnothing_b = 6 \times 10^{-4}$ Wb, $\varnothing_c = 2 \times 10^{-4}$ Wb in the direction shown.

Determine the electrical current in each coil, giving its magnitude and direction.

Answer: i_1 = 0.46 A up; i_2 = 50 mA down.

2.6 If the magnetic flux in legs A and B of magnetic structure of figure 2.29 is $\varnothing_a = \varnothing_b = 4 \times 10^{-4}$ Wb in the counter clockwise direction and the

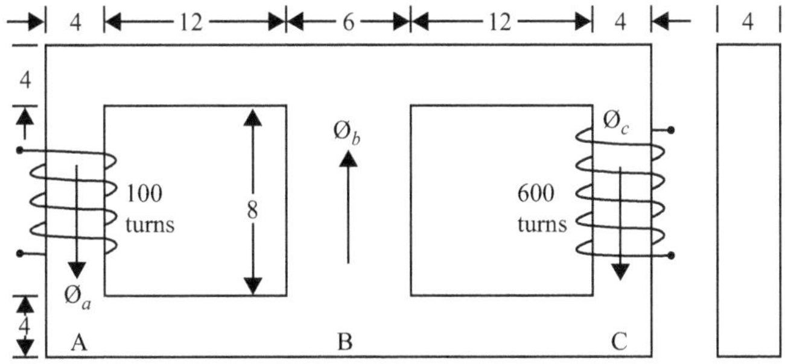

Figure 2.32. Magnetic structure of problem 2.5.

magnetic flux in leg C is zero, determine the magnitude and direction of the electric current in the two exciting coils.

Answer: $i_1 = 0.43$ A up; $i_2 = 1$ mA up.

2.7 In the magnetic circuit of figure 2.33 that is manufactured with sheet steel the magnetic flux density in the air-gap is 0.4 Wb m^{-2}. Assuming: $N = 40$ turns; cross-section = 80 cm^2; stacking factor = 0.95; air-gap = 0.8 mm; $l_{\text{bad}} = l_{\text{bcd}} = 75$ cm; $l_{\text{bd}} = 25.4$ cm. Determine the electric current in the coil.

Answer: 20A.

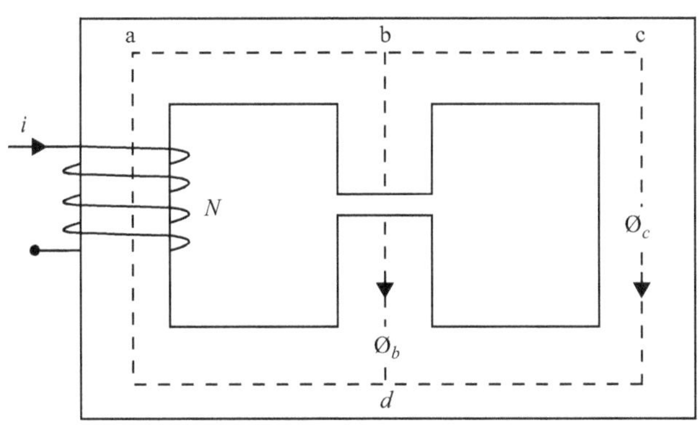

Figure 2.33. Magnetic circuit of problem 2.7.

2.8 The cast steel magnetic core shown in the figure 2.34 has a uniform 8 cm × 8 cm cross-section. There are two exciting coils, one on leg A and the other on leg B. Coil A has 1000 turns and carries a current of 0.5 A in the direction shown. Determine the electric current in coil B in the direction shown to make the magnetic flux in the center leg zero. Coil B has 200 turns. All dimensions are in centimeters.

Answer: $i = 1.25$A.

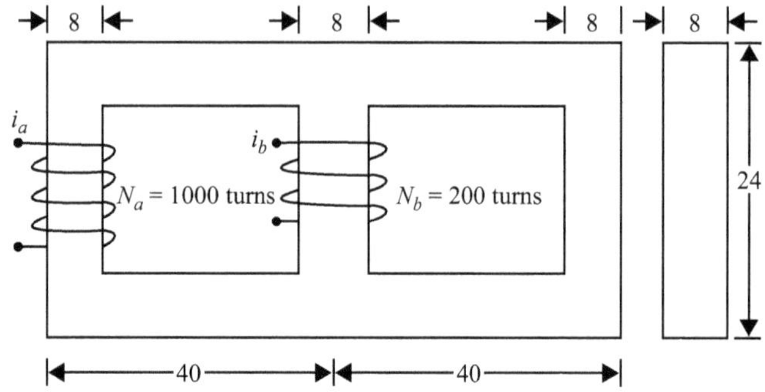

Figure 2.34. Magnetic structure of problem 2.8.

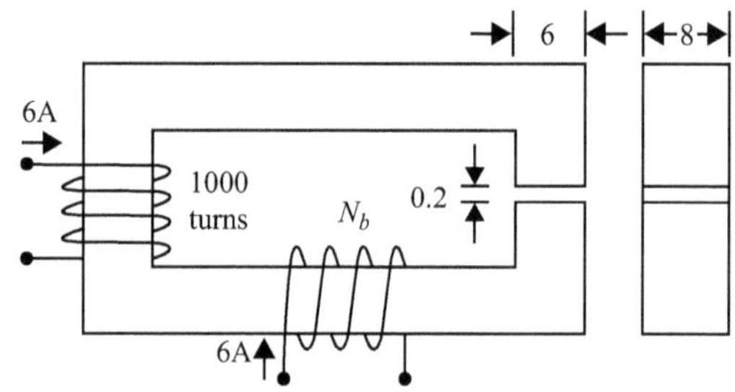

Figure 2.35. Magnetic structure of problem 2.9.

2.9 The magnetic structure shown in figure 2.35 is made of sheet steel laminations. The stacking factor is 0.95. The mean length of magnetic path is 0.75 m in the steel portion. The cross-sectional dimensions are 6 cm × 8 cm. The length of air-gap is 0.2 cm. The flux in the air-gap is 4×10^{-3} Wb. Coil A has 1000 turns, and both coils A and B carry 6 A. Determine the number of turns of coil B. The mean length of the magnetic circuit is 180 cm. Neglect leakage fluxes but allow for fringing. All dimensions are in centimeters.

Answer: 17 turns.

2.10 The magnetic core shown in figure 2.36 consists of two sections made of two hypothetical magnetic materials (figure 2.37). Find the flux in the air-gap if the mmf in the exciting coil is 800 Aesp. Area of cross-section of the core is uniform and 8 cm × 8 cm. The mean lengths of magnetic paths are:

- Material (1): 40 cm
- Material (2): 50 cm
- Neglect leakage and fringing effect

(Hint: Construct the saturation characteristics of the magnetic structure)

Answer: ~30×10^{-4} Wb.

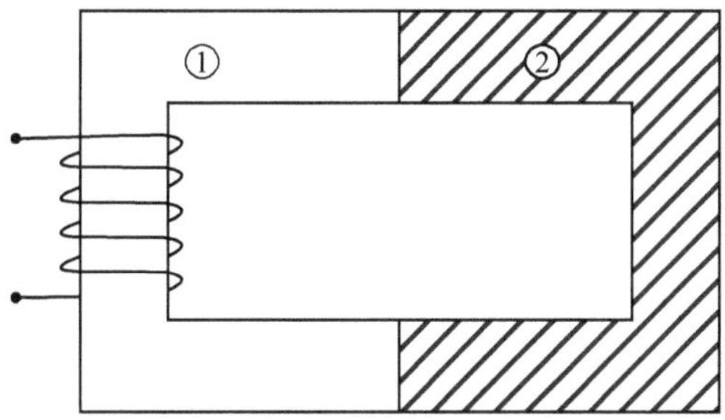

Figure 2.36. Magnetic structure of problem 2.10.

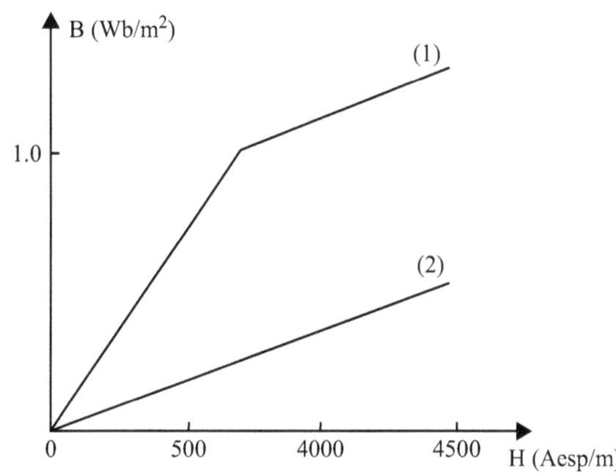

Figure 2.37. Magnetization characteristics of hypothetical materials.

Further reading

[1] Gourishankar V 1966 *Electromechanical Energy Conversion* (Scanton: International Textbook Company)
[2] Skilling H H 1962 *Electromechanics: A First Course in Electromechanical Energy Conversion* (New York: Wiley)
[3] Bansal R (ed) 2004 *Handbook of Engineering Electromagnetics* (New York: Marcel Decker)
[4] Bozorth R M 1993 *Ferromagnetism* (Piscataway, NJ: IEEE Press)
[5] Fitzgerald A E, Kingsley C Jr and Umans S D 1992 *Electric Machinery* 5th edn (New York: McGraw-Hill)

IOP Publishing

Electromechanical Energy Conversion Through Active Learning

J R Cardoso, M B C Salles and M C Costa

Chapter 3

Transformers

3.1 Introduction

If you were asked to name an electrical device that changed the world, this would likely be the *transformer*.

In the late 19th century the DC current versus AC current or Edison versa Tesla battles achieved their climax.

The most important inventor of all time, Thomas Alva Edison, advocated that DC current was both the most suitable and safest for distributing electric energy in cities. On the other hand, the Serbian Nikola Tesla who worked for Edison at the beginning of his career was fighting to prove that his AC current system was better.

As we know, the DC current is only viable for small distances between source and load due to the impossibility of compensating the voltage drop as the current flows in the feeder. In the AC system, this challenge was overcome with the transformer not only compensating for the voltage drop in the feeder, but also allowing the use of high voltage with low current.

With these properties, the AC system spread quickly in all USA cities, while the DC current became the electric system for public transportation due to its easy control.

The transformer operation is based on the *magnetic induction law* discovered by Michael Faraday on 29 August 1831, which established that an electromotive force appears in any electric circuit subjected to time-dependent magnetic flux, and it can induce an electric current that flows for avoiding the magnetic flux variation.

Some years later, James Clerk Maxwell translated it mathematically, as:

$$e = -\frac{d\lambda}{dt}(\text{V}) \tag{3.1}$$

with λ being *concatenate magnetic flux* with the circuit.

doi:10.1088/978-0-7503-2084-9ch3

Remarks: When the electric circuit has only one closed path the concatenate magnetic flux is identical to the magnetic flux (chapter 1). Therefore, if the electric circuit has N closed paths like a coil of N turns, the concatenate magnetic flux with the circuit is:

$$\lambda = N\varnothing \text{ (Wb.turn)} \tag{3.2}$$

Therefore, the 'magnetic induction law' is often written as:

$$e = -N\frac{d\varnothing}{dt}\text{(V)} \tag{3.3}$$

3.2 The ideal transformer

The simpler ideal transformer consists of a single reactor built with an ideal magnetic material whose magnetic permeability is very high, so we can consider its permeability infinite, i.e., ($\mu \rightarrow \infty$).

There are two ideal coils (no electrical resistance) around it with N_1 and N_2 turns normally called *primary* and *secondary*, respectively.

Figure 3.1 shows such a device. As the sense of winding is important, we must introduce the concept of *polarity* of the coils.

Polarity: if injecting electrical current into two different terminals (one of each coil) and the magnetic flux produced by each current is concordant, we state that both terminals of different coils have the same 'polarity'. The terminals of the same polarity are indicated by the same symbol, i.e. points, triangles and so on.

3.2.1 Conventions

The positive quantities of the ideal transformer are established according the *receptor* convention in both coils as shown in figure 3.2. In the receptor convention the *positive* power is the one introduced in the transformer.

The quantities involved in the transformer operation are:

$\varnothing$: magnetic flux in the core;

v_1 and v_2: voltages at the primary and secondary sides;

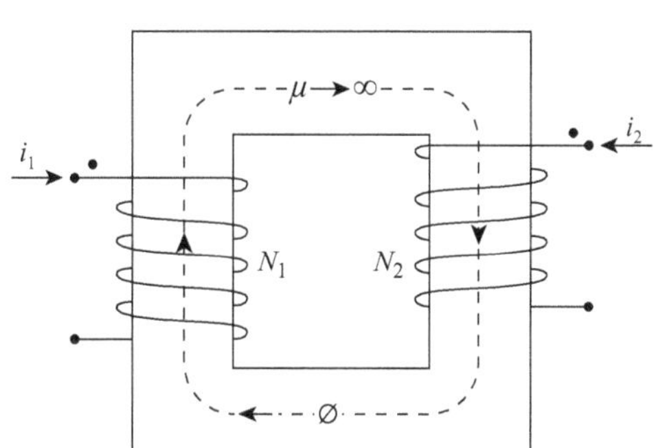

Figure 3.1. The ideal transformer $\mu \rightarrow \infty$ and $r_1 = r_2 = 0$. Terminals with the same polarity have the same symbol.

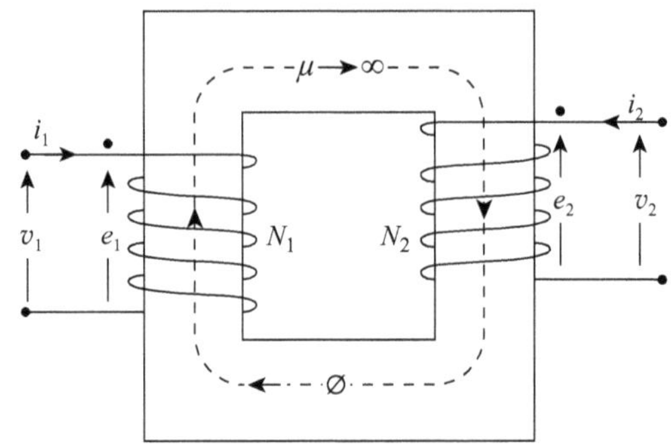

Figure 3.2. Receptor convention for v and i. Input power is positive. Output power is negative.

i_1 and i_2: electric currents at the primary and secondary sides;
e_1 and e_2: emf induced by Ø in both coils.

3.2.2 The electromotive force relation

According to Faraday's Law, we can write the following relations for each coil:

$$e_1 = -N_1 \frac{d\varnothing}{dt} \text{(V)} \tag{3.4}$$

$$e_2 = -N_2 \frac{d\varnothing}{dt} \text{(V)} \tag{3.5}$$

In (3.4) and (3.5) the magnetic flux is the same because the magnetic permeability is infinite that results in all magnetic flux being confined in the core and no leakage magnetic flux should be considered.

From (3.4) and (3.5) we get:

$$\frac{e_1}{e_2} = \frac{N_1}{N_2} = a \tag{3.6}$$

where $a = \frac{N_1}{N_2}$ is called the transformer relation.

3.2.3 The voltage relation

As the resistance of the coils is zero, on each side of the transformer we can write:

$$v_1 = e_1 \text{ and } v_2 = e_2$$

therefore:

$$\frac{v_1}{v_2} = \frac{e_1}{e_2} = \frac{N_1}{N_2} = a \tag{3.7}$$

3.2.4 The electric current relation

The relation between the electric current in both coils is obtained by solving the magnetic circuit of the core.

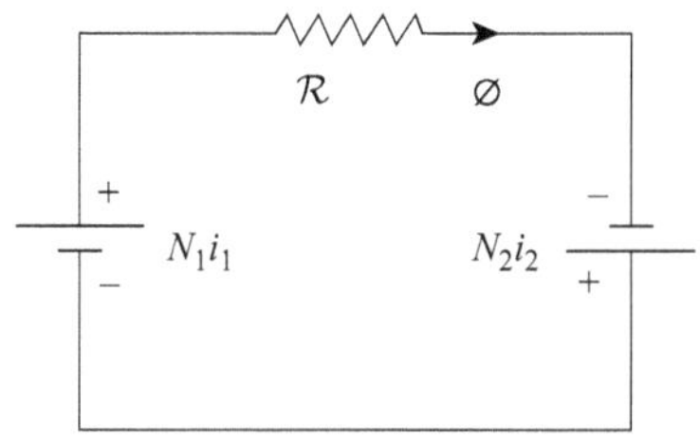

Figure 3.3. Analog electric circuit. $\mathcal{R} = \frac{l_m}{\mu S}$ is the reluctance of the magnetic circuit.

Figure 3.3 shows the analog electric circuit of the magnetic structure of the transformer. Applying the right-hand rule, the sense of mmfs are given by the polarity of fictitious batteries $N_1 i_1$ and $N_2 i_2$.

Solving the analog electric circuit of figure 3.3 we get:

$$N_1 i_1 + N_2 i_2 = \mathcal{R}\varnothing \tag{3.8}$$

The magnetic permeability is infinite $\mathcal{R} = 0$, therefore:

$$\frac{i_1}{i_2} = -\frac{N_2}{N_1} = -\frac{1}{a} \tag{3.9}$$

3.2.5 The electric power relation

The instant power developed on each side of the transformer is given by:

$$p_1 = v_1 i_1 \tag{3.10}$$

$$p_2 = v_2 i_2 \tag{3.11}$$

From (3.7) and (3.9), we get:

$$v_1 = a v_2$$
$$i_1 = -\frac{i_2}{a}$$

Substituting in (3.10) results in:

$$p_1 = a v_2 \left(-\frac{i_2}{a}\right) = -p_2 \tag{3.12}$$

Based on the basic electrical relations of the ideal transformer:

$$\begin{cases} \dfrac{v_1}{v_2} = \dfrac{e_1}{e_2} = \dfrac{N_1}{N_2} = a \\ \dfrac{i_1}{i_2} = -\dfrac{N_2}{N_1} = -\dfrac{1}{a} \\ p_1 = -p_2 \end{cases}$$

we conclude:

(1) The voltages v_1 and v_2 always have the same signals.
(2) The electric currents i_1 and i_2 have the opposite signals, so it is impossible to have both currents entering or leaving the terminals of the same polarity at the same instant.
(3) The power flowing on each coil has the inverse sense, so no energy is stored or dissipated in an ideal transformer.

3.2.6 The transformer under load

Let us suppose that an impedance is connected on the secondary side of the transformer. Immediately afterwards, an electric current (i_L) will flow through the load. The sense of this current is opposite to the positive sense chosen for the transformer model of figure 3.2, as shown in figure 3.4.

It is convenient to change the convention of the secondary side from the receptor to the generator. For adapting the secondary side to the generator convention is enough to change the sense of the current. For that we must to work with (i_L) instead of (i_2).

As $i_L = -i_2$ the main relations of the ideal transformer become:

$$\begin{cases} \dfrac{v_1}{v_2} = \dfrac{e_1}{e_2} = a \\ \dfrac{i_1}{i_L} = \dfrac{1}{a} \\ p_1 = p_L \end{cases}$$

Applying Ohm's Law to the load, we have:

$$v_2 = z_L i_L \tag{3.13}$$

Using the basic relations in an ideal transformer, Ohm's Law can be rewritten as:

$$\frac{v_1}{a} = z_L a i_1$$

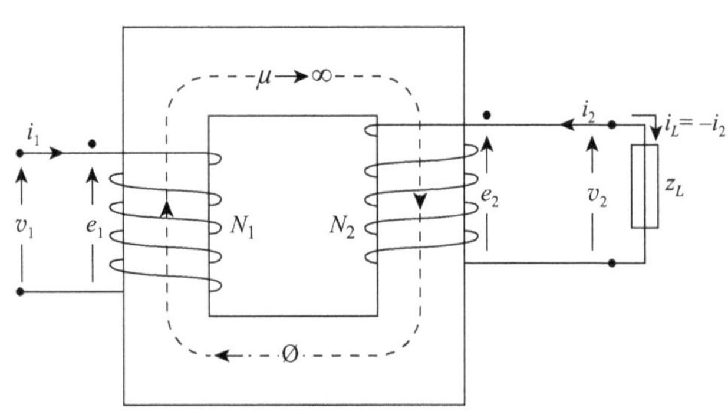

Figure 3.4. Transformer under load. The electric current of load is opposite of i_2 – ($i_L = -i_2$).

or,

$$v_1 = a^2 z_L i_1 \tag{3.14}$$

The relation,

$$\frac{v_1}{i_1} = a^2 z_L$$

is Ohm's Law applied to the primary side. That is, the ideal transformer feeding an impedance z_L is equivalent to an impedance:

$$Z_L' = a^2 Z_L \tag{3.15}$$

This impedance is called the *referred impedance* to the primary (figure 3.5).

Example 3.1

The transformer shown in figure 3.6 is ideal ($\mu \to \infty$). It is wound with three ideal coils ($r_1 = r_2 = r_3 = 0$) of $N_1 = 100$ turns, $N_2 = 50$ turns and $N_3 = 150$ turns, respectively. The polarity terminal of a 100 turns coil is given. A time-dependent voltage is applied to the 100 turns coil and at instant t_1 its value is $v_1(t_1) = 150$ V. Determine:

(a) the polarity terminal of both 50 turns *and* 150 turns' coils;
(b) the voltages $v_2(t_1)$ *and* $v_3(t_1)$;
(c) the current $i_1(t_1)$ when switch S_1 is closed and S_2 is open;

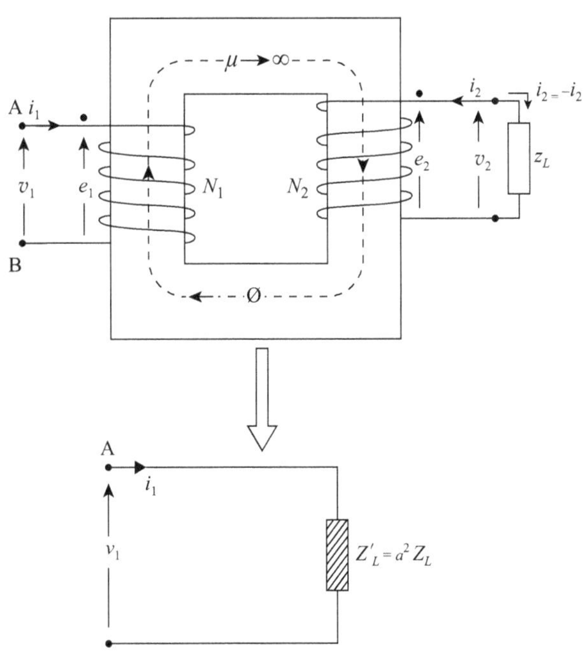

Figure 3.5. Referred impedance to the primary. Referred impedance is equivalent to *ideal transformer + impedance of load.*

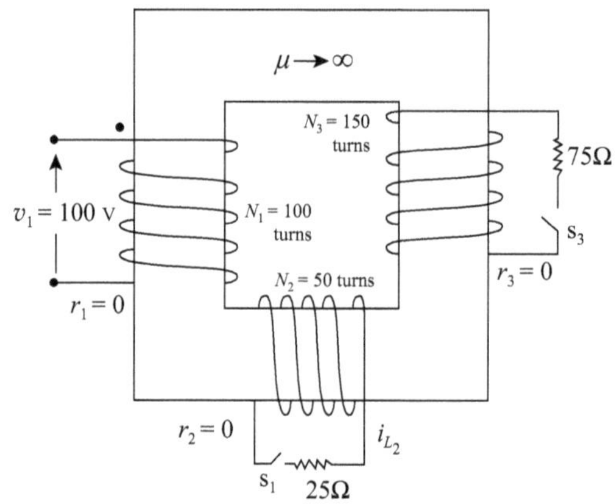

Figure 3.6. Three coils ideal transformer.

(d) the current $i_1(t_1)$ when switch S_2 is closed and S_1 is open;
(e) the current $i_1(t_1)$ when both switches are closed;
(f) the power supplied to the transformer in the last three cases.

Solutions

(a) The polarities of the coils are determined by injecting the electrical current in any terminal of the other coils and observing if the sense of magnetic flux is according to the one produced by the reference coil (figure 3.7).

(b) The voltages of coils are obtained for this transformation by applying the basic relations in an ideal transformer:

$$\frac{v_1}{v_2} = \frac{100}{50}$$

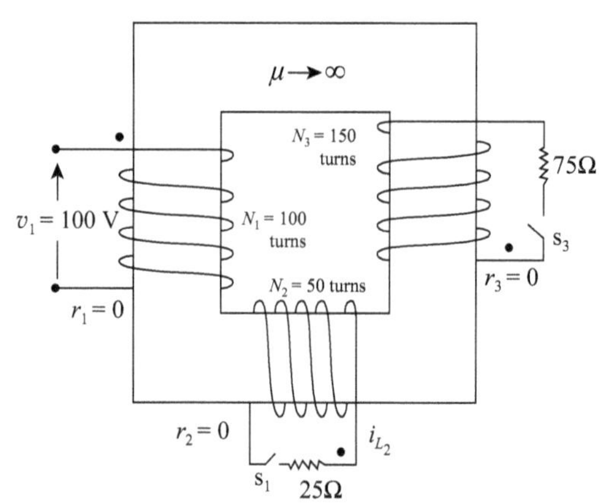

Figure 3.7. Polarities.

As $v_1 = 150$ V results $v_2 = 75$ V.

For the other coil:

$$\frac{v_1}{v_3} = \frac{100}{150}$$

As $v_1 = 150$ V results $v_3 = 225$ V.

(c) With the switch S_1 closed the electrical current in the coil is:

$$i_{L2} = \frac{v_2}{25} = \frac{75}{25} = 3\text{A}$$

Applying the basic relations in an ideal transformer:

$$\frac{i_1}{i_{L2}} = \frac{50}{100} \rightarrow i_1 = 1.5\text{A}$$

(d) With the switch S_2 closed the electrical current in the coil is:

$$i_{L3} = \frac{v_3}{75} = \frac{225}{75} = 3\text{A}$$

Applying the basic relations in an ideal transformer:

$$\frac{i_1}{i_{L3}} = \frac{150}{100} \rightarrow i_1 = 4.5\text{A}$$

(e) With both switches closed, the total current in the primary is:

$$i_1 = 1.5 + 4.5 = 6\text{A}$$

Remarks: The total mmf in an ideal transformer is always zero.

(f) The power supplied to the transformer is:

Case b: $p_1 = v_1 i_1 = 150 \times 1.5 = 225$ W

Case c: $p_1 = v_1 i_1 = 150 \times 4.5 = 675$ W

Case d: $p_1 = v_1 i_1 = 150 \times 6 = 900$ W

3.3 A non-ideal linear transformer

This kind of transformer has the following *imperfections* compared to an ideal transformer:

(a) The magnetic permeability is finite and constant;
(b) the coil's resistance is non-null.

Consequently, two effects should be considered:

(i) As the magnetic permeability is finite, the total amount of magnetic flux produced by the coils is not confined in the core. Some parcels of it will be closed in the air.

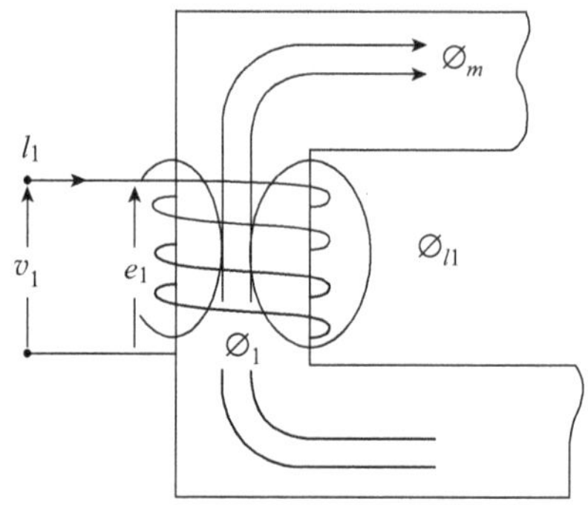

Figure 3.8. The magnetic fluxes on the primary coil. $\varnothing_1$: the total amount of magnetic flux that cross the coil 1. $\varnothing_m$: the mutual magnetic flux. $\varnothing_{l1}$: the primary leakage magnetic flux.

(ii) As the resistances are non-null the induced electromotive force is different to the terminal voltage of the coil.

Firstly, let us discuss the different kind of magnetic fluxes that are involved in the transformer operation. Figure 3.8 shows the primary coil of the transformer excited by an external source of voltage v_1. The total magnetic flux produced in the coil is $\varnothing_1$. As the magnetic flux leaves the coil it is divided in two parts $\varnothing_m$ and $\varnothing_{l1}$, where:

$\varnothing_m$: is the mutual magnetic flux, practically confined in the core as we will see later;

$\varnothing_{l1}$: is the primary leakage magnetic flux. This parcel of $\varnothing_1$ has no magnetic coupling with the secondary coil.

Concerning the emf induced in the primary we can write:

$$e_1 = N_1\frac{d\varnothing_1}{dt} = N_1\frac{d\varnothing_m}{dt} + N_1\frac{d\varnothing_{l1}}{dt} \tag{3.16}$$

The total emf could be supposed as composed of two emfs. The first one is due to the mutual magnetic flux $\varnothing_m$ given by:

$$e_{m1} = N_1\frac{d\varnothing_m}{dt} \tag{3.17}$$

called the induced emf on the primary due to the mutual magnetic flux.

The other one is due to the leakage magnetic flux $\varnothing_{l1}$ given by:

$$e_{l1} = N_1\frac{d\varnothing_{l1}}{dt} \tag{3.18}$$

called the induced emf on the primary due to the leakage magnetic flux.

Based on this consideration, we can magnetically decouple both emfs supposing that $\varnothing_m$ is the only magnetic flux in the core and the effect of $\varnothing_{l1}$ is emulated by an external N_1 turns coil of the transformer as shown in figure 3.9.

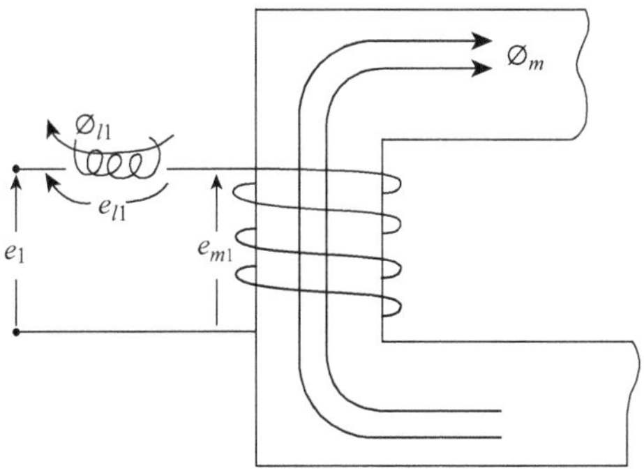

Figure 3.9. The induced emf in the coil 1. e_{m1}: the induced emf due to the mutual magnetic flux. e_{l1}: the induced emf due to the primary leakage magnetic flux.

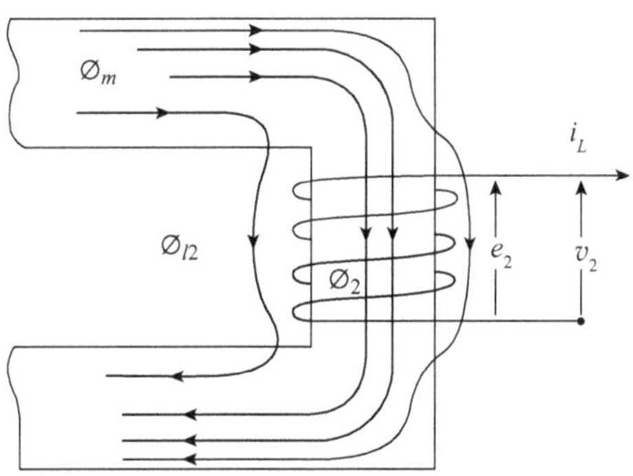

Figure 3.10. The magnetic fluxes on the secondary coil. $\varnothing_2$: the total amount of magnetic flux that cross the coil 2. $\varnothing_m$: the mutual magnetic flux. $\varnothing_{l2}$: the secondary leakage magnetic flux.

Figure 3.10 shows the secondary coils of the transformer that are crossed by the magnetic flux $\varnothing_2$. The magnetic flux $\varnothing_2$ is a parcel of $\varnothing_m$ because the effect of the secondary reaction of the load current spreads the flux lines outside the coil creating the secondary magnetic leakage flux $\varnothing_{l2}$. So, we can write:

$$\varnothing_2 = \varnothing_m - \varnothing_{l2} \tag{3.19}$$

Concerning the emf induced in the secondary coil we can write:

$$e_2 = N_2\frac{d\varnothing_2}{dt} = N_2\frac{d\varnothing_m}{dt} - N_2\frac{d\varnothing_{l2}}{dt} \tag{3.20}$$

The total emf could be supposed to be composed of the difference in the two emfs. The first is due to the mutual magnetic flux $\varnothing_m$ given by:

$$e_{m2} = N_2\frac{d\varnothing_m}{dt} \tag{3.21}$$

called the induced emf on the secondary coil due to the mutual magnetic flux.

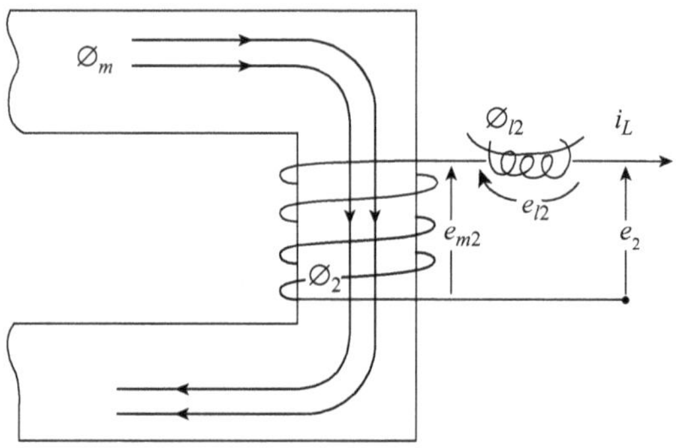

Figure 3.11. The induced emf in the coil 2. e_{m2}: the induced emf due to the mutual magnetic flux. e_{l2}: the induced emf due to the secondary leakage magnetic flux.

The other one is due to the leakage magnetic flux $\varnothing_{l2}$ given by:

$$e_{l2} = N_2 \frac{d\varnothing_{l2}}{dt} \tag{3.22}$$

called the induced emf on the secondary due to the leakage magnetic flux.

Using the same approach, we can magnetically decouple both emfs supposing that $\varnothing_m$ is the only magnetic flux in the core and the effect of $\varnothing_{l2}$ is emulated by an external N_2 turns coil of the transformer, as shown in figure 3.11.

3.3.1 The leakage inductances

As the leakage magnetic flux is produced by the electrical current in the coil, it is possible to associate the emf due to the leakage magnetic flux to the drop voltage in an inductance defined as:

$$e_{l1} = N_1 \frac{d\varnothing_{l1}}{dt} = L_{d1} \frac{di_1}{dt} \tag{3.23}$$

and

$$e_{l2} = N_2 \frac{d\varnothing_{l2}}{dt} = L_{d2} \frac{di_L}{dt} \tag{3.24}$$

where:

$$L_{d1} i_1 = N_1 \varnothing_{l1} \tag{3.25}$$

and

$$L_{d2} i_L = N_2 \varnothing_{l2} \tag{3.26}$$

Figure 3.12 shows a partial equivalent electric circuit of a non-ideal linear transformer that includes the representation of the electrical resistances of each coil.

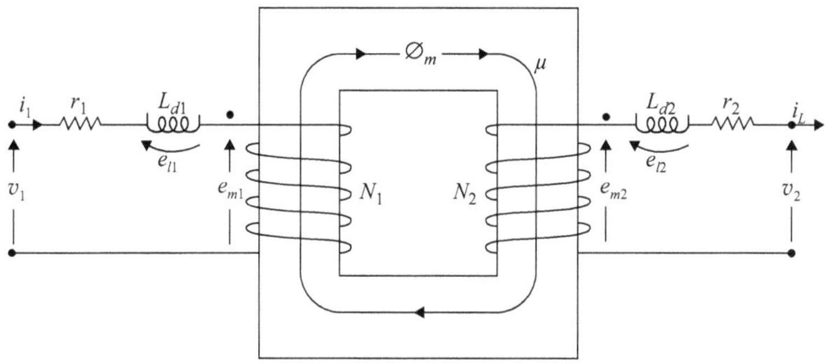

Figure 3.12. The partial equivalent electric circuit of a non-ideal linear transformer. L_{d1}: leakage inductance of the primary. L_{d2}: leakage inductance of the secondary. r_1 and r_2: electrical resistance of both primary and secondary.

This electric circuit represents some imperfections of a non-ideal linear transformer compared to an ideal one through the voltage drop in electrical parameters, vis-a-vis, L_{d1}, L_{d2}, r_1 and r_2.

3.3.2 The magnetization inductance

The only imperfection of a non-ideal linear transformer compared to an ideal one that we need to consider is the magnetic permeability. In the ideal transformer the magnetic permeability is such that $\mu \to \infty$.

In this case, the total amount of mmf required to produce a magnetic flux in the core is null.

$$N_1 i_1 - N_2 i_L = 0 \tag{3.27}$$

As the magnetic permeability of the core is finite, the total amount of mmf required to produce $\varnothing_m$ is non-null. Therefore, we can write:

$$N_1 i_1 - N_2 i_L = \mathcal{R}\varnothing_m \tag{3.28}$$

where $\mathcal{R} = \frac{1}{\mu}\frac{l}{S}$ is the magnetic reluctance of the magnetic circuit.

As we did with the leakage magnetic flux, we will emulate the effect of the finite permeability of the core by an electric parameter outside the transformer. For that, we will split the primary electric current in two parcels, such that:

$$i_1 = i_m + i_2' \tag{3.29}$$

Therefore, we can rewrite (3.28) as:

$$N_1 i_m + N_1 i_2' - N_2 i_L = \mathcal{R}\varnothing_m$$

Choosing i_2' such that:

$$N_1 i_2' - N_2 i_L = 0 \tag{3.30}$$

results in:

$$N_1 i_m = \mathcal{R}\varnothing_m \tag{3.31}$$

or,

$$i_m = \frac{\mathcal{R}\varnothing_m}{N_1} = \frac{N_1\varnothing_m}{\frac{N_1^2}{\mathcal{R}}}$$

Finally,

$$i_m = \frac{\lambda_m}{L_m} \tag{3.32}$$

where,

$$L_m = \frac{N_1^2}{\mathcal{R}} \tag{3.33}$$

is called the *magnetization inductance* of the transformer.

Locating this inductance outside the transformer to emulate the imperfection of finite magnetic permeability, the equivalent electric circuit is shown in figure 3.13.

After representing all imperfections of the non-ideal linear transformer through electrical parameters outside the magnetic structure, the resulting transformer became an ideal transformer. Therefore, the secondary parameters could be referred (reflected) to the primary as usual. So, the final equivalent electric circuit of a non-ideal linear transformer is the one shown in figure 3.14.

3.3.3 The sinusoidal steady-state—the AC system

The transformer was created at the end of 19th century to operate under the alternating current system which had recently been invented by Nikola Tesla. In an AC system, all voltages and currents vary sinusoidally.

The analysis of the transformer operation under sinusoidal steady-state conditions can be done by using complex notation.

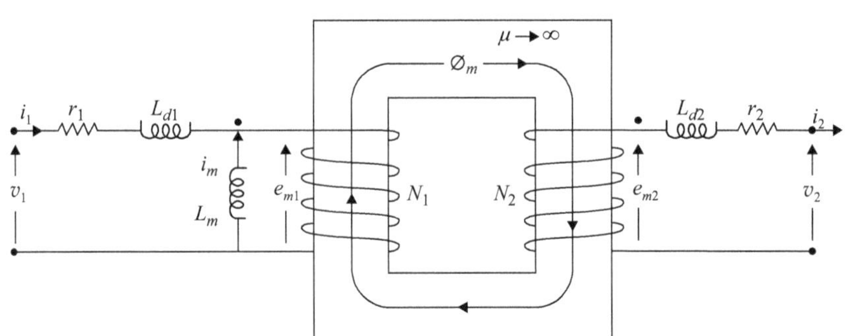

Figure 3.13. Emulating the magnetic permeability. L_m: the magnetization inductance. The transformer in this figure is an ideal transformer.

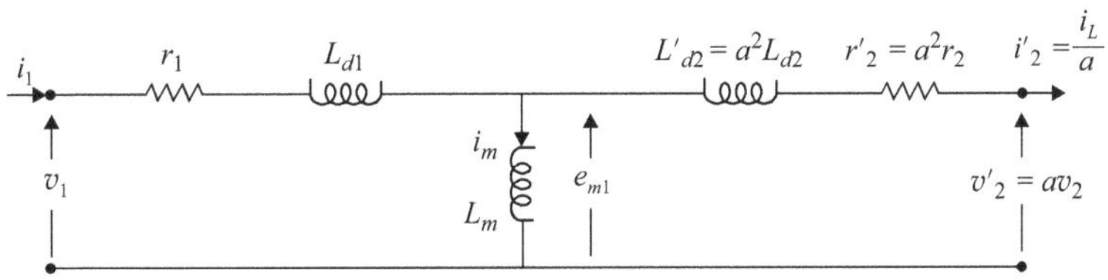

Figure 3.14. Equivalent electric circuit of a non-ideal linear transformer. All parameters referred to the primary.

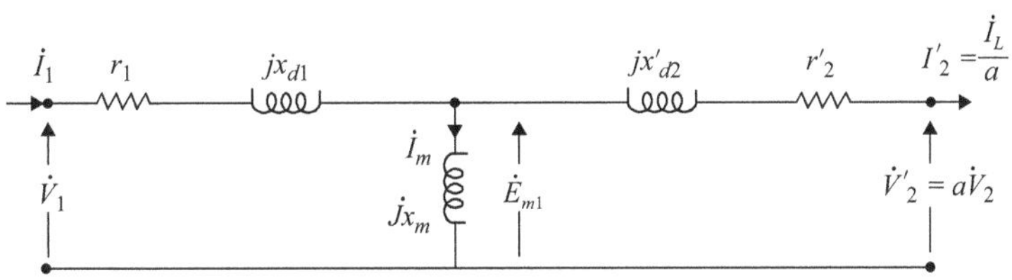

Figure 3.15. Equivalent electric circuit of a non-ideal linear transformer—complex notation. All parameters are represented by their impedance. All voltages and currents are represented by their phasor.

In complex notation, both voltages and currents are represented by their 'phasor' and all electrical parameters by their 'impedance'.

Applying complex notation to the electric circuit of figure 3.14, the equivalent electric circuit of the transformer is shown in figure 3.15.

The parameters of the equivalent electric circuit in complex notation are:

r_1: resistance of primary;
$r_2' = a^2r_2$: resistance of the secondary coil *referred* to the primary;
$x_{d1} = \omega L_{d1}$: leakage reactance of the primary;
$x_{d2}' = a^2\omega L_{d2}$: leakage reactance of the secondary *referred* to the primary;
$x_m = \omega L_m$: magnetization reactance;
$\dot{V}_1$, $\dot{I}_1$: phasors of both voltage and current of the primary;
$\dot{V}_2' = a\dot{V}_2$: phasor of the secondary voltage *referred* to the primary;
$\dot{I}_2' = \frac{\dot{I}_L}{a}$: phasor of the load current *referred* to the primary;
$\dot{E}_{m1}$: phasor of induced emf of the primary.

3.3.4 The induced electromotive force

The induced emf E_{m1} is evaluated by the application of Faraday's Law.

As the mutual magnetic flux is a sinusoidal function, we can write:

$$\varnothing_m(t) = \varnothing_{\max} \text{ sen } \omega t \tag{3.34}$$

Applying Faraday' law:

$$e_{m1} = -N_1\frac{d\varnothing_m}{dt} \tag{3.35}$$

results in:

$$e_{m1} = \omega N_1 \varnothing_{max} \cos \omega t \tag{3.36}$$

whose rms value is:

$$E_{m1} = \frac{\omega N_1 \varnothing_{max}}{\sqrt{2}} = 4.44 f N_1 \varnothing_{max} \tag{3.37}$$

3.3.5 Approximate equivalent electric circuit

Depending on the study involving the existence of a transformer we can simplify the equivalent electric circuit.

As both the resistance and leakage reactance are very small compared to the magnetization reactance, we can move it to the primary terminal as shown in figure 3.16. This model is commonly called the 'constant flux model'.

In this modified equivalent electric circuit, there are two new parameters:

$r_{sc} = r_1 + r_2'$: the short-circuit resistance;

$x_{sc} = x_{d1} + x_{d2}'$: the short-circuit reactance.

Another approach is to neglect the magnetization current comparing it with the load current because the magnetization reactance can reach more than 100 times the total series impedance of both the primary and secondary. The equivalent electric circuit is shown in figure 3.17.

In power system analysis, it is very common to use a simpler equivalent electric circuit where only the short-circuit reactance is considered, as shown in figure 3.18.

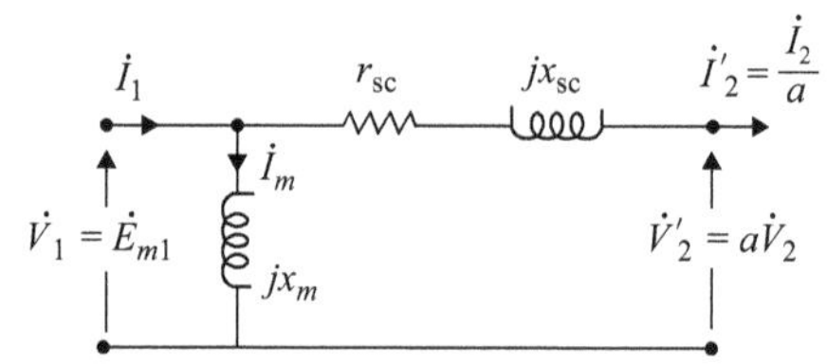

Figure 3.16. Constant flux model. r_{sc}: the short-circuit resistance. x_{sc}: the short-circuit reactance.

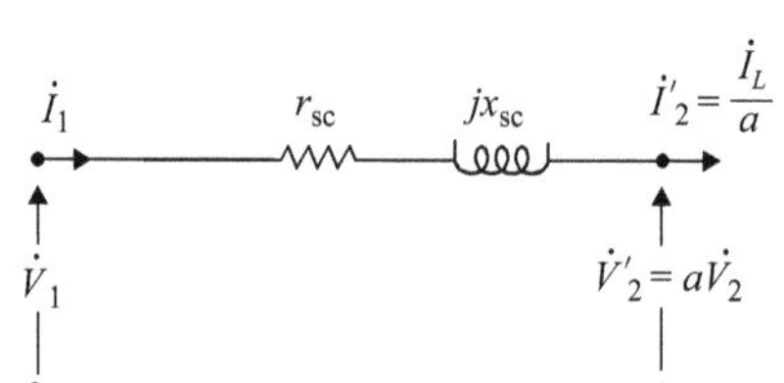

Figure 3.17. The modified equivalent electric circuit. The magnetization current is very small compared to the load current.

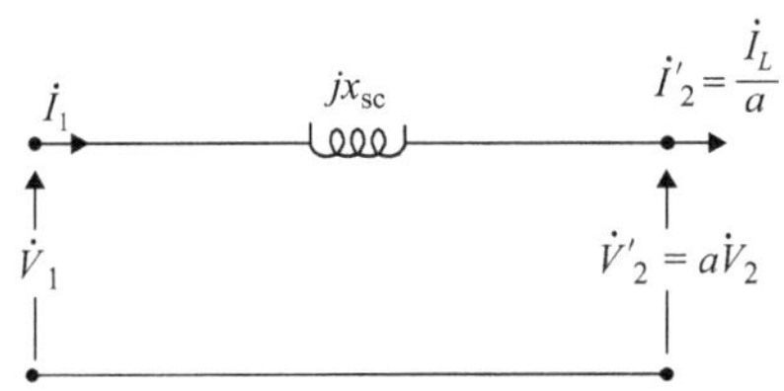

Figure 3.18. Equivalent electric circuit for power system analysis.

3.3.6 The rated values of a transformer

All transformers are designed to operate under specific conditions. Those conditions are commonly called the 'rated values' of the transformer. The most important rated quantities are:

(a) Rated voltage: this is the maximum voltage of both the primary and secondary at which the transformer can work without reducing its life expectancy.
(b) Rated current: this is the maximum current of both the primary and secondary at which the transformer can work without reducing its life expectancy.
(c) Transformer volt–ampere rating: this is the maximum apparent power that the transformer can deliver without reducing its life expectancy.

The overload operation is limited to a few percent of its rated condition accompanied with an important reduction of its life expectancy.

Example 3.2

The non-ideal linear transformer of figure 3.19 is excited by a 60 Hz AC voltage source. It is wound with three coils of $N_1 = 100$ turns, $N_2 = 50$ turns and $N_3 = 150$ turns, respectively. The electrical parameters of this transformer are:

Side 1	Side 2	Side 3
$r_1 = 0.001\ \Omega$	$r_2 = 0.00025\ \Omega$	$r_3 = 0.00225\ \Omega$
$L_m = 0.40$ H	$L_{d2} = 0.0005$ H	$L_{d3} = 0.0045$ H
$L_{d1} = 0.002$ H		

A mutual magnetic flux of magnitude 5.6×10^{-3} Wb is produced when the coil 1 is fed by the source. Determine:

(a) the induced electromotive force in all coils;
(b) the equivalent electric circuit when the primary is coil 1 and the secondary is coil 2;
(c) the equivalent electric circuit when the primary is coil 2 and the secondary is coil 3;
(d) using the first arrangement both the voltage and the current in the primary when a resistive load of 10 Ω is connected in the secondary.

Solution

(a) The induced emf

According to equation (3.37), the rms values of the electromotive forces are:

$$E_{m1} = 4.44fN_1\varnothing_{max} = 4.44 \times 60 \times 100 \times 5.6 \times 10^{-3} = 149.2 \text{ V}$$
$$E_{m2} = 4.44fN_2\varnothing_{max} = 4.44 \times 60 \times 50 \times 5.6 \times 10^{-3} = 74.6 \text{ V}$$
$$E_{m3} = 4.44fN_3\varnothing_{max} = 4.44 \times 60 \times 150 \times 5.6 \times 10^{-3} = 223.8 \text{ V}$$

(b) The parameters of the equivalent electric circuit when $N_1 = 100$ turns and $N_2 = 50$ turns are:

As the transformer relation is

$$a = \frac{100}{50} = 2$$

Results (figure 3.20):

$r_1 = 0.001\ \Omega$: resistance of the primary;
$r_2' = a^2 r_2 = 2^2 \times 0.0025 = 0.001\ \Omega$: resistance of the secondary referred to the primary;
$x_m = \omega L_m = 377 \times 0.4 = 150.8\ \Omega$: magnetization reactance;
$x_{d1} = \omega L_{d1} = 377 \times 0.002 = 0.754\ \Omega$: leakage reactance of the primary;

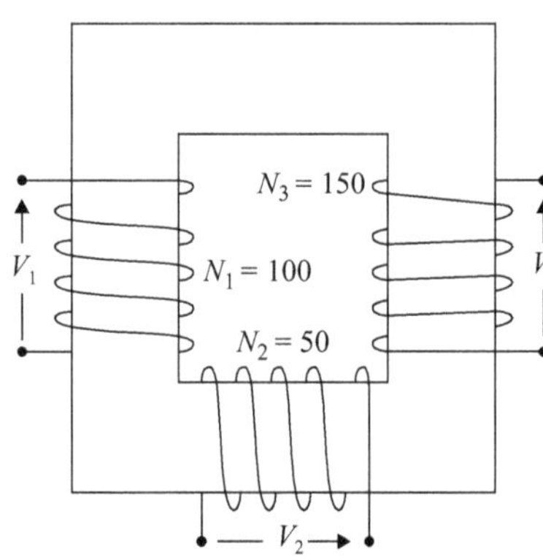

Figure 3.19. Transformer of example 3.3.7.

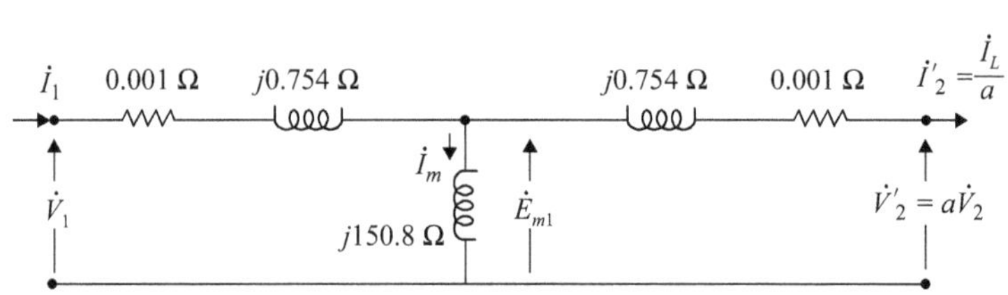

Figure 3.20. The equivalent electric circuit referred to side 1.

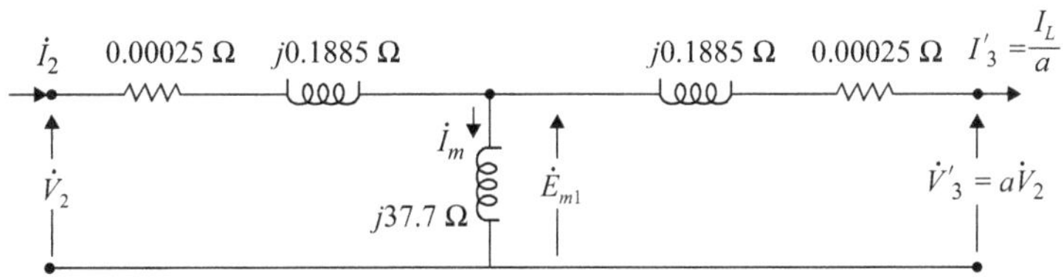

Figure 3.21. The equivalent electric circuit referred to side 2.

$x'_{d2} = a^2\omega L_{d2} = 2^2 \times 377 \times 0.0005 = 0.754\ \Omega$: leakage reactance of the secondary referred to the primary.

(c) The parameters of the equivalent electric circuit when $N_2 = 50$ turns and $N_3 = 150$ turns are:

As the transformer relation is

$$a = \frac{50}{150} = 0.3333$$

Results (figure 3.21):

$r_2 = 0.000\ 25\ \Omega$: resistance of the primary;

$r'_3 = a^2 r_3 = (0.333)^2 \times 0.002\ 25 = 0.000\ 25\ \Omega$: resistance of the secondary referred to the primary;

As $L_m = 0.40$ H referred to the coil of $N_1 = 100$ turns, the same inductance referred to the coil of $N_2 = 50$ turns will be:

$$L_{m2} = \left(\frac{50}{100}\right)^2 \times 0.40 = 0.10\ \text{H}$$

Therefore:

$x_{m2} = \omega L_{m2} = 377 \times 0.1 = 37.7\ \Omega$: magnetization reactance;

$x_{d2} = \omega L_{d2} = 377 \times 0.0005 = 0.1885\ \Omega$: leakage reactance of the primary;

$x'_{d3} = a^2\omega L_{d3} = (0.333)^2 \times 377 \times 0.0045 = 0.1885\ \Omega$: leakage reactance of the secondary referred to the primary.

(d) The voltage and current in the primary when a resistive load of 10 Ω is connected in the secondary for the first arrangement.

As $a = 2$ the resistive load of 10 Ω referred to the primary is:

$$Z'_L = a^2 Z_L = 2^2 \times 10 = 40\ \Omega$$

Therefore, the equivalent electric circuit is given in figure 3.22.

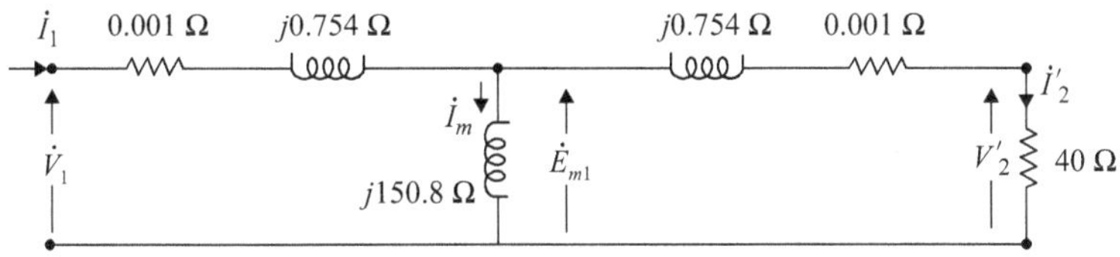

Figure 3.22. Equivalent electric circuit with secondary load.

As E_{m1} = 149.2 V results:

$$\dot{I}_2' = \frac{149.2\angle 0^\circ}{0.001 + j0.754 + 40} = 3.73\angle -1.08^\circ(\text{A})$$

$$\dot{I}_m = \frac{149.2\angle 0^\circ}{j150.8} = 0.99\angle -90^\circ$$

$$\dot{I}_1 = \dot{I}_m + \dot{I}_2' = 3.73\angle -1.08^\circ + 0.99\angle -90^\circ = 3.88\angle -15.9^\circ(\text{A})$$

The primary voltage is given by:

$$\dot{V}_1 = (0.001 + j0.754) \times 3.88\angle -15.9^\circ + 149.2\angle 0^\circ$$

$$\dot{V}_1 = 150\angle 1.07^\circ(\text{V})$$

3.4 The real transformer

3.4.1 Constructive aspects

The real transformer has a core made of a very thin sheet of iron–silicon alloy (~95% Fe and ~0.5% Si) (figure 3.23). The presence of Si is important for increasing the resistivity of the sheet steel that minimizes the induced current due to the magnetic flux time variation as we will see later.

Both coils from the primary and the secondary are normally wound on the same leg with the high-voltage coil enclosing the low-voltage one. This kind of assemblage is very suitable for minimizing the leakage magnetic flux.

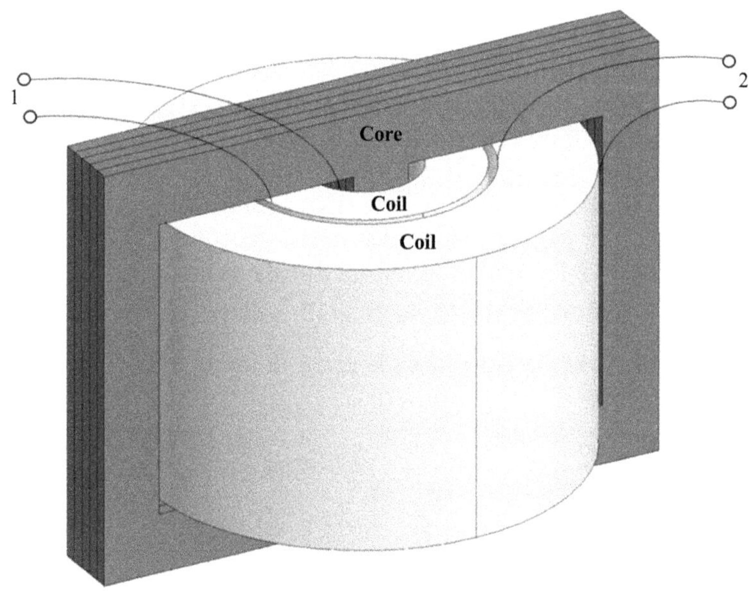

Figure 3.23. Typical monophasic real transformer.

Figure 3.24. The real power transformer. Source: Courtesy of M S Wilerson Calil, from a personal collection.

The power transformer with more than tens of kVA is placed inside an oil tank whose function is to make the power loss dissipation easier (figure 3.24).

3.4.2 The iron losses

The iron loss is divided into hysteresis loss, eddy-current loss and excess loss. The excess loss is beyond the scope of this book and corresponds to the anomalous loss component, which is influenced by intricate phenomena, such as microstructural interactions, magnetic anisotropy, and nonhomogeneous locally induced eddy-currents.

3.4.2.1 The hysteresis losses

As we discussed in chapter 2, the iron sheet is composed of a very expressive number of magnetic domains. When a magnetic field is impressed by an alternative electric current, the movement and rotations of the magnetic domains' walls are observed.

Figure 3.25 shows two different orientations of magnetic domains for two different values of electrical current.

This movement promotes the losses of energy due to the friction in the wall of the magnetic domain.

The evaluation of hysteresis power loss was firstly proposed by the German engineer Charles Steinmetz who encouraged the use of the alternating electrical current in the USA.

He observed that the variation of the magnetic flux density related to magnetic field is described by hysteresis curve, as shown in figure 3.26.

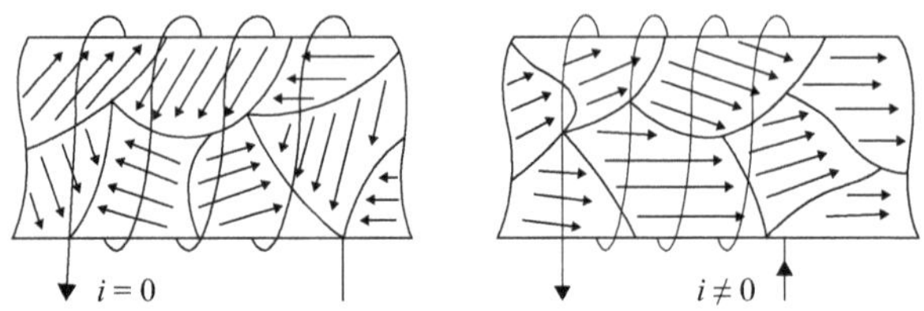

Figure 3.25. The movement and rotation of magnetic domain walls under time variation of the electrical current excitation.

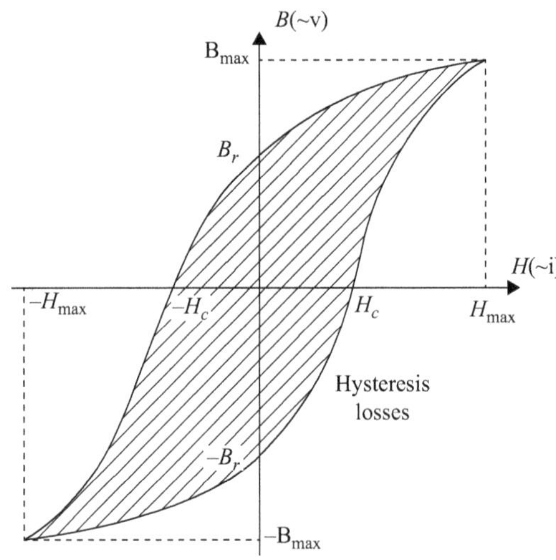

Figure 3.26. The hysteresis curve of sheet steel for transformer at sinusoidal excitation. The loss energy per m^3 at each cycle is the inside area of the curve. B is associated with V–H is associated with i.

The well-known Steinmetz formula for the hysteresis power loss evaluation is:

$$P_{\text{hist}} = K_{\text{hist}} \times \text{Vol} \times f \times B_{\max}^{n} \tag{3.38}$$

where:

K_{hist}: coefficient that depends on the material;
Vol: the volume of the magnetic core;
f: the excitation frequency;
$B_{\max}$: the magnitude of the magnetic flux density.

The range of application for this expression is limited by $1.2 < B_{\max} < 1.4$ for $1.5 < n < 2.5$.

Neglecting both the resistance of coil and the leakage reactance, the applied voltage on the primary of the transformer at no-load can be expressed as:

$$v(t) \cong \omega N_1 \varnothing_{\max} \cos \omega t \tag{3.39}$$

As $\varnothing(t) = \varnothing_{max} \text{sen}\, \omega t$, the wave form of $v(t)$ has the same wave form of $\varnothing(t)$. On the other hand, the wave form of $\varnothing(t)$ has the same wave form of $B(t)$ because $\varnothing(t) = B(t) \times S$.

Based on the constitutive relation $B(t) = \mu H(t)$, the wave form of $H(t)$ can be represented by:

First case: μ = constant
The wave shape of $H(t)$ is the same as $B(t)$. As $N_1 i(t) = H(t)l$, we concluded that the wave shape of $i(t)$ has the same form as $v(t)$.

Second case: the relationship $B(t) \times H(t)$ is defined by the hysteresis curve.
In this case, the wave shape of the primary electrical current $i(t)$ of a transformer at no-load is completely different from the wave form of $v(t)$.

Figure 3.27 shows the no-load current of both types of transformer, the non-ideal (linear) and the real one (non-linear).

This effect on a real transformer is normally neglected because the magnitude of the no-load current is a few percent (1%–3%) of the rated current.

3.4.2.2 The eddy-current losses
The eddy-current (also called Foucault's current) phenomenon is associated with the induced current in a conductor material subjected to time-dependent magnetic flux density according to Faraday's Law.

The time variation of the magnetic flux density $B(t)$ inside the ferromagnetic material produces an electrical current distribution comprised of concentric loops as shown in figure 3.28.

A Joule loss associated with this electrical current distribution is responsible for heating the core. To minimize this, very thin sheet steel—tenths of millimeters thick—is used. This procedure limits the establishment of current loops, reducing the eddy-current intensity. Another precaution for minimizing the eddy-currents is

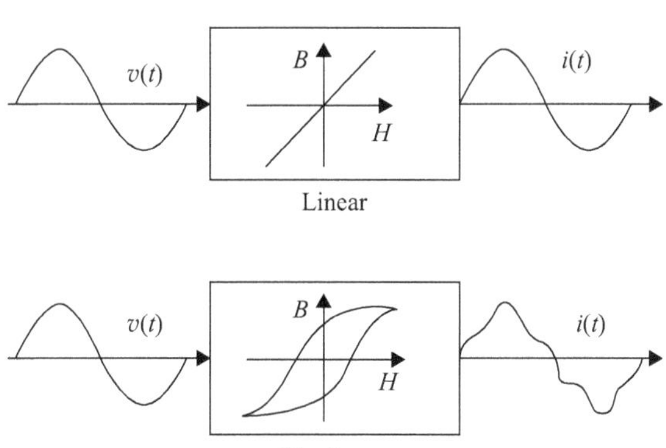

Figure 3.27. Comparison of both $v(t)$ and $i(t)$ wave form at no-load operation. Top: non-ideal linear transformer. Bottom: real transformer.

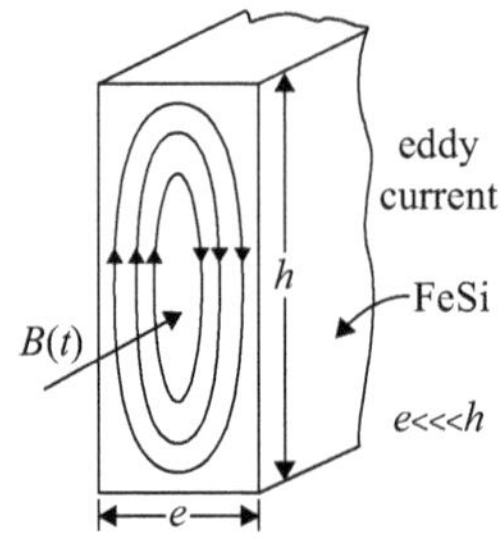

Figure 3.28. The eddy-currents in a ferromagnetic material.

doping the sheet steel with silicon creating the alloy iron–silicon sheet steel which increases the resistivity of the material.

From the electromagnetism, it is possible to demonstrate that the eddy-current losses can be expressed by:

$$P_f = K_f \times \mathrm{Vol} \times (f \times B_{\max} \times e)^2 \tag{3.40}$$

where:

K_f: coefficient that depends on the material;
Vol: the volume of the magnetic core;
f: the excitation frequency;
$B_{\max}$: the magnitude of the magnetic flux density;
e: the sheet steel thickness.

The total amount of iron losses is given by:

$$P_{\mathrm{Fe}} = P_{\mathrm{hist}} + P_f \tag{3.41}$$

As the real transformer normally works at a magnetic load of $1.2 < B_{\max} < 1.4\ \mathrm{Wb\ m^{-2}}$, it is possible to consider the exponent of the hysteresis losses closed to $n \cong 2$. On the other hand, the real transformer also works at constant frequency. Considering those approaches, we can write:

$$P_{\mathrm{hist}} = K_1 \times B_{\max}^2 \tag{3.42}$$

where:

$$K_1 = K_{\mathrm{hist}} \times \mathrm{Vol} \times f$$

and,

$$P_f = K_2 \times B_{\max}^2 \tag{3.43}$$

where:

$$K_2 = K_f \times \mathrm{Vol} \times (f \times e)^2$$

Therefore, the real transformer iron losses at a constant frequency excitation can be expressed by:

$$P_{\mathrm{Fe}} = K_3 B_{\max}^2 \qquad (3.44)$$

where:

$$K_3 = K_1 + K_2$$

3.4.3 Considering the iron loss in the equivalent electric circuit

We are now able to represent the iron losses by a circuit element introduced in the equivalent electric circuit based on both equation (3.44) and Faraday's Law.

From (3.37), we extracted from Faraday's Law that the induced electromotive force by the sinusoidal magnetic flux is given by:

$$E_{m1} = 4.44 f N_1 \varnothing_{\max} \qquad (3.45)$$

As $\varnothing_{\max} = B_{\max} S_{fe}$ and supposing that the transformer is operating at constant frequency, from (3.45) we obtain:

$$B_{\max} = \frac{E_{m1}}{K_4} \qquad (3.46)$$

Substituting (3.36) in (3.44) and after some mathematical manipulation, we get:

$$P_{\mathrm{Fe}} = \frac{E_{m1}^2}{K_5} \qquad (3.47)$$

We can identify (3.47) with a dissipated power in a suitable resistor R_{Fe} fed by a voltage E_{m1} such that:

$$P_{\mathrm{Fe}} = \frac{E_{m1}^2}{R_{\mathrm{Fe}}} \qquad (3.48)$$

For that, it is enough to impose:

$$R_{\mathrm{Fe}} = K_5 \; (\Omega) \qquad (3.49)$$

As E_{m1} is the voltage drop in the magnetization reactance X_m, we must locate R_{Fe} parallel to X_m as shown in figure 3.29.

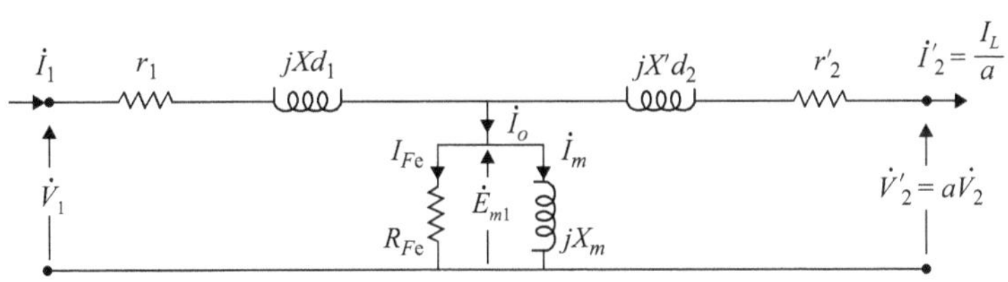

Figure 3.29. The complete equivalent electric circuit for a real transformer. R_{Fe}: the iron loss resistance.

Remarks: the same considerations regarding the simplification of the equivalent electric circuit for non-ideal linear transformers can also be adopted in the case of real transformers.

3.4.4 Obtaining parameters from tests

The parameters of a real transformer are obtained from two main tests and a measurement carried out by instruments.

We can divide the parameters into two classes. The first is the series parameters that are where the load current flows, like the coil's resistances and leakage reactance. The second one is the parallel parameters that are submitted to the full voltage like the iron loss resistance and the magnetization reactance.

During the tests the series parameters should be fed by the rated current, while the parallel ones by the rated voltage.

The short-circuit test: the low-voltage side in this test is short-circuited and the high-voltage side is fed by the controlled voltage source large enough to impose the rated current of the coil (figure 3.30). This procedure is adopted because it is easier to have an AC source available for the low current of the high-voltage side than the high current of the low-voltage side.

As the applied voltage is few percent of the rated voltage of the HV coil and considering the reaction of the secondary electrical current on the magnetic flux, it is possible to neglect both the iron losses and the magnetization current. Therefore, the parallel impedance can be suppressed as is shown in figure 3.31.

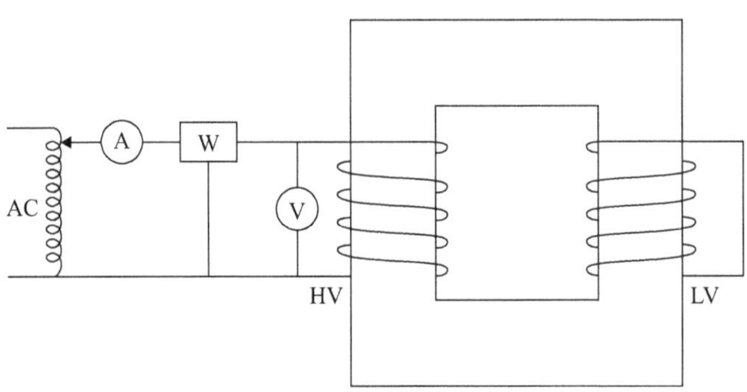

Figure 3.30. Short-circuit test. LV side short-circuited. HV fed by a controlled AC voltage source.

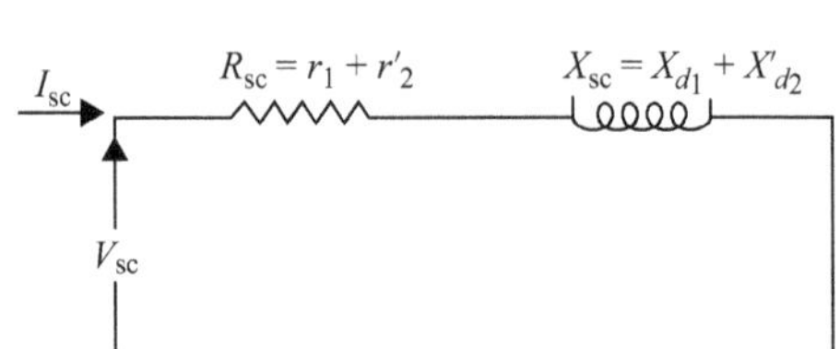

Figure 3.31. Equivalent electric circuit of the short-circuited transformer. The parallel impedance effect is suppressed.

The short-circuit resistance R_{sc} is obtained from the measured active power P_{sc} delivered to the circuit through the relation:

$$R_{sc} = \frac{P_{sc}}{I_{sc}^2} \tag{3.50}$$

as,

$$Z_{sc} = \frac{V_{sc}}{I_{sc}}$$

results in:

$$X_{sc} = \sqrt{Z_{sc}^2 - R_{sc}^2} \tag{3.51}$$

Considering that the path of the leakage magnetic flux lines of both the primary and the secondary coils are the same, we can consider that:

$$X_{d1} = X'_{d2} = \frac{X_{sc}}{2} \tag{3.52}$$

Normally, the resistances are commonly measured by instruments at the end of the tests to consider the heating process of the tests, if not we can also consider that:

$$R_1 = R'_2 = \frac{R_{sc}}{2} \tag{3.53}$$

The open-circuit test: the open-circuit test is suitable for evaluating the iron losses and the magnetization current level of the transformer (figure 3.32). The main requirement for it is to impose the rated magnetic flux in the core. As the rated magnetic flux is due to the rated voltage of the transformer, the LV side should be chosen for this test because it is safer handling low voltages than the high voltages in a laboratory.

As the electric current is a few percent of the rated current of the transformer, the effects of series impedance could be neglected. Therefore, the series impedance effects can be suppressed in the equivalent circuit as is shown in figure 3.33.

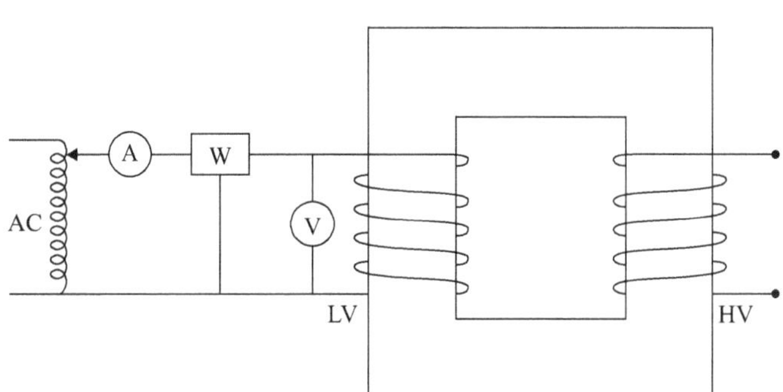

Figure 3.32. Open-circuit test. HV side is open-circuited. LV fed by a controlled AC voltage source.

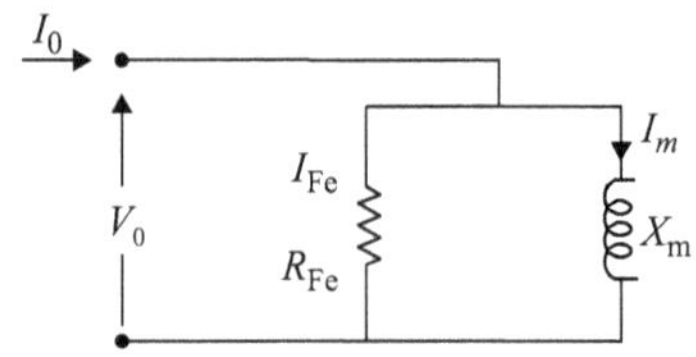

Figure 3.33. Equivalent electric circuit of the open-circuit test. The series impedance effect is suppressed.

The iron resistance R_{Fe} is obtained from the measured active power P_0 delivered to the circuit through the relation:

$$R_{Fe} = \frac{V_0^2}{P_0} \tag{3.54}$$

as,

$$I_{Fe} = \frac{V_0}{R_{Fe}}$$

results in:

$$I_m = \sqrt{I_0^2 - I_{Fe}^2} \tag{3.55}$$

therefore:

$$X_m = \frac{V_0}{I_m} \tag{3.56}$$

3.4.5 Efficiency

The efficiency of a transformer is defined by:

$$\eta = \frac{P_{del}}{P_{del} + P_j + P_{Fe}} \times 100\% \tag{3.57}$$

where:

$P_{del} = V_2 \times I_L \times \cos\varphi$: active power delivered to the load (W);
$P_j = r_1 \times I_1^2 + r_2' \times I_2'^2$: joule losses (W);
$P_{Fe} = \frac{E_{m1}^2}{R_{Fe}}$: iron losses (W).

3.4.6 Voltage regulation

The measure of how well a transformer maintains constant secondary voltage over a range of load currents is called the transformer's voltage regulation. It can be calculated using the following formula:

$$\mathcal{R} = \frac{V_{20} - V_2}{V_2} \times 100\% \tag{3.58}$$

where:

V_2: secondary voltage at load (V);

V_{20}: secondary voltage at no-load (V).

The voltage regulation is strongly affected by load nature. It can be positive under both resistive and inductive load and negative under a capacitive one.

3.4.7 The phasor diagrams

The phasor diagram is a fast and powerful tool for analyzing the behavior of the transformer under all types of load. Although there are several values involved in the transformer's operation, it enables us to solve almost all steady-state problems of the device.

We normally divide the phasor diagram drawing into two steps:

3.4.7.1 The secondary side

Figure 3.34 shows part of the equivalent electric circuit of the transformer representing only the secondary parameters and associated values.

As the load is specified by its apparent power, voltage and power factor, we start the phasor diagram tracing both the secondary voltage reflected to the primary side and the secondary current reflected to the primary side lagged of the power factor angle φ. Under inductive load, the current is lagging of the voltage, so $\varphi < 0$, while under capacitive load, the current is in advance of the voltage, in this case $\varphi > 0$. Finally, under resistive load, both the current and the voltage have the same phase, so $\varphi = 0$.

Let us suppose that an inductive load is located at the secondary side. Figure 3.35 shows the phasor diagram of the secondary values reflected to the primary.

In the secondary, we have:

$$\dot{E}_{m1} = \dot{V}_2' + r_2'\dot{I}_2' + jX_{d2}'\dot{I}_2' \tag{3.59}$$

The phasor diagram represents this equation by the sum of those complex numbers represented by its polar representation.

As both $\dot{V}_2'$ and $\dot{I}_2'$ are given by the load requirements, we draw the product $r_2'\dot{I}_2'$ by the phasor indicated in figure 3.35 which has the same phase of $\dot{I}_2'$ (parallel).

The product $jX_{d2}'\dot{I}_2'$ is a phasor 90° ahead of the phasor of the current $\dot{I}_2'$. Note that in figure 3.35 the phasors of $\dot{I}_2'$ and $jX_{d2}'\dot{I}_2'$ are perpendicular.

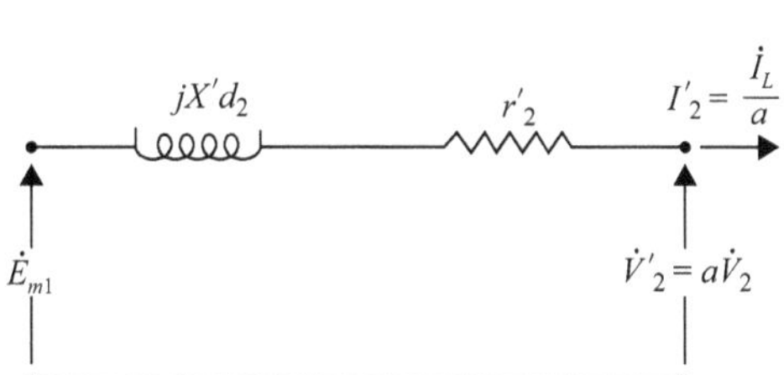

Figure 3.34. Equivalent electric circuit of secondary.

The sum of complex numbers in polar representation is made of the sum of the vectors, so the vectorial sum of $(\dot{V}_2' + r_2'\dot{I}_2' + jX_{d2}'\dot{I}_2')$ is $\dot{E}_{m1}$, as shown in figure 3.35.

3.4.7.2 The primary side

The first step for drawing the phasor diagram, referred to as the primary side, represents the phasor of both the loss current $\dot{I}_{\mathrm{Fe}}$ and the magnetization current $\dot{I}_m$.

Figure 3.36 shows the other part of the equivalent electric circuit of the transformer representing only the primary parameters and associated values.

From the circuit of figure 3.36, we obtain:

$$\dot{I}_{\mathrm{Fe}} = \frac{\dot{E}_{m1}}{R_{\mathrm{Fe}}} \tag{3.60}$$

In figure 3.37, the phasor $\dot{I}_{\mathrm{Fe}}$ is represented by the vector in the same direction and sense of $\dot{E}_{m1}$.

The magnetization current is given by:

$$\dot{I}_m = \frac{\dot{E}_{m1}}{jX_m} \tag{3.61}$$

which is represented in figure 3.37 by a phasor 90° lagged from $\dot{E}_{m1}$.

The vectorial sum of both currents gives:

$$\dot{I}_0 = \dot{I}_{\mathrm{Fe}} + \dot{I}_m \tag{3.62}$$

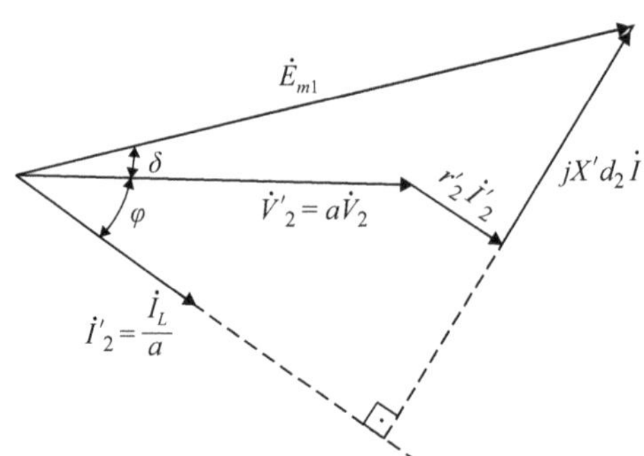

Figure 3.35. Phasor diagram of an inductive load at the secondary side.

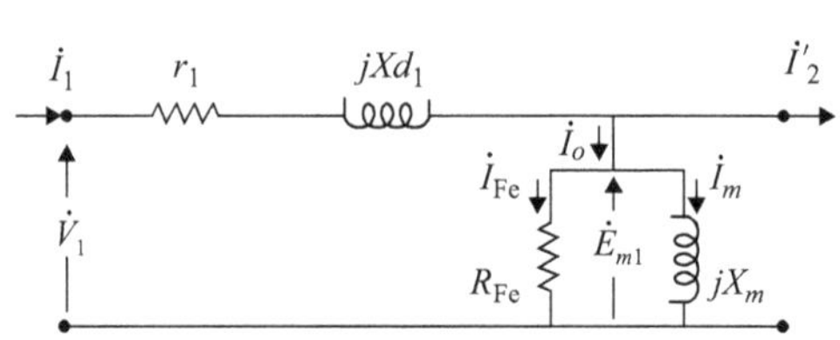

Figure 3.36. Equivalent electric circuit of primary.

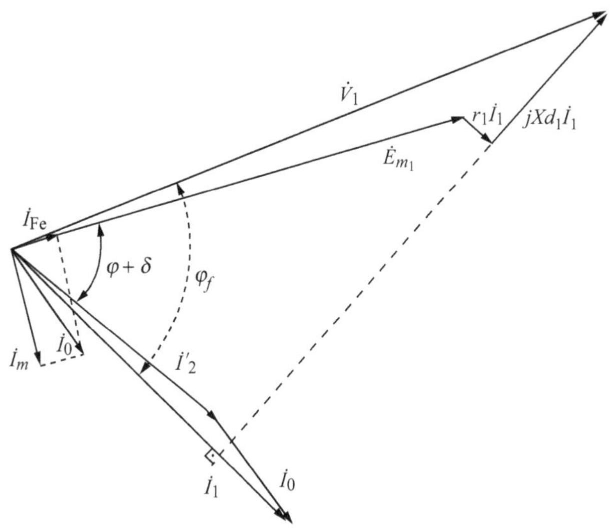

Figure 3.37. Primary phasor diagram of the transformer. $\dot{I}_{Fe}$ has the same phase of $\dot{E}_{m1}$. $\dot{I}_m$ is 90° behind $\dot{E}_{m1}$.

This is the current commonly called the 'no-load' current of the transformer.

As $\dot{I}_{Fe} \ll \dot{I}_m$, the 'no-load current' is lagged almost 90° from $\dot{E}_{m1}$, as figure 3.37 shows.

The primary current $\dot{I}_1$ is obtained by the vectorial sum of $\dot{I}_2'$ and $\dot{I}_0$ as:

$$\dot{I}_1 = \dot{I}_2' + \dot{I}_0 \tag{3.63}$$

Remark: The vectorial sum of $\dot{I}_2'$ and $\dot{I}_0$ was performed by the polygon rule.

On the primary side, we have:

$$\dot{V}_1 = \dot{E}_{m1} + r_1\dot{I}_1 + jx_1\dot{I}_1 \tag{3.64}$$

Since both $\dot{E}_{m1}$ and $\dot{I}_1$ are known, we obtain $\dot{V}_1$ by the vector sum indicated in (3.64), remembering that $r_1\dot{I}_1$ has the same phase of $\dot{I}_1$ and $jx_1\dot{I}_1$ is 90° in front of $\dot{I}_1$.

As a final remark, the cosine of the angle between $\dot{V}_1$ and $\dot{I}_1$ ($\cos\varphi_f$) is the power factor 'seen' by the AC source and can be considered for evaluating the efficiency of the transformer as:

$$\eta = \frac{V_2 I_L \cos\varphi}{V_1 I_1 \cos\varphi_f} \times 100\% \tag{3.65}$$

3.5 Summary

The discovery of the magnetic induction phenomenon by Michael Faraday on 29 August 1831 was one of the most important achievements made in universal history. The understanding of magnetic induction enabled humankind to develop all the electromechanical equipment that opened the world to a new era of development.

The transformer was a device that enabled not only the possibility of transforming both voltage and electric current but also the possibility of long-distance transmission of electric energy.

The chapter began by introducing the main concepts applied in an ideal transformer. Although these are basic concepts, they are widely used for evaluating the performance of a real transformer when we have not enough information about its parameters.

The non-ideal transformer was introduced to identify some transformer parameters associated with the leakage flux issued when the magnetic permeability is a finite number.

As the real transformer is built with ferromagnetic material and operated with alternating current, especially with the sinusoidal time variation shape, the chapter ended by introducing the iron losses issued from both the hysteretic phenomenon and eddy-current effects.

Some rates like regulation and efficiency were also introduced in this chapter together with the equivalent circuit approximations that we can adopt in specific studies of power electric systems. The main experimental tests used to determine such electrical parameters, i.e., the no-load and the short-circuit tests, were also presented.

3.5.1 Transformer design guidelines: a project-based learning approach

It is suggested that this activity is developed by five teams composed of five students each. The main objective is promoting competition among them and rewarding the team whose design was the most efficient transformer or the one that reaches the maximum power-to-weight ratio or any other indicator suggested by the teacher.

The design of a transformer starts imposing the maximum value to both the magnetic flux density (B_{max}) in the core and the current density (J_{max}) in the conductors.

The transformer chosen for this exercise is a single-phase, two limb transformer, whose general design is shown in figure 3.38. Both windings are mounted on the same limb and the other limb is used as a return limb. For the rated apparent power range of table 3.1, we can adopt $B_{max} = 1.6\ T$ and $J_{max} = 3\ \text{A mm}^{-2}$.

In this figure the high-voltage (HV) winding is enclosing the low-voltage (LV) one. We also suggest you choosing one of the options offered in table 3.1 where the rated apparent power and the voltages are specified.

The kick start is the Faraday's Law application for an AC excitation given by:

$$V_{ind} = 4.44 f N B_{max} A_{core} \tag{3.66}$$

We can define the induced voltage per turn as:

$$E_t = \frac{V_{ind}}{N} = 4.44 f B_{max} A_{core} \tag{3.67}$$

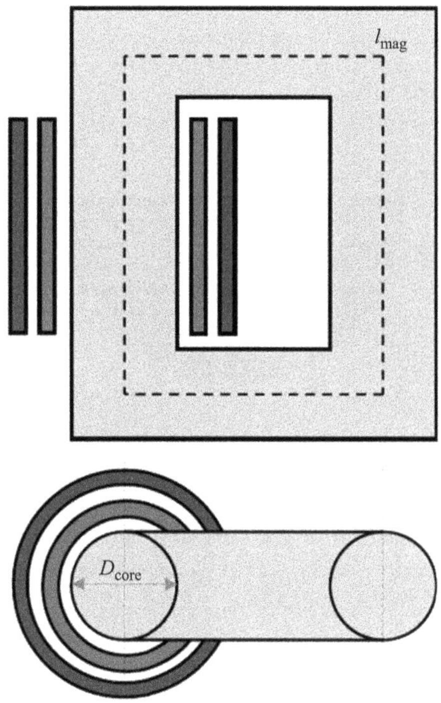

Figure 3.38. Single-phase transformer layout.

Table 3.1. Single-phase transformer specifications—frequency 60 Hz.

Option	S(kVA)	V_h(kV)	V_l(V)
1	50	13.8	3000
2	50	13.8	4500
3	25	13.8	4500
4	25	13.8	3000
5	25	13.8	1380

The induced voltage per turn is a specification dependent on the past history of the manufacturer. As an orientation, we suggest adopting it by using the following empirical equation:

$$E_t = K\sqrt{S} \tag{3.68}$$

where S is the rated apparent power in kVA and K is a factor that depends on the conductor material type. For copper windings, the K factor is in the range between 0.37 and 0.45 and for aluminum windings, K is between 0.32 and 0.35. In this work, we suggest considering copper windings with $K = 0.45$.

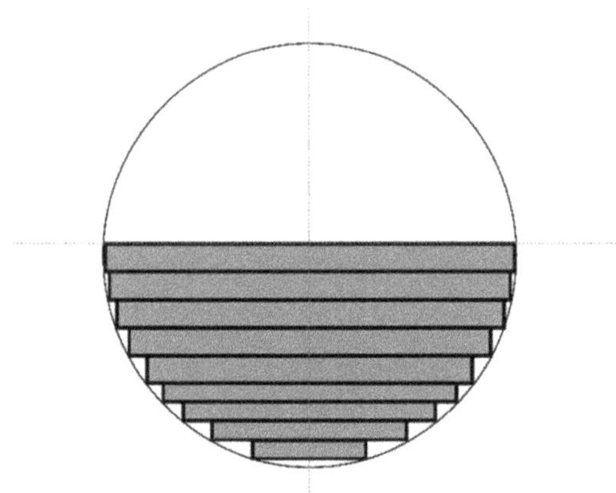

Figure 3.39. Magnetic core made of packets of laminations.

As E_t is evaluated, not only the number of turns of both windings (N_{hv} and N_{lv}) but also the area of the cross-section of the core can be found from equation (3.67).

As the core cross-section is not exactly a circle, as figure 3.39 shows, it is convenient to define the block package k_l such that:

$$k_l = \frac{A_{\text{effcore}}}{A_{\text{core}}} \tag{3.69}$$

where;

A_{effcore}: effective area of laminated steel;
A_{core}: geometric area of the involved circle.

We also should consider the stacking factor k_s as defined in the previous chapter, when the core is made of laminated Fe–Si steel. As a result, the effective core cross-section area is given by:

$$A_{\text{effcore}} = k_l k_s A_{\text{core}} \tag{3.70}$$

We suggest for this range of rated power choose $k_l = 0.97$ and $k_s = 0.98$.

The cross-section of the conductors is defined from the relation:

$$J_{\text{max}} = \frac{I_{\text{hv}}}{A_{\text{hv}}} = \frac{I_{\text{lv}}}{A_{\text{lv}}} \tag{3.71}$$

where:

I_{hv}: rated current of high-voltage winding;
I_{lv}: rated current of low-voltage winding;
A_{hv}: Cross-section of high-voltage conductor;
A_{lv}: Cross-section of low-voltage conductor.

As there is limited space for accommodating both HV and LV windings to achieve not only a core that is as light as possible but also as safe as possible to avoid insulating problems, some distance should be imposed, as shown in figures 3.40 and 3.41.

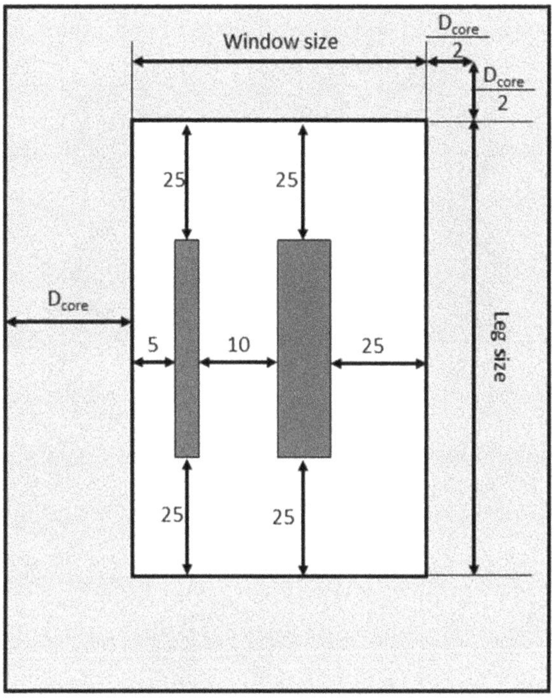

Figure 3.40. Window size and insulation distances in mm.

Concerning the low-voltage winding, for an industrial transformer whose rated power is higher than tens of kVA and the low voltage is hundreds of volts, the coil can be made by rectangular conductors specified by the manufacturer. In this situation the dimension of b_2 is given in figure 3.42.

Remarks: The use of non-standard conductors increases the cost of transformer manufacture.

Using the average radius of the coil it is possible to evaluate the average length of the conductor of each winding by:

$$l_{hv} = 2\pi R_{a1} N_{hv} \tag{3.72}$$

$$l_{lv} = 2\pi R_{a2} N_{lv} \tag{3.73}$$

As a result, the total of copper mass is such that:

$$M_{Cu} = d_{Cu}(A_{hv} l_{hv} + A_{lv} l_{hv}) \tag{3.74}$$

where d_{Cu} is the copper density in kg m^{-3}.

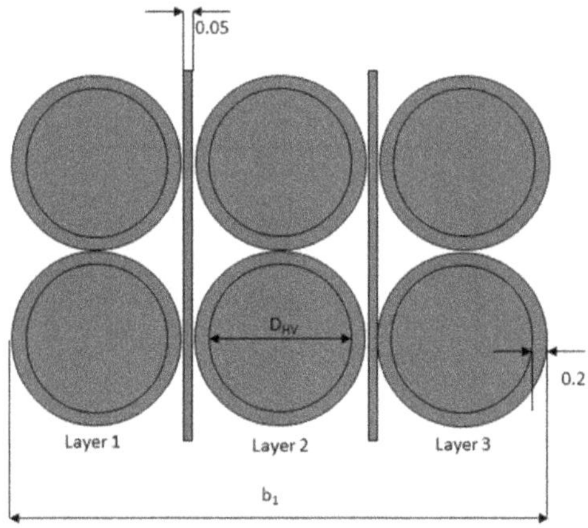

Figure 3.41. HV winding in detail with insulations in mm. $b_1 = n_{\mathrm{hv}}(0.2 + D_{\mathrm{hv}} + 0.2) + (n_{\mathrm{hv}} - 1)0.05$. n_{hv} is the number of layers of HV winding, D_{hv} diameter of the conductor.

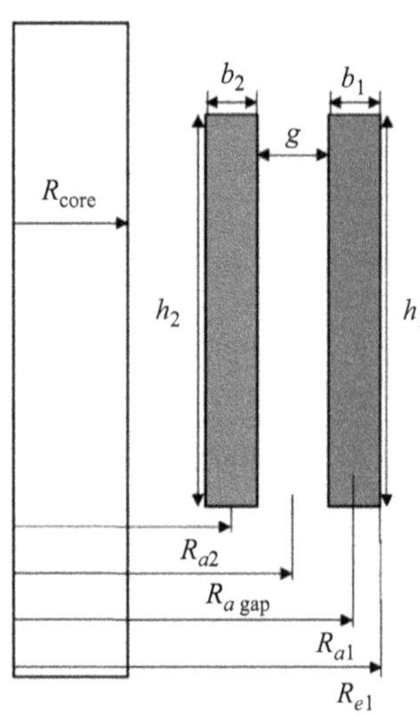

Figure 3.42. LV winding in detail with insulations in mm. $b_2 = n_{\mathrm{lv}}(0.2 + L_{l\mathrm{hv}} + 0.2) + (n_{\mathrm{lv}} - 1)0.05$. n_{lv} is the number of layers of LV winding. L_{hv} diameter of the conductor.

We are now able to evaluate the winding resistances that are given by:

$$r_{\mathrm{hv}} = \rho \frac{l_{\mathrm{hv}}}{A_{\mathrm{hv}}} \tag{3.75}$$

$$r_{\mathrm{lv}} = \rho \frac{l_{\mathrm{lv}}}{A_{\mathrm{lv}}} \tag{3.76}$$

The core volume is evaluated from the winding dimension and the minimum distances from the coil to the core. From figure 3.42 we get:

$$\text{Window Size} = 5 + b_2 + 10 + b_1 + 25 \tag{3.77}$$

The leg size is such that:

$$\text{Leg Size} = 25 + \max[h_1; h_2] + 25 \tag{3.78}$$

l_{mag} (figure 3.37) is given by:

$$l_{mag} = 2(\text{Leg size} + \text{Window Size}) + 4D_{core} \tag{3.79}$$

The volume of the core yields:

$$V_{core} = l_{mag} D_{core} \tag{3.80}$$

The transformer losses are evaluated from its components:
Joule losses:

$$P_{joule} = r_{hv} i_{hv}^2 + r_{lv} i_{lv}^2 \tag{3.81}$$

Iron losses:

$$P_{iron} = d_{iron} V_{core} w_{iron} \tag{3.82}$$

where:
d_{iron}: density of iron—kg m^{-3}
w_{iron} : specific iron losses—W kg^{-1}

Remarks: The specific losses of iron are dependent on the maximum value of B_{max} and frequency. For $B_{max} = 1.6\ T$ and frequency $f = 60$ Hz the specific iron loss is about 4.5 W kg^{-1}.

Activity: The teams should develop a project of the chosen transformer from table 3.1, evaluating not only all dimensions of both windings but also the dimensions of the core including the mass of both copper and iron. The approximate efficiency of the transformer should also be evaluated considering that it is supplying its rated apparent power under 0.8 inductive power factor.

Problems

3.1 Find a transformer and learn all you can about it from observation. Any kind of transformer may be used: power, audio-frequency, doorbell, impedance-matching, power-supply, etc. Write about two pages of description, including a diagram. Write as if for an engineering report. Give its purpose, its general size, shape, and construction, provision for cooling. Give the information about its electrical operation that is available from the name plate, with any deductions that you are able to make. Possible data to include are voltage, current, power ratings, number of windings, number of turns, size of wire, number and thickness of core laminations, estimated weight, insulation, turn ratio, input and output impedance, etc.

3.2 Describe all components of an ideal transformer and outline the physical laws involved in its operation.

Answer: magnetic core with infinite magnetic permeability. Two windings with different turns coil. Induction magnetic law. Balance of magneto-motive force. Energy conservations law.

3.3 Explain the electromagnetic phenomena associated with:
   (a) leakage and magnetization inductances;
   (b) resistances of coils;
   (c) iron losses resistance.

Answer: (a) leakage inductance is associated with the percentage of flux lines that is not closed through the magnetic core and the magnetization inductance is associated with the mutual flux of both coils.

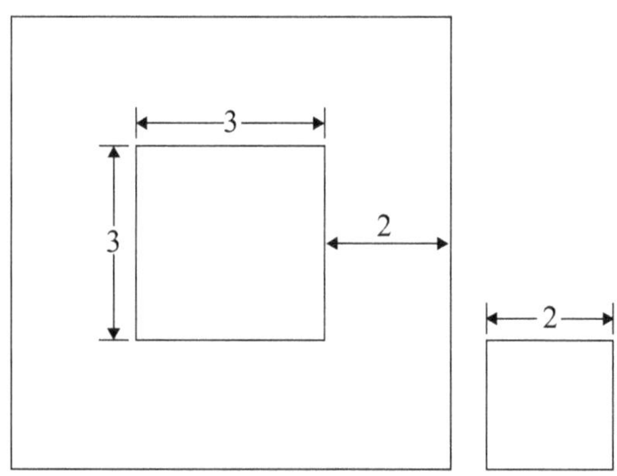

Figure 3.43. Problem 3.3. Dimensions in cm.

3.4 Figure 3.43 shows a linear magnetic core for a transformer designed for transforming from 120 V (rms) to 6 V output, 60 Hz. Use maximum $B = 0.6\ \text{Wb m}^{-2}$ and relative magnetic permeability of $\mu_r = 5000$.

The leakage inductances of both sides are 6.7 mH and 0.268 mH and the resistances of windings are 1 Ω and 0.04 Ω. Compute:
   (a) the number of turns in each winding;
   (b) the magnetization reactance referred to the primary (120 V);
   (c) input impedance with 6 Ω resistive load on the secondary (the magnetization current can be neglected);

(d) compare the magnetizing current to the total primary current at load.

Answers: (a) 1877, 94; (b) 97.4 Ω, (c) 152 Ω, (d) 1%.

3.5 An AC voltage source can be represented by a constant back emf of 5 V in series with an internal resistance of 2000 Ω. This source is feeding a headphone represented by a 50 Ω constant resistance through an ideal transformer. Sketch the power delivered (in mW) to the load as a function of the transformer relation highlighting the relation that the power delivered to the headphone is maximum. Extract from the plotted curve the maximum value of the power delivered to the load.

Answer: 6.32; 3.125 mW.

3.6 A small industrial business is fed by a power system that provides 2400 V–60 Hz in a substation located 2 km away. The unifilar diagram of the system is shown in figure 3.44. The feeder that connects the substation to the business has a *distributed impedance* given by $0.15 + j0, 8\ \Omega\ \text{km}^{-1}$. The rated power of the transformer is 50 kVA, 2400/240 V, 60 Hz and its *primary series impedance* is $0.72 + j0.92\ \Omega$ and the *secondary* $0.007 + j0.009\ \Omega$. The magnetization current can be neglected.

All electrical installations of the plant are represented by an impedance connected at the low-voltage winding of the transformer given by $0.92 + j0.69\ \Omega$.

(a) sketch the electrical circuit of the system referred to the high-voltage side of transformer indicating the values of all electrical parameters;
(b) compute the electrical current in the feeder;
(c) compute both the voltage at the load and the drop voltage in the feeder;
(d) the electric power delivered to the load;
(e) the electrical efficiency of the system.

Answers: (b). 20.4 A; (c) 2340 V, 33 V.

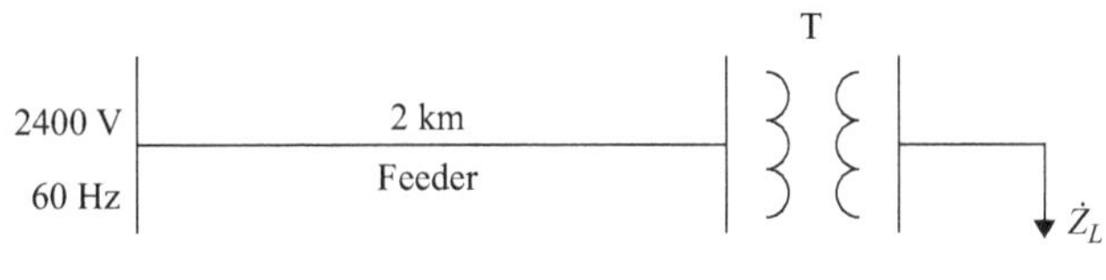

Figure 3.44. Problem 3.4.

3.7 Two different transformers present the characteristics shown in table 3.2. both transformers are used for delivering active power to a 1.5 Ω load resistance connected to 230 V side. Compute:

(a) the primary and secondary currents of each transformer;
(b) the total current supplied to the transformer group by the AC electrical system;

Table 3.2. Problem 3.7. Voltages and leakage reactances of the transformers.

Rated power (kVA)	Primary voltage (V)	Secondary voltage (V)	Primary leakage reactance (Ω)	Secondary leakage reactance (Ω)
10	2300	230	6	0.06
30	2300	230	2	0.02

(c) the active power delivered to the load for each transformer;
(d) both the active power and the power factor that the AC electrical system supplies to each transformer

Answers: (a) Primary side: 3.83 A / 11.5 A; Secondary side: 38.33 A / 115 A; (b) 15.33 A; (c) 8.82 kW / 26.45 kW; (d) 35.27 kW / 0.98.

3.8 The energy consumption of an electrical plant is not constant throughout the day. It is variable dependent on the goods being produced during the day. Figure 3.45 shows the energy demand curve applied in a typical work day of a hypothetical plant. The problem that all companies face is to choose a suitable transformer to operate the electrical system at its maximum efficiency. That is not so simple, because all transformers are designed to operate at their maximum efficiency when delivering their rated power. As the power consumption is not constant throughout the day the rated efficiency of the transformer is not enough to define it as the best choice for a plant.

Two proposals of a 20 MW transformer issued from different manufacturer were offered to the hypothetical company. Table 3.3 highlights the efficiency of the transformer as a function of the load.

Imagine you are the engineer in charge of recommending which transformer should be chosen; prepare a report analyzing the problem

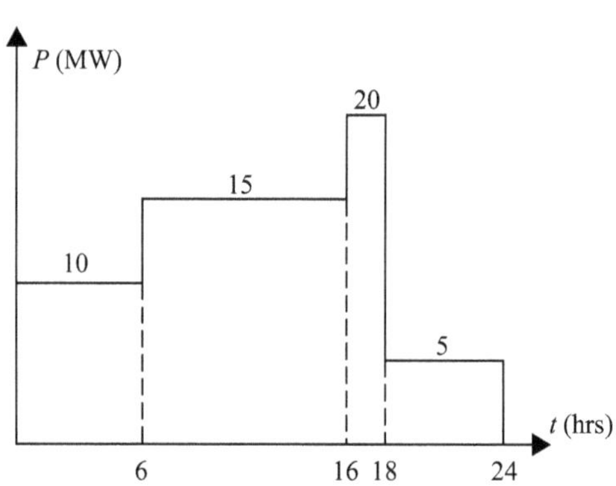

Figure 3.45. Problem 3.5.

Table 3.3. Efficiency based on the rated power of a 20 MVA transformer.

% of power delivered	Transformer 1	Transformer 2
Rated	80%	95%
75%	75%	72%
50%	70%	60%
25%	50%	50%

justifying (if so) why the transformer with better efficiency at the rated power was not the most suitable for this application.

3.9 Assume that a 100-kVA transformer, 11 000/2200 V–60 Hz is feeding an impedance of 50 Ω under a leading power factor 0.7 connected to the 2200 V side. The circuit parameters referred to the HV side are given by:

$r_1 = 6.1\ \Omega$	$r_2' = 7.21\ \Omega$
$x_1 = 31.2\ \Omega$	$x_2' = 31.2\ \Omega$
$X_m = 57\ 300\ \Omega$	$R_{Fe} = 124\ 000\ \Omega$

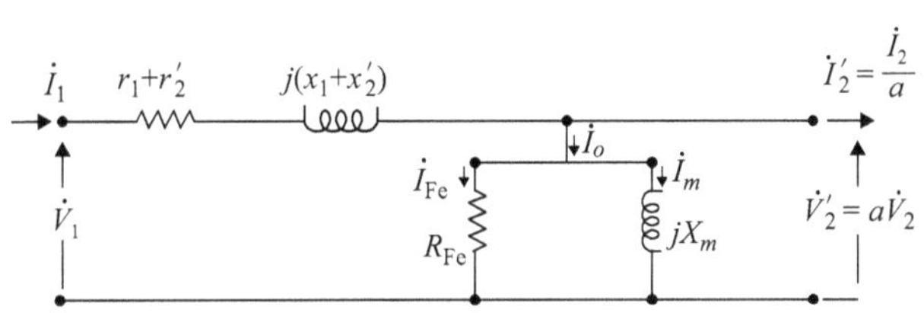

Figure 3.46. Approximate equivalent circuit for problem 3.8.

Supposing that it can be represented by the approximate equivalent circuit of figure 3.46 determine all voltages and currents involved in this operation and draw the phasor diagram (not to scale) showing the various phasor magnitudes and angles.

Hint: Adopt $\dot{V}_2 = 2000\angle 0°$ V as a reference.

Determine also the total losses of the transformer

(a) by subtracting output active power from input active power;

(b) by calculating the power dissipated the in the resistive elements of the approximate equivalent circuit.

3.10 The transformer shown in figure 3.46 is to operate with 115 V– 60 Hz applied to its primary winding. The secondary winding is to produce an output at 500 V. All core dimensions are given in mm. The magnetic core is to operate at a maximum flux density of 1.4 T. The core stacking factor k_s is 0.95. The winding can be operated at current density of 2 A mm^{-2}. The space used by winding should be 45% maximum of the window area:

(a) determine the required numbers of turns on both the primary and secondary winding;
(b) determine the kVA rating of the transformer;
(c) at the operating temperature the resistivity of the copper is $2 \times 10^{-8}\,\Omega \cdot m$, Assuming that the operating ends of the coil are semicircular in shape, estimate the Joule loss in the windings when operated at rated current;
(d) determine the iron losses in the core if it was made of sheet steel laminations that exhibit the characteristics losses of 3.5 W kg^{-1} at 1.4 T–60 Hz and mass density of 7800 kg m^{-3}.

Answers: (a) $N_1 = 90$ turns; $N_2 = 391$ turns; (b) 1 kVA; (c) 26 W; (d) 27 W.

Further reading

[1] Gourishankar V 1966 *Electromechanical Energy Conversion* (Scranton: International Textbook Company)
[2] Skilling H H 1962 *Electromechanics: A First Course in Electromechanical Energy Conversion* (New York: Wiley)
[3] Bansal R (ed) 2004 *Handbook of Engineering Electromagnetics* (New York: Marcel Decker)
[4] Bozorth R M 1993 *Ferromagnetism* (Piscataway, NJ: IEEE Press)
[5] Fitzgerald A E, Kingsley C Jr and Umans S D 1992 *Electric Machinery* 5th edn (New York: McGraw-Hill)
[6] Falcone A G 1979 *Eletromecanica* (São Paulo: Editora Edgard Blucher Ltda) (in Portuguese)
[7] Jordao R G 2002 *Transformadores* (São Paulo: Editora Edgard Blucher Ltda) (in Portuguese)
[8] Slemon G R and Straughen A 1980 *Electric Machines* (Reading MA: Addison-Wesley)

IOP Publishing

Electromechanical Energy Conversion Through Active Learning

J R Cardoso, M B C Salles and M C Costa

Chapter 4

The elementary electromechanical energy conversion

4.1 Introduction

All electromechanical systems are composed of three subsystems whose properties develop several different functions depending on the *trajectory* we want to be done by the electromechanical device.

Figure 4.1 shows the blocks where the characteristics of each module are highlighted.

The ***electric circuit*** is responsible for the electromechanical system control when the device is operating as an electric motor. When the device is operating as a generator the electric circuit is necessary for making the generator output voltage compatible with the one required by the load or by the power system.

A typical equation system of the electric circuit connected to the electromechanical device is:

$$V_i = r_i i_i + L_i \frac{di_i}{dt} + e_i(t); i = 1, 2 \ldots m$$

The *mechanical subsystem* is described by the force (or torque) balance equations in the motor operation case or by the prime move equations in the generator operation one.

A typical equation system for a translational movement issuing from the mechanical subsystem is given by:

$$F_{\text{mec}} - F_{\text{mag}} = m \frac{dv}{dt} + Kv + C \int_{t_0}^{t} v dt$$

For a rotating electromechanical system, we have:

$$T_{\text{mec}} - T_{\text{mag}} = J \frac{d\omega}{dt} + K\omega + C \int_{t_0}^{t} \omega dt$$

doi:10.1088/978-0-7503-2084-9ch4

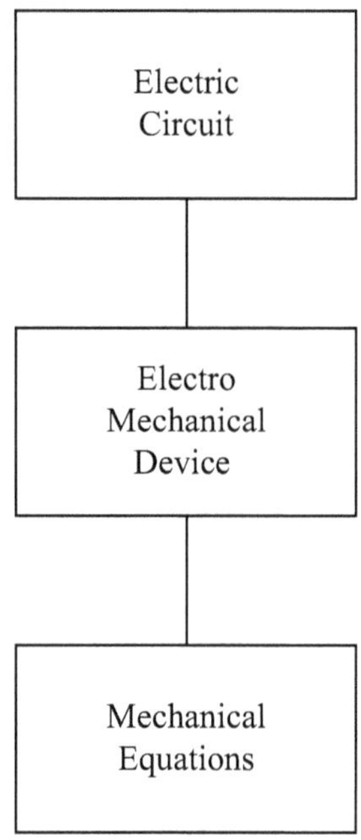

Figure 4.1. Flow chart of electromechanical device analysis.

where:

m : mass of the moving parts
K: damping constant
C: spring constant
J: moment of inertia

The ***block of the electromechanical device*** is represented by the equations that link both the electric circuit and mechanical equations blocks. Therefore, the equations of the electromechanical device present values of both the electric and mechanical subsystems.

As the ordinary electromechanical devices are made using non-complex geometries, the laws of electromagnetism involving both the electromotive force and force (or torque) are very simple.

4.2 The electromotive force in a straight conductor

When a moving electric conductor is placed in a region that has a magnetic flux density distribution, an electromotive force (emf) appears immediately after it at its end. This phenomenon is reflected in an external voltage drop that can be measured by instruments.

If the conductor is placed with its axis perpendicular to both the magnetic flux density and the speed direction, as shown in figure 4.2, the induced electromotive force is given by:

$$e = Blv \tag{4.1}$$

The direction of this emf is obtained using the left-hand rule as shown in figure 4.3.

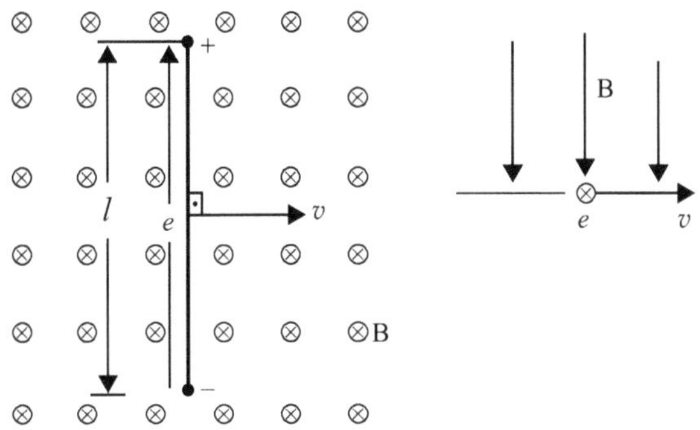

Figure 4.2. Moving electric conductor embedded in a magnetic flux density. ⊗ Direction of magnetic flux density. The left image shows the top view, the right shows the view from the front.

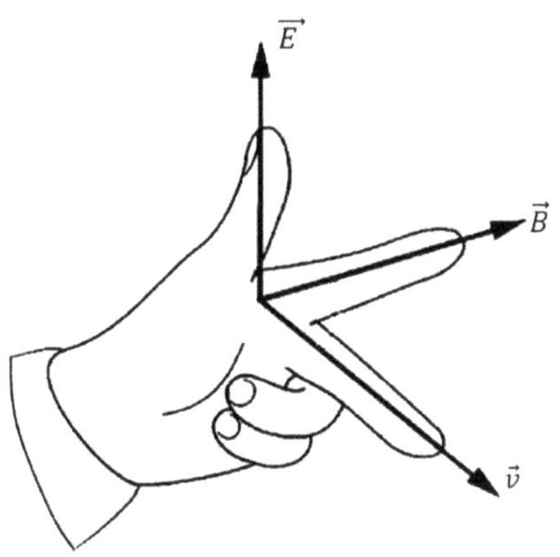

Figure 4.3. The left-hand rule for emf direction.

Example 4.1

A magnetic flux density distribution is composed of an infinite uniform sequence of N–S poles. A single conductor traveling with constant speed of 50 m s^{-1} perpendicular to both the magnetic flux density and the conductor is shown in figure 4.4. The active length of the conductor is 50 cm. The magnitude of the magnetic flux density is 1.0 Wb m^{-2}.

Determine:

(a) the induced emf in the conductor in function of the x, sketch the function;
(b) The induced emf in the conductor in function of the t and the correspondent time-frequency, sketch the function.

Solution

(a) As we discussed earlier, the induced emf is given by (figure 4.5):

$$e = Blv \text{ (V)}$$

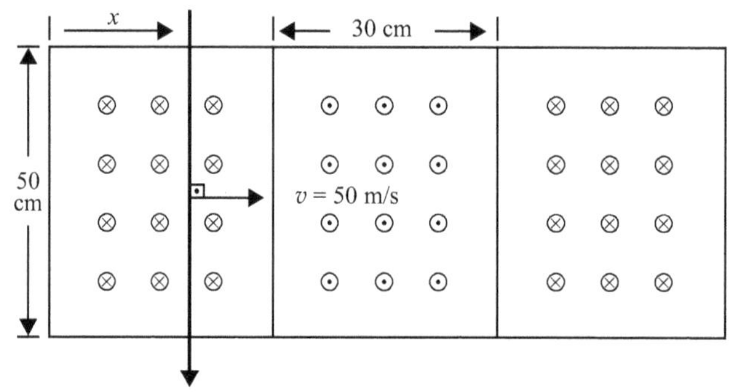

Figure 4.4. Straight conductor immersed in a magnetic flux density distribution.

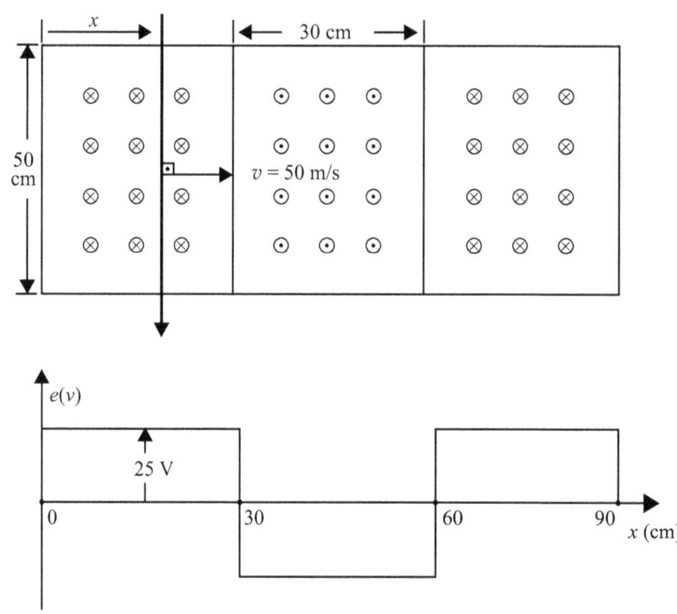

Figure 4.5. The emf as a function of x. The polarity of the emf changes at each pole pitch. Pole pitch: length of the structure where B has same polarity N or S.

where:

(i) B: perpendicular magnetic flux density in the conductor;
(ii) l: active length of the conductor (the length of conductor under the field);
(iii) v: speed of the conductor whose direction is perpendicular to the conductor.

Resulting in:

$$e = Blv = 1.0 \times 50 \times 10^{-2} \times 50 = 25 \text{ V}$$

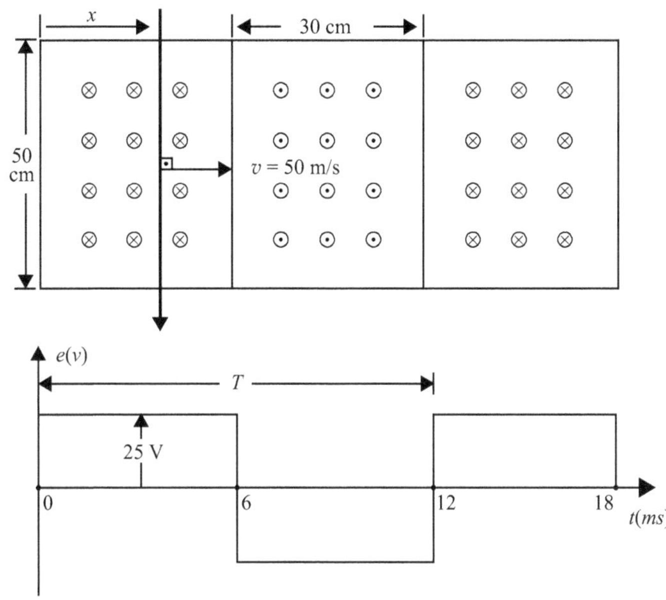

Figure 4.6. The emf as a function of t. The polarity of the emf changes at each half period because of the changing of the magnetic flux density polarity.

(iv) To get the induced emf in the conductor as function of (t), it is necessary to find the time for completing a pole pitch (t_p) as:

$$\Delta t = \frac{t_p}{v} = \frac{30 \times 10^{-2}}{50} = 6 \text{ ms}$$

As the period $T = 2t_p = 12$ms, the resulting frequency is:

$$f = \frac{1}{T} = 83.3 \text{ Hz}$$

Figure 4.6 shows the emf as a function of t.

Example 4.2

A magnetic flux density distribution (figure 4.7) is composed of an infinite uniform sequence of N–S poles. A five turns coil is traveling with a constant speed of 50 m s^{-1} perpendicular to both the magnetic flux density and the conductor. The gap between the two sides of the coil is the same as the pole pitch (30 cm). The magnitude of the magnetic flux density is 1.0 Wb m^{-2} and the active length of the coil is 50 cm. Determine:

(a) the induced emf in the coil in function of (x) sketch the function;
(b) the induced emf in the coil in function of (t) and the correspondent time-frequency, sketch the function.

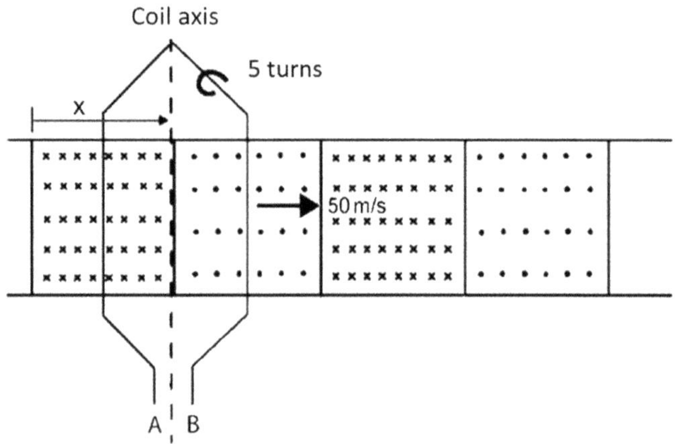

Figure 4.7. Coil in magnetic flux density distribution.

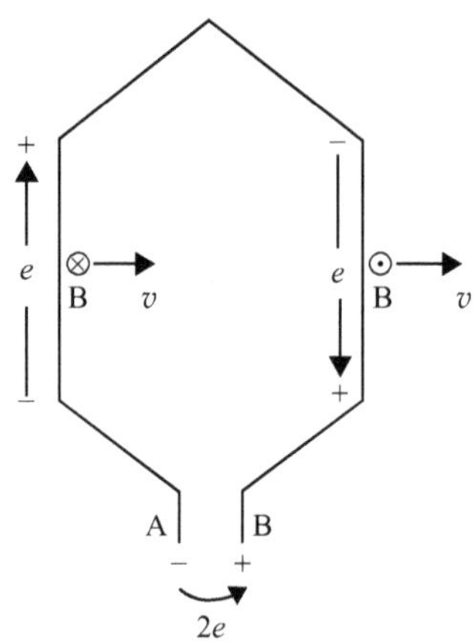

Figure 4.8. Polarities in the function of x. The polarity of A is (−), the polarity of B is (+).

Solution

(a) Due to the symmetry, both sides of the coil are subjected to the same magnetic flux density with the opposite directions. The emf of each side is the same, given by:

$$e = NBlv = 5 \times 50 \times 10^{-2} \times 50 = 125 \text{ V}$$

As both sides are connected in series with concordant polarities the resulting emf in the coil is (figure 4.8):

$$e_{\text{coil}} = 2 \times e = 250 \text{ V}$$

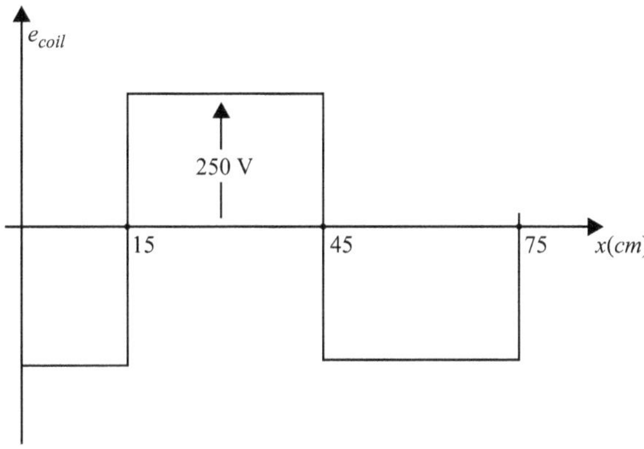

Figure 4.9. Induced emf in function of (x) The polarity of emf changes at each half period because the changing of magnetic flux density polarity.

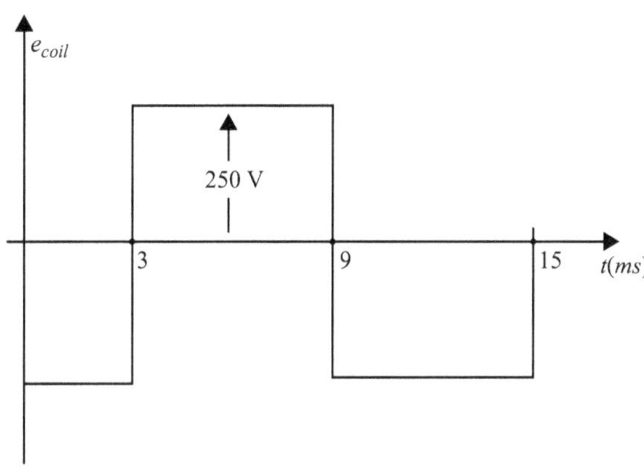

Figure 4.10. The emf as a function of (t) The polarity of the emf changes at each half period because the changing of the magnetic flux density polarity.

From $x = 15_{+}$cm to $x = 45_{-}$cm the terminal A is *negative,* and the terminal B is *positive* and from $x = 45_{+}$cm to $x = 75_{-}$cm, the terminal A is *positive,* and the terminal B is *negative* (figure 4.9).

(b) To get the induced emf in the conductor as a function of (t), it is necessary to find the time for completing a pole pitch (t_p) as:

$$\Delta t = \frac{t_p}{v} = \frac{30 \times 10^{-2}}{50} = 6 \text{ ms}$$

As the period is $T = 2t_p = 12$ ms, the resulting frequency is:

$$f = \frac{1}{T} = 83.3 \text{ Hz}$$

Figure 4.10 shows the emf as a function of t.

4.3 The magnetic force in a straight conductor

When a current-carrying conductor is placed in a region that has a magnetic flux density distribution, immediately a magnetic force appears in the conductor reacting from the source that produces the electric current.

If the conductor is placed with its axis perpendicular to the magnetic flux density as shown in figure 4.11, the force direction is perpendicular to both the magnetic flux density and the current, and its value is given by:

$$F = Bli \tag{4.2}$$

The direction of this force is obtained by the same left-hand rule as shown in figure 4.12.

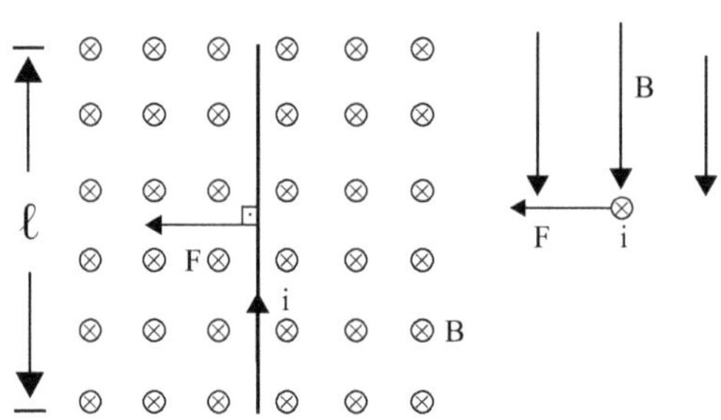

Figure 4.11. Force in a current-carried conductor embedded in a magnetic flux density. ⊗ Direction of magnetic flux density and current. Left—top view Right—frontal view.

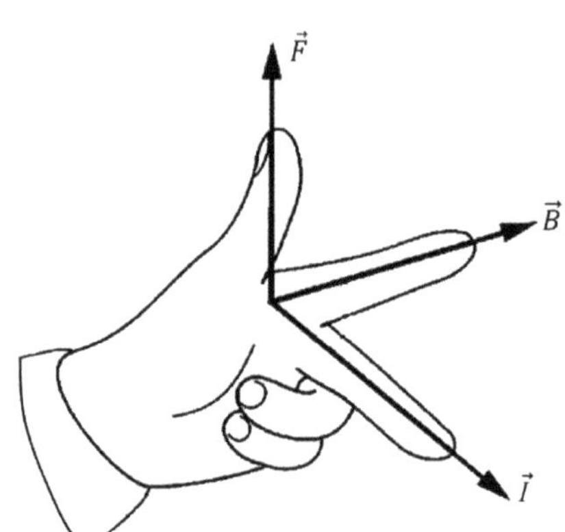

Figure 4.12. The left-hand rule for force direction.

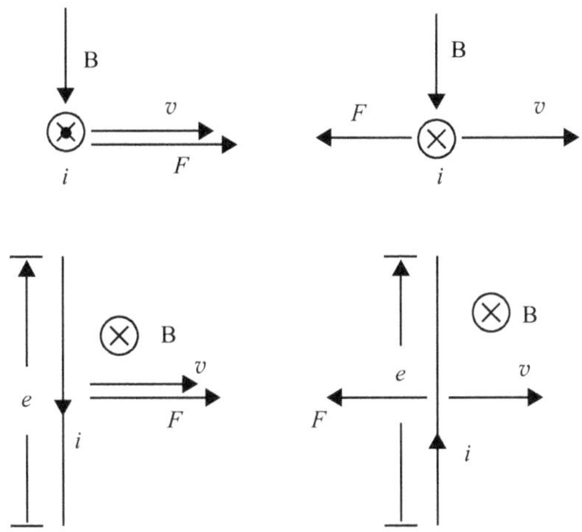

Figure 4.13. Motor (left) and generator operation (right). Motor: emf and i in opposite direction—speed and force in same direction. Generator: emf and i in the same direction—speed and force in the opposite direction.

If the current-carried conductor is moving under a magnetic flux density, it is possible to identify its operation as a motor or as a generator analyzing the sense of both force and speed.

The motor operation is characterized when both force and speed are concordant. Otherwise, in generator operation the sense of both force and speed are opposite.

We can also identify the generator operation when the direction of emf and current i are concordant. Otherwise, motor operation is identified when the emf and i are in the opposite sense. Figure 4.13 shows both types of operation.

Steady state × transient analysis

Two kinds of analysis are applied when an engineer needs to design a drive system using an electromagnetic device. The simple one is the steady state analysis when we want to evaluate all values involved in the electromagnetics device operating continuously. In the steady state, all values are constant and normally reproduce the rated values of the device.

Otherwise, the transient analysis is applied when we need to evaluate the time-dependence of the main values during the start process or during an electrical or mechanical fault. This kind of study is very important, not only for defining protection devices, but also for specifying electrical and/or mechanical components.

Example 4.3

A 25 Ω resistor is connected at the A and B ends of the coil of example 4.2.2 as shown in figure 4.14. Both the inductance and the resistance of the coil should be neglected. Determine:

(a) the steady state electrical current in the coil;
(b) the steady state force on the coil;

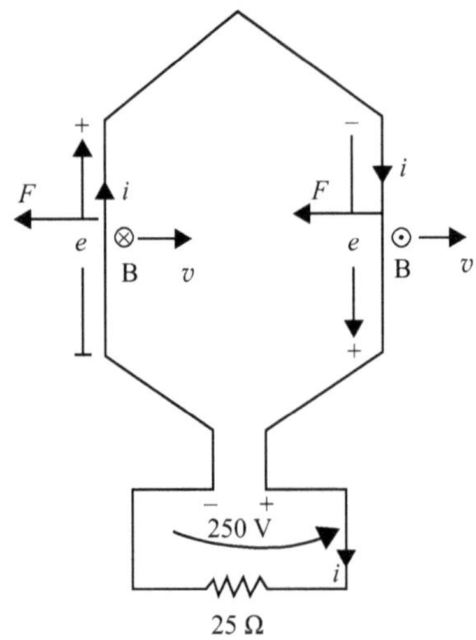

Figure 4.14. Coil of example 4.2.2 feeding a 25 Ω resistor.

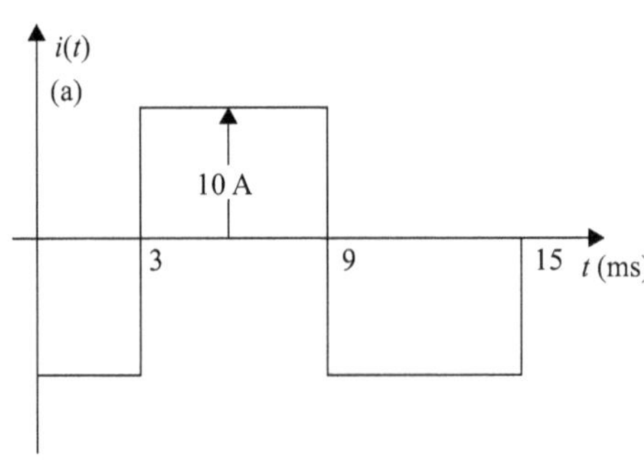

Figure 4.15. The electrical current as a function of *t*.

(c) the mechanical power developed by the force evaluated at (b);
(d) the Joule losses in the resistor;
(e) compare the results obtained in both item (c) and (d).

Solution

(a) The steady state electric current

Figure 4.15 shows the time-dependent emf in the coil. Neglecting both the resistance and the inductance of the coil, this emf is the voltage at its ends.

Applying Ohm's law, the electric current is given by:

$$i = \frac{\text{emf}}{R} = \frac{250}{25} = 10\ \text{A}$$

(b) Applying (4.2), the magnetic *force in the coil* is given by:

$$F = 2NBli = 2 \times 5 \times 1.0 \times 50 \times 10^{-2} \times 10 = 50 \text{ N}$$

The sense of force does not change with the movement.

(c) *The mechanical power* is obtained by:

$$P_{mec} = Fv = 50 \times 50 = 2500 \text{ W}$$

(d) *The Joule losses in the resistor* are given by:

$$P_{Joule} = Ri^2 = 25 \times 10^2 = 2500 \text{ W}$$

(e) Comparing the results obtained from both item (c) and (d), we see that the power transferred to the resistor is the electric power converted from mechanical to electrical in the coil.

4.4 The elementary DC machine

The electromechanical device we have studied so far is a kind of synchronous machine because the period of the induced voltage should be synchronized with the speed.

The synchronization effect requires that the coil should spend half the period of the induced emf wave developing forces or torques. To overcome this requirement, a suitable accessory called a commutator is used.

Figure 4.16 shows the moving coil in three different positions depending on the direction of the magnetic flux density. On the left, the polarity of the ends A and B are (−) and (+), respectively. These polarities are the same at the commutator ends C and D.

When both sides of the coil are crossing the border that divides both poles, a short-circuit is created by the commutator switching as shown in the central position of figure 4.16.

This operation does not cause any electrical damage because at this position not only the magnetic flux density, but also the emf in these lines are null. This position is called the *neutral line* (NL in figure 4.16) of the machine.

Immediately after, the commutator switches again to its third position shown in the right-hand figure 4.16. As the magnetic flux density has moved from N to S, the polarity of the A and B ends have also changed to A (+) and B (–), but with the switching no change is observed in the polarities of both the C and D ends of the commutator.

Figure 4.17 shows the time-dependent representation of both the emf and the commutator voltage.

Using the commutator, no constraints are required between speed and the period of the induced emf during motor operation. The only requirement is to install a sensor to indicate the NL position for the commutator switching.

These kinds of characteristics allow us to easily control the speed of the motor.

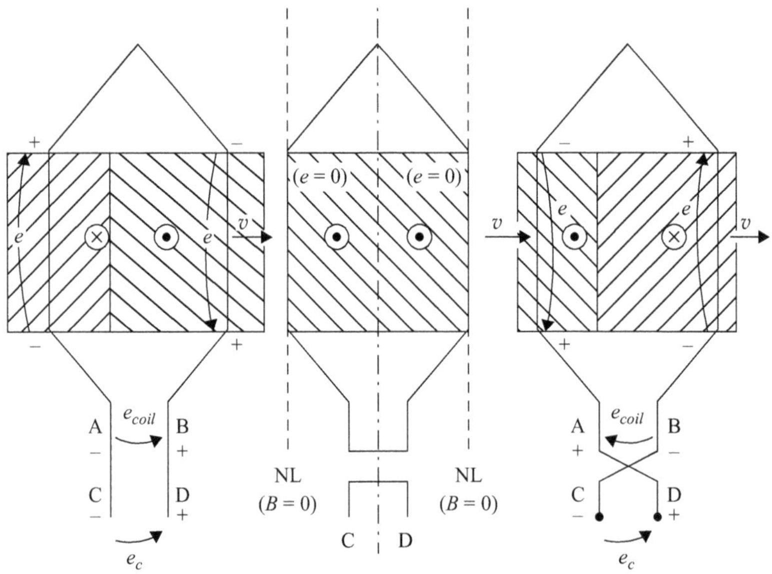

Figure 4.16. The commutator operation. e_{coil}: The polarity of emf changes each half period. e_c: The polarity of the commutator remains constant. B = 0 at the neutral line NL.

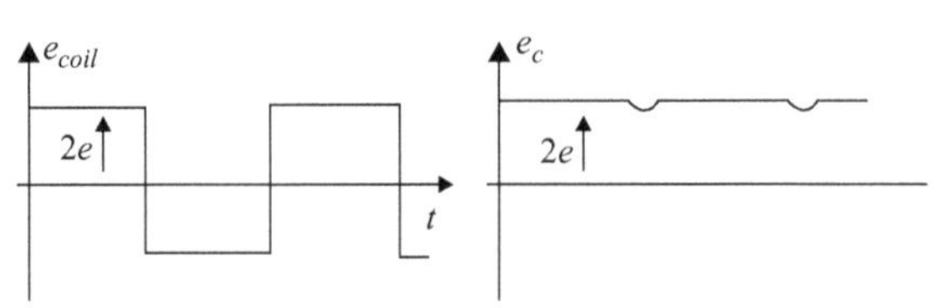

Figure 4.17. The emf and commutator voltage as a function of time. Left: emf × time is an alternate electric voltage. Right: e_C × time is a DC electric voltage with a small perturbation at the commutation point.

To illustrate the transient analysis of an elementary DC generator, example 4.4.1 was created to evaluate the speed time evolution of this kind of machine from the initial instant where the coil is at rest to the final steady state speed.

Example 4.4

A commutator is installed at the ends of the coil of example 4.3.2 allowing it to work as a DC machine. Its weight is 1 kg. Determine the time evolution of the speed when an external force of 50 N is suddenly applied to the coil at rest.

Solution
This is a case of a time-domain analysis. To find the solution, we must identify the equations for each part of the system.

Concerning the electric subsystem, we have:

$$e = Ri \tag{4.3}$$

Neglecting both the friction ($K = 0$) and the spring effect ($C = 0$) related to the mechanical part, we have:

$$F_{\text{ext}} - F_{\text{mag}} = m\frac{dv}{dt} \tag{4.4}$$

Finally, for the electromechanical part we have:

$$e = 2NBlv \tag{4.5}$$

$$F_{\text{mag}} = 2NBli \tag{4.6}$$

Combining these equations, we get:

$$F_{\text{ext}} - 2NBl\frac{2NBl}{R}v = m\frac{dv}{dt}$$

Substituting numerically by its value, the final differential equation becomes:

$$\frac{dv}{dt} + v = 50 \tag{4.7}$$

There are two solutions for (4.7). The first is a constant that we can write as:

$$v_1 = V_o$$

Substituting V_o in (4.7), we obtain:

$$V_o = 50$$

The second is the response free solution where:

$$\frac{dv_2}{dt} + v_2 = 0$$

Whose general solution is:

$$v_2 = k_1 e^{k_2 t}$$

Substituting in the previous equation, we obtain:

$$k_2 k_1 e^{k_2 t} + k_1 e^{k_2 t} = 0$$

The result is:

$$k_2 = -1$$

The final solution for v is:

$$v = v_1 + v_2 = 50 + k_1 e^{-t}$$

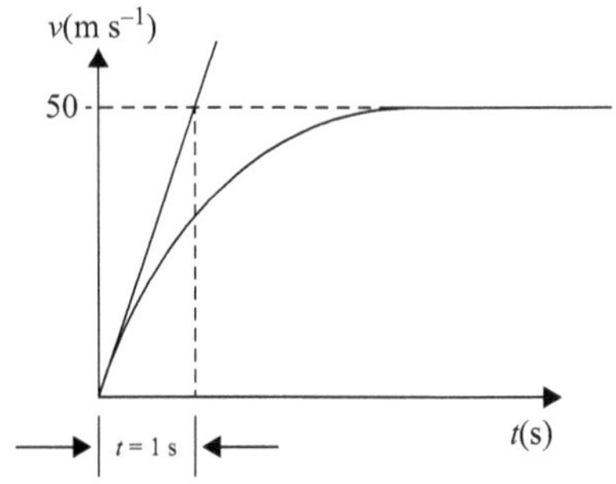

Figure 4.18. Speed × time—example 4.4.1.

As the coil is at rest in $t = 0$, the result is:

$$0 = 50 + k_1$$

Or:

$$k_1 = -50$$

The result is:

$$v = 50(1 - e^{-t}) \tag{4.8}$$

Figure 4.18 shows the time-dependent representation curve of the speed.

The harmonic steady state

Since the electrical energy distribution was made by AC current at the end of the 19th century, most electromechanical devices were designed to operate under this condition. Not only electrical motors and generators, but also several electromechanical accessories were adapted for working with AC excitation.

Example 4.5

An N turns coil is moving under an infinite sequence of static sinusoidal magnetic flux density distribution with speed v. This kind of magnetic field distribution is normally created by a set of permanent magnets that are not shown in figure 4.19.

Each magnet pole has a rectangular shape of dimensions $t_p \times L$ (figure 4.20). Determine:

(a) the rms value of the induced emf in the coil;
(b) the frequency of the induced emf;
(c) the electric current carried by a resistor R connected to the coil ends;
(d) the force F developed by the coil as a function of t and its average value, sketch the function $F(t)$;
(e) the mechanical power developed by coil, compare it with the Joule losses in the resistor.

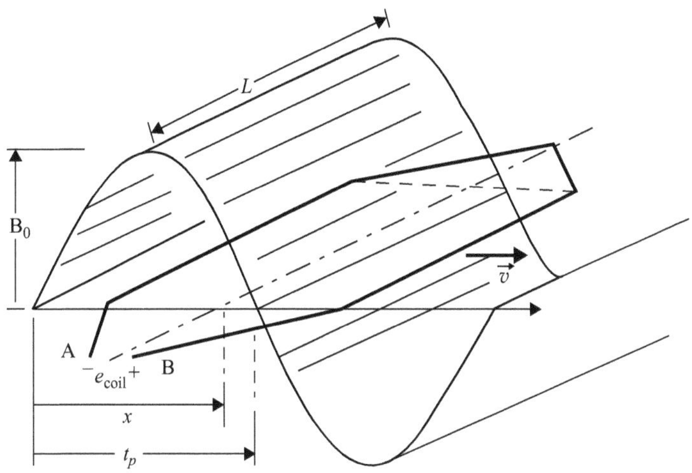

Figure 4.19. Moving coil under sinusoidal magnetic flux density— three-dimensional view.

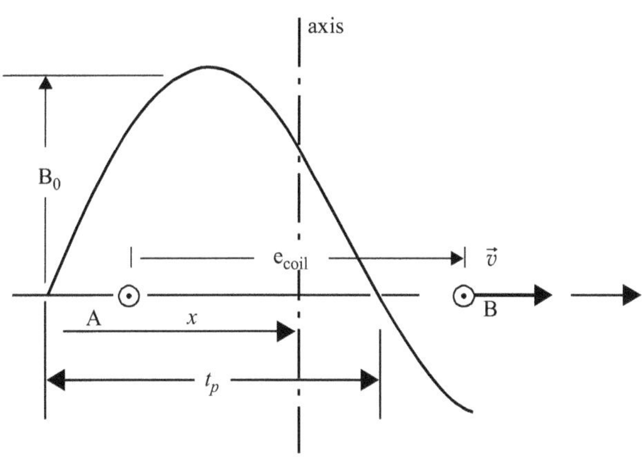

Figure 4.20. Frontal view of a moving coil under sinusoidal magnetic flux density distribution.

Solution

(a) The induced emf in the moving coil is given by the algebraic sum of the induced emf of each side of the coil:

$$e_{coil} = e_A - e_B$$

where:

$$e_A = NB\left(x - \frac{t_p}{2}\right)Lv$$

$$e_B = NB\left(x + \frac{t_p}{2}\right)Lv$$

as:

$$B(x) = B_0 \operatorname{sen}\left(\frac{\pi}{t_p}x\right) \tag{4.9}$$

results:

$$e_A = -e_B = -NB_0 \cos\left(\frac{\pi}{t_p}x\right)Lv$$

Therefore:

$$e_{\text{coil}} = -2NB_0 \cos\left(\frac{\pi}{t_p}x\right)Lv$$

As the coil is moving at speed v, its position can be written as:

$$x = x_0 + vt$$

So, the induced emf in the coil is given by:

$$e_{\text{coil}}(t) = -2NLvB_0 \cos(\omega t + \alpha)$$

where:

$$\omega = \frac{\pi}{t_p}v \text{ and } \alpha = \frac{\pi}{t_p}x_0$$

Applying the right-hand rule, the polarities of the coil ends are B (+) and A (−).

Using the complex notation, the phasor of $e_{\text{coil}}(t)$ is written as:

$$\dot{E} = -E\angle\alpha$$

where:

$$E = \frac{2NLvB_0}{\sqrt{2}}(V) \tag{4.10}$$

is the rms value of the induced emf, while its phase is:

$$\alpha = \frac{\pi}{t_p}x_0 \tag{4.11}$$

(b) Concerning the frequency, we have:

$$\omega = 2\pi f \text{ and } \omega = \frac{\pi}{t_p}v$$

Comparing both expression, results in:

$$f = \frac{v}{2t_p}(\text{Hz}) \tag{4.12}$$

(c) The current carried by the resistor R is obtained applying Ohm's law

$$\dot{I} = -\frac{E}{R}\angle\alpha$$

Using a time-domain representation, results in:

$$i(t) = -\sqrt{2}\frac{E}{R}\cos(\omega t + \alpha)\ (\text{A}) \tag{4.13}$$

(d) The interaction between both electric current in the coil and the magnetic flux density distribution makes the sense of the exerted force opposite to the movement and given by:

$$F = 2NBLi$$

Due to the symmetry, the force on both sides of the coil is the same. Therefore, we can write:

$$F_{\text{coil}} = 2NB\left(x - \frac{t_p}{2}\right)Li(t)$$

as:

$$B\left(x - \frac{t_p}{2}\right) = -B_0\cos(\omega t + \alpha)$$

resulting:

$$F_{\text{coil}} = \frac{2\sqrt{2}NB_0LE}{R}\cos^2(\omega t + \alpha)$$

Applying the identity:

$$\cos^2\theta = \frac{1}{2}(1 + \cos 2\theta)$$

The final expression of the force in the coil is:

$$F_{\text{coil}} = \frac{\sqrt{2}NB_0LE}{R} + \frac{\sqrt{2}NB_0LE}{R}\cos(2\omega t + 2\alpha) \tag{4.14}$$

The first term of (4.14) is independent of the time. The second is a sinusoidal time-dependent function (figure 4.21) with double-frequency. As the average value of the last term is null, the average value of the function (4.14) is given by:

$$F_{\text{av}} = \frac{\sqrt{2}NB_0LE}{R}\ (\text{N}) \tag{4.15}$$

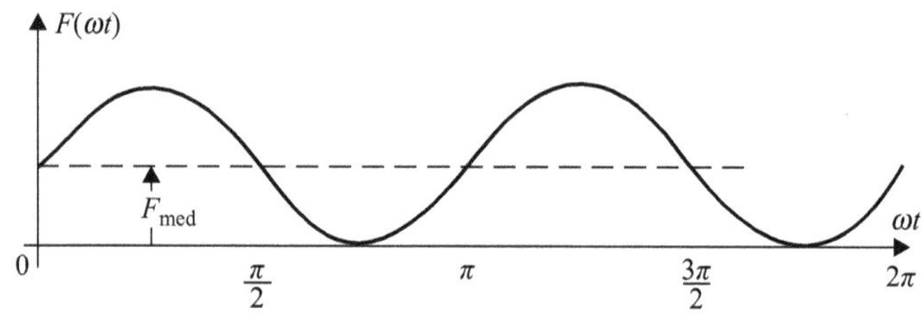

Figure 4.21. The force in the coil as a function of time.

Remarks: The average force F_{av} is the force that the prime mover should apply in the coil for transfer electric power to the resistor R.

(e) The mechanical power applied to the coil by the prime mover is:

$$P_{mec} = F_{av} \times v = \frac{\sqrt{2}\, NB_0LE}{R} v \tag{4.16}$$

as:

$$E = \frac{2NLvB_0}{\sqrt{2}} \text{ (V)}$$

results:

$$NLB_0 = \frac{\sqrt{2}\, E}{2v}$$

Substituting this value in (4.16), we get:

$$P_{mec} = \frac{E^2}{R} \tag{4.17}$$

Remarks: As we did not consider any losses in the system, all mechanical power is converted into electrical power that is transferred to the resistor. This is a typical generator operation.

The rotating electrical machine fundamentals

We have discussed the behavior of a moving coil under an infinite sequence of planar N–S magnetic poles. It is only possible to build this kind of device if the magnetic poles are distributed on a cylindrical surface. This is the origin of a rotating electrical machine.

Figure 4.22 shows a cylindrical surface divided into $2p$ sections. Each section covers a pole pitch N or S and has one or more coils associated with it.

The coils are connected in series or parallel depending on the voltage being used.

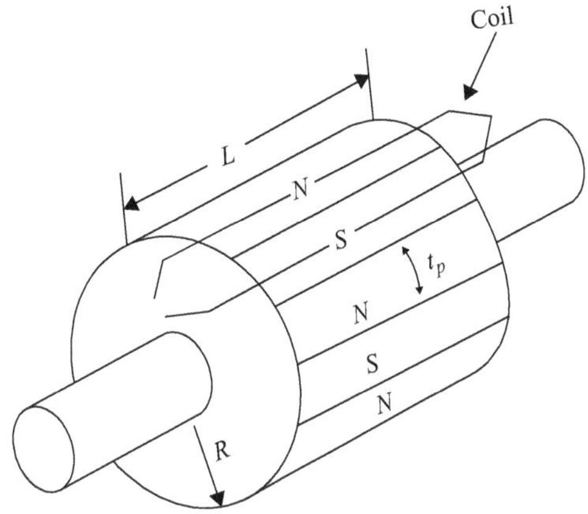

Figure 4.22. Cylindrical surface with $2p$ poles (p poles N and p poles S).

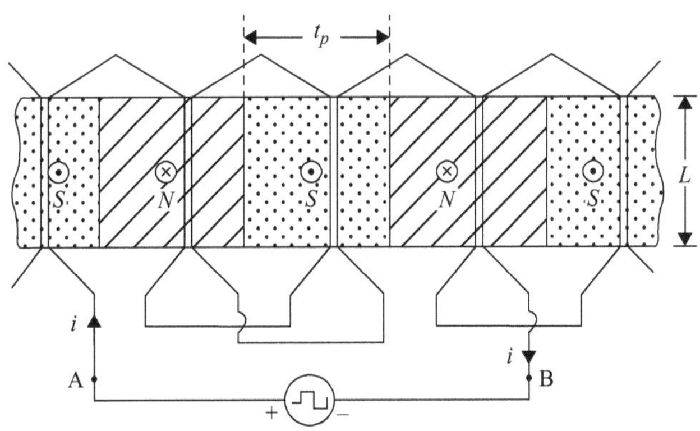

Figure 4.23. Winding of four poles.

Example 4.6

A cylinder (R = 20 cm and length L = 40 cm) of an electric machine presents four magnetic poles uniformly distributed on its surface. A set of four coils of 10 turns each are associated in series via its commutator as shown in figure 4.23. A square wave source voltage of 450 V and frequency of 50 Hz is feeding the winding. The resistance of the winding is 0.96 Ω and its inductance can be neglected. The magnitude of the magnetic flux density is 0.4 Wb m^{-2} and can be considered constant in the pole pitch.

Determine:

(a) the speed of the winding;
(b) the magnetic force developed by each coil;
(c) the magnetic torque developed by the winding.

Solution

This is a steady state analysis with a constant speed synchronized with the frequency of the source. In the synchronized operation, the time for turning a pair of poles is the period of the voltage source wave. If this condition is not obeyed, no force (or torque) is developed.

(a) As the cylinder has four poles on its surface, the time for completing two poles—or half circumference—is $T = \frac{1}{f} = 20$ ms. Therefore, the speed of the winding is given by:

$$v = \frac{2t_p}{T} = 2t_p f \tag{4.18}$$

where:

$$t_p = \frac{\text{perimeter}}{2p} = \frac{\pi R}{p} = 0.10\pi \text{ m}$$

That results:

$$v = \frac{2 \times 0.1\pi}{20 \times 10^{-3}} = 10\pi \text{ m s}^{-1}$$

The angular speed is given by:

$$\omega = \frac{v}{R} = \frac{10\pi}{0.2} = 50\pi \text{ rad s}^{-1}$$

As $\omega = 2\pi\, n$ results:

$$n = \frac{\omega}{2\pi} = \frac{50\pi}{2\pi} = 25 \text{ rps}(1500 \text{ rpm})$$

(b) From the electric subsystem, we can write:

$$V = E + ri \tag{4.19}$$

From the electromechanical device, we can write:

$$F_{\text{mag}} = 8NBLi \tag{4.20}$$

Remarks: Each coil has $2N$ conductors. The winding has four coils associated in series. So, the total number of conductors is $4 \times 2N = 8N$.

From the mechanical subsystem:

$$v = 10\pi \text{ m s}^{-1}$$
$$\omega = 50\pi \text{ rad s}^{-1}$$

The induced emf in the coil is extracted from:

$$E = 8NBLv = 8 \times 10 \times 0.4 \times 0.4 \times 10\pi = 402 \text{ V}$$

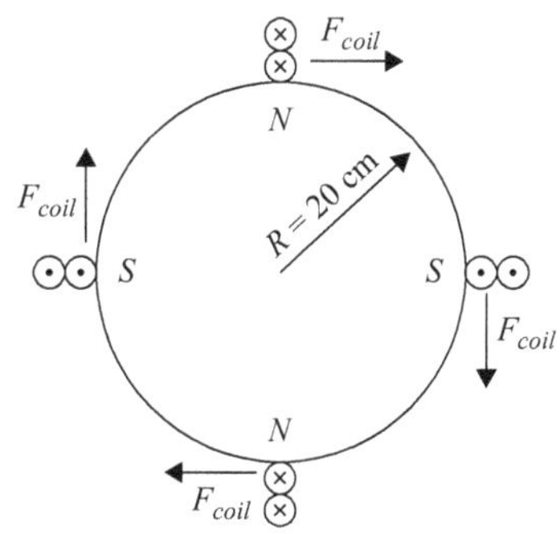

Figure 4.24. The torque developed by the winding.

From (4.19), the electric current is given by:

$$i = \frac{V - E}{r} = \frac{450 - 402}{0.96} = 50 \text{ A}$$

Therefore, the total magnetic force developed by the winding is given by:

$$F_{\text{mag}} = 8NBLi = 8 \times 10 \times 0.4 \times 0.4 \times 50 = 640 \text{ N}$$

(c) The developed magnetic torque is obtained from (figure 4.24):

$$T_{\text{mag}} = F_{\text{mag}} \times R = 640 \times 0.2 = 128 \text{ Nm}$$

Remarks: The force developed by each coil is $F_{\text{coil}} = \frac{F_{\text{mag}}}{4} = 160$ N.

4.5 The Faraday disc

The monopolar generator is the easiest electric generator we can build. Known as the Faraday disc, this generator is made by a conductor disc crossed by a magnetic flux density produced by a coil connected to a DC source or by a permanent magnet.

As the disc is turning, an emf is induced in two contact points, one located at the center of the disc ($r = a$) and the other in its external surface ($r = b$), as shown in figure 4.25.

A constant magnetic flux density B is normal to the disc surface which is turning at angular speed ω.

The emf is collected by two brushes located at $r = a$ and $r = b$ and can be evaluated isolating a radial differential strip of length dr as shown in figure 4.26.

The induced emf in this differential strip is given by:

$$de = B \times dr \times v \tag{4.21}$$

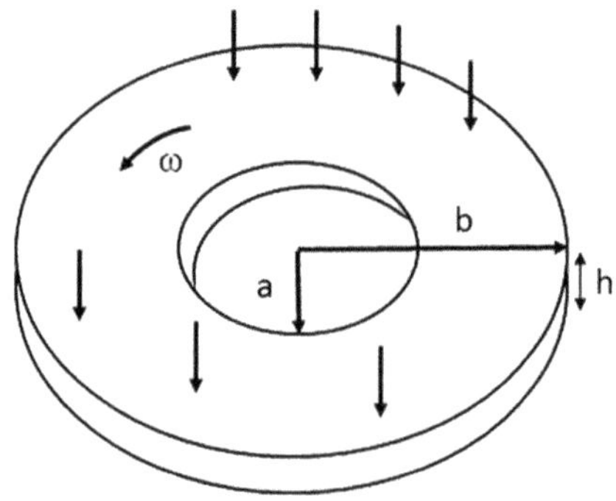

Figure 4.25. Properties of a Faraday disc generator. Two contact points are located at $r = a$ and $r = b$. h = thickness, σ = conductivity (S m^{-1}).

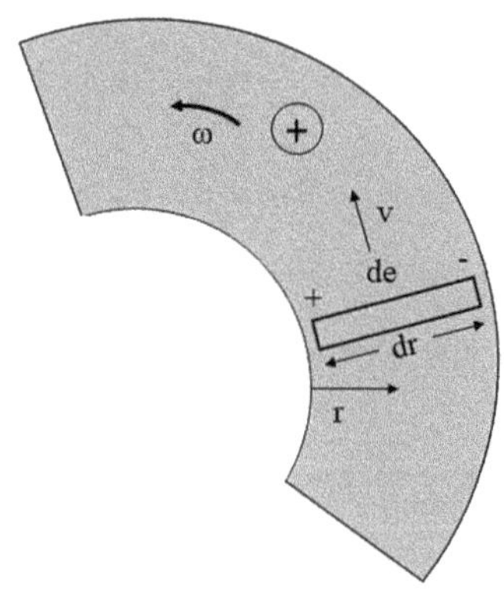

Figure 4.26. The differential strip, dr = differential strip. The polarity of de is obtained by the right-hand rule.

Since $v = \omega \times r$:

$$de = B \times dr \times \omega \times r \tag{4.22}$$

The resulting emf between both the internal and external radius is obtained from:

$$e = \omega B \int_a^b r dr$$

Therefore:

$$e = \frac{\omega B}{2}[b^2 - a^2](V) \tag{4.23}$$

To complete the equivalent electric circuit of a Faraday Disc, we need to evaluate the electrical resistance. We can evaluate this parameter through the differential ring extracted from the disc whose dimensions are:

- r: the internal radius
- $r + dr$: the external radius
- h: the thickness

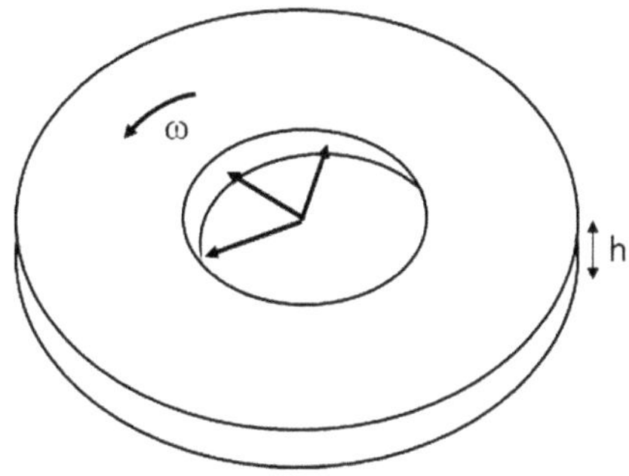

Figure 4.27. Differential ring. The electrical current line has a radial symmetry. The electrical current is coming through the internal surface and comes out in the external one.

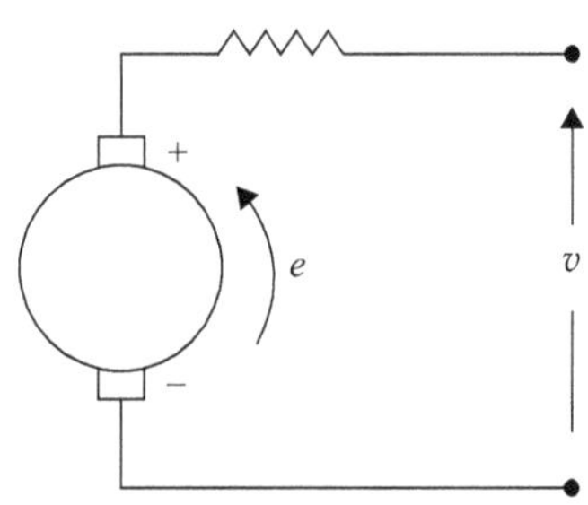

Figure 4.28. Equivalent electric circuit of the Faraday disc. $e = \frac{\omega B}{2}[b^2 - a^2]$ = back emf (V). $R = \frac{1}{2\pi h\sigma} \times \ln\frac{b}{a}$ = internal resistance (Ω).

As the electric current lines have a radial distribution as shown in figure 4.27, the differential ring's electrical resistance is:

$$dR = \frac{1}{\sigma} \times \frac{dr}{2\pi rh}$$

The disc's resistance is obtained from the mathematical integration of the following expression:

$$R = \frac{1}{2\pi h\sigma} \times \int_a^b \frac{\mathrm{d}r}{r}$$

Therefore,

$$R = \frac{1}{2\pi h\sigma} \times \ln\frac{b}{a}(\Omega) \tag{4.24}$$

The equivalent electric circuit of the Faraday Disc could be represented by the electric circuit shown in figure 4.28.

To use a Faraday disc as a brake, we need to short-circuit its ends. Doing this, an electric current will flow from the internal to the external surface where the magnitude of the surface is given by:

$$I = \frac{e}{R} = \frac{\pi\omega B\sigma h}{\ln\frac{b}{a}}[b^2 - a^2](A) \tag{4.25}$$

Once an electrical current interacts with the magnetic flux density, an electromagnetic torque is developed.

The differential electromagnetic torque is given by:

$$dC = dF \times 2r \tag{4.26}$$

Based on figure 4.29, the differential electromagnetic force will be:

$$dF = B \times dr \times dI \tag{4.27}$$

As the current lines have a radial symmetry, the differential current could be evaluated by the proportionality:

$$I \rightarrow 2\pi r$$
$$dI \rightarrow rd\theta$$

Therefore:

$$dI = \frac{I}{2\pi}d\theta$$

As a result, the electromagnetic torque could be evaluated by the mathematical integration:

$$C = B \times \frac{I}{2\pi}\left[\int_a^b rdr\right]\left[\int_0^{2\pi} d\theta\right]$$

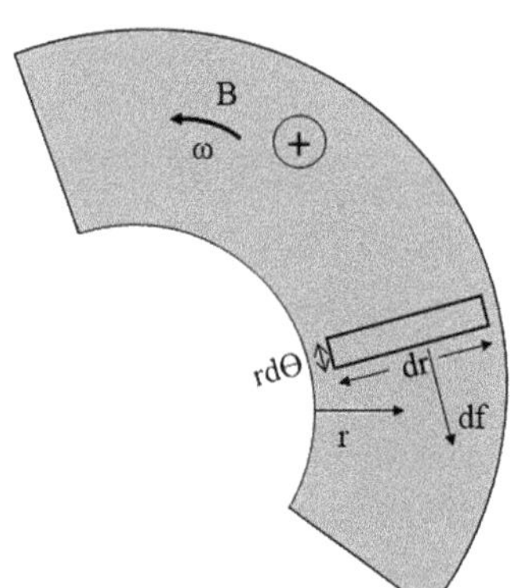

Figure 4.29. Geometry for electromagnetic torque evaluation.

Therefore:

$$C = BI[b^2 - a^2] \tag{4.28}$$

Substituting the current with its value extracted from (4.25), results:

$$C = \frac{\pi\omega\sigma h}{ln\frac{b}{a}}[b^2 - a^2]^2 B^2 (Nm) \tag{4.29}$$

Remarks: The sense of the electromagnetic torque is opposite to the angular speed, characterizing the brake action.

Another kind of Faraday disc is the one shown in figure 4.30. In this case, the magnetic flux density is acting in a small portion of the disc generating an electrical current distribution on its surface.

No brushes are necessary to promote the circulation of electrical current because the current lines are confined in a small portion of the disc. This type of current line distribution is normally called eddy current distribution.

The eddy current brake is used by racing cars to minimize friction brake wear.

As the eddy current Faraday disc is excited by a concentrated magnetic pole that covers only a small part of the disc's surface, the expression of the developed torque should be improved by the introduction of a correcting factor $f_c < 1.0$, as shown in equation (4.30).

$$C = f_c \frac{\pi\omega\sigma h}{ln\frac{b}{a}}[b^2 - a^2]^2 \, B^2 (\mathrm{Nm}) \tag{4.30}$$

The correcting factor f_c could be evaluated by both tests or electromagnetic numerical simulation.

Example 4.7

The eddy current Faraday disc of figure 4.31 has the following characteristics:
conductivity $\sigma = 5.61 \times 10^7$ S m^{-1}
internal radius $a = 30$ mm

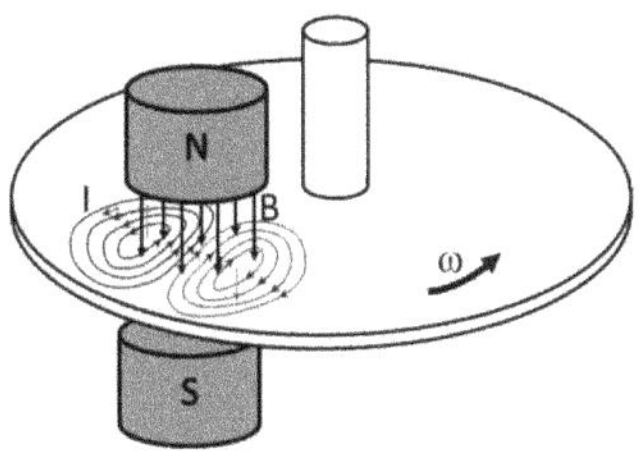

Figure 4.30. The eddy current distribution in a Faraday disc.

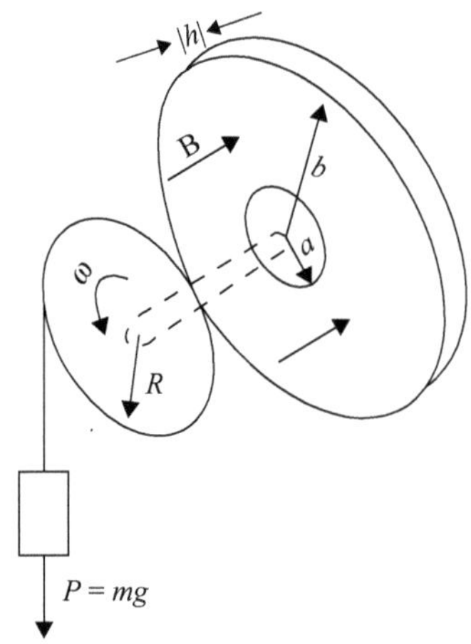

Figure 4.31. System of example 4.5.1.

external radius b = 350 mm
thickness h = 12.7 mm

Such a disc is excited by a normal magnetic flux density of $B = 0.4$ Wb m^{-2}. As the magnetic excitation covers only a small surface of the disc, the correcting factor $f_c = 0.15$ should be considered.

This disc is part of the system used for limiting the angular speed of the suspending mass $m = 1000$ kg. The total moment of inertia of the moving part is $J = 5000$ kgm^2. All friction losses should be neglected. Consider $g = 9.8$ m s^{-2} and $R = 0.5$ m:

(a) write the differential equation that governs the angular speed of the system;
(b) determine the maximum speed of the system;
(c) plot the curve $\omega \times t$.

Solution

(a) The balance torque equation of the system is:

$$mgR - C_{\text{mag}} = J\frac{d\omega}{dt} \tag{4.31}$$

where C_{mag} is given by (4.30) that could be written as:

$$C_{\text{mag}} = K_1\omega \tag{4.32}$$

with

$$K_1 = f_c\frac{\pi\sigma h}{\ln\frac{b}{a}}[b^2 - a^2]^2 B^2$$

Substituting in (4.31), we get:

$$mgR - K_1\omega = J\frac{d\omega}{dt}$$

or

$$\frac{d\omega}{dt} + \frac{K_1}{J}\omega = \frac{mgR}{J} \tag{4.33}$$

The constant solution of (4.33) should satisfy:

$$\frac{K_1}{J}\omega_1 = \frac{mgR}{J}$$

or

$$\omega_1 = \frac{mgR}{K_1} \tag{4.34}$$

The second one should satisfy:

$$\frac{d\omega_2}{dt} + \frac{K_1}{J}\omega_2 = 0 \tag{4.35}$$

whose solution is:

$$\omega_2 = k_1 e^{k_2 t} \tag{4.36}$$

From (4.35), we get:

$$k_2 k_1 e^{k_2 t} + \frac{K_1}{J} k_1 e^{k_2 t} = 0$$

That results in:

$$k_2 = -\frac{K_1}{J}$$

The general solution of (4.31) is:

$$\omega = \omega_1 + \omega_2 = \frac{mgR}{K_1} + k_1 e^{-\frac{t}{\tau}} \tag{4.37}$$

where:

$$\tau = \frac{J}{K_1}\ (s)$$

is the mechanical time constant of the system.

Supposing that for $t = 0$ the angular speed is $\omega(0) = 0$, from (4.37) we obtain:

$$k_1 = -\frac{mgR}{K_1}$$

Finally, the solution for (4.33) is:

$$\omega = \frac{mgR}{K_1}(1 - e^{-\frac{t}{\tau}}) \tag{4.38}$$

(b) The maximum speed of the system is the one obtained imposing $t \to \infty$ in (4.38), that results:

$$\omega_f = \frac{mgR}{K_1} \tag{4.39}$$

(c) Figure 4.32 shows the angular speed × time curve of example 4.5.1.

The numerical values of the solution are:

$$K_1 = f_c \frac{\pi \sigma h}{ln\frac{b}{a}}[b^2 - a^2]^2 B^2 = 323 \text{ Nm. s rad}^{-1}$$

$$\tau = \frac{J}{K_1} = 15.5 \text{ (s)}$$

$$\omega_f = \frac{mgR}{K_1} = 15.2 \text{ rad s}^{-1}$$

$$n_f = \frac{\omega_f}{2\pi} = 2.4 \text{ rps (144 rpm)}$$

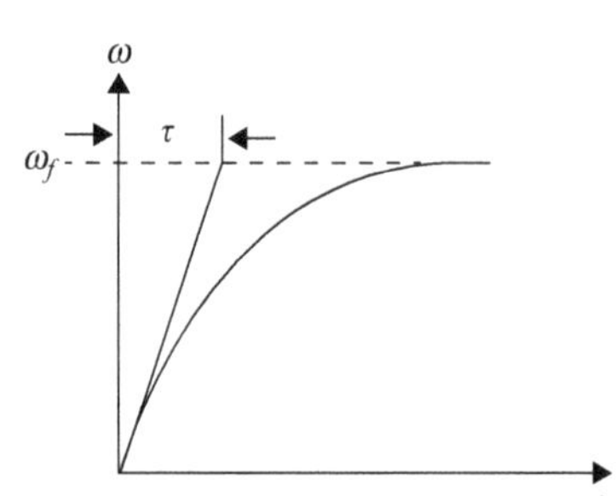

Figure 4.32. Angular speed × time.

4.6 The axisymmetric symmetry

Several electromechanical devices are designed to develop not only forces but also working as a position (or speed) transducer with axisymmetric (or cylindrical) geometry.

A loudspeaker is an example, which is built with a high permeability magnetic structure embedded inside a permanent magnet and an N turns coil located in the air-gap. The magnetic excitation is made by a cylindrical PM that creates a radial magnetic flux density distribution in the air-gap.

The coil is fed by an electrical current issuing from the receiver. As the coil is attached to a complex elastic moving frame, the dynamic equation of the system should include several parameters like the spring constant, friction (or damping) coefficient and mass. Both the spring constant and the friction coefficient could be non-linear parameters that depend on the speed, temperature, humidity and others that require a complex mathematical formulation for optimizing the design of this kind of device.

Figure 4.33 presents a cylindrical core of loudspeaker showing the cylindrical coil located at the air-gap.

All components of a loudspeaker are presented in figure 4.34.

Figure 4.35 shows a sketch of the loudspeaker system. The electrical source is composed of a back emf $e(t)$ and an internal resistance r_1.

The electromechanical side is composed of the magnetic core considered built of an ideal magnetic material and an N turns coil that is modeled by a single resistance r_c. As the inductance of the coil is very small, it could be neglected in this approach.

The mechanical side is the moving frame of the loudspeaker.

The equation system for each side of the loudspeaker system is:

$$\text{Electric circuit: } e(t) = (r_1 + r_c)i(t) \tag{4.40}$$

$$\text{Electromechanical side: } F_{\text{mag}} = 2\pi RNBi(t) \tag{4.41}$$

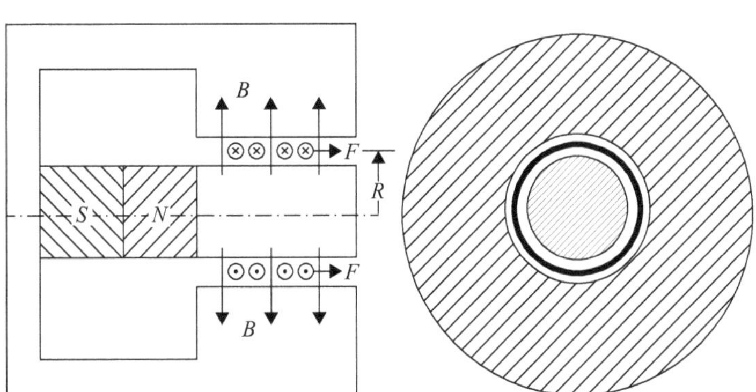

Figure 4.33. The core of a loudspeaker. The magnetic flux density distribution is radial.

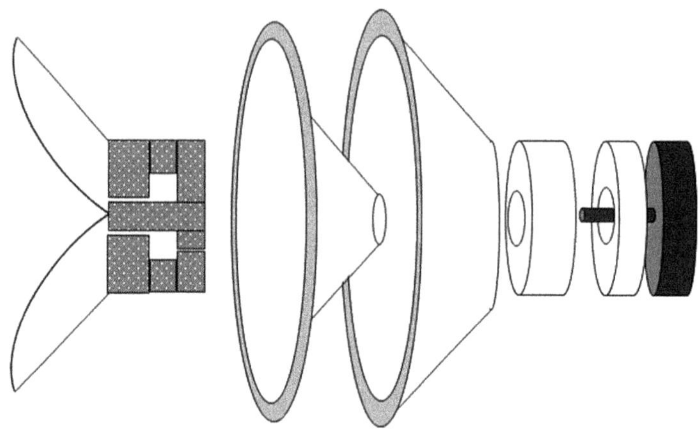

Figure 4.34. Components onside a loudspeaker.

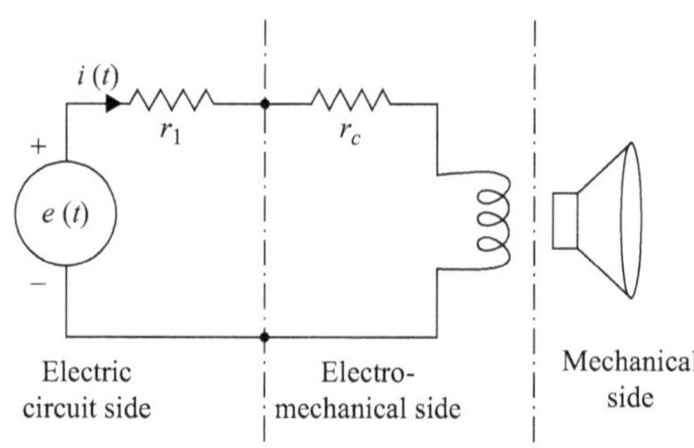

Figure 4.35. The loudspeaker electromechanical system. The inductance of the coil is very small and could be neglected.

$$\text{Mechanical side: } F_{\text{mag}} = m\frac{d^2x}{dt^2} + c\frac{dx}{dt} + Kx \tag{4.42}$$

After some mathematical manipulation, we get:

$$K_l e(t) = m\frac{d^2x}{dt^2} + c\frac{dx}{dt} + Kx \tag{4.43}$$

where:

x: coil position measured from a fixed frame;
$K_l = \frac{2\pi RNB}{r_1 + r_c}$: electromechanical constant (N V^{-1});
m: mass of the moving parts of the loudspeaker;
c: damping constant (Ns m^{-1});
k: elastic constant (N m^{-1});

The general solution of (4.42) can be written as:

$$x(t) = x_1(t) + x_2(t)$$

where $x_1(t)$ is the first solution of (4.43), also called the natural response of the system, issuing from the solution of the following differential equation:

$$m\frac{d^2x_1}{dt^2} + c\frac{dx_1}{dt} + Kx_1 = 0 \tag{4.44}$$

and $x_2(t)$ is the forced response whose solution is linked with the excitation. This solution is also called the steady state solution of the dynamic system.

$$m\frac{d^2x_2}{dt^2} + c\frac{dx_2}{dt} + Kx_2 = K_l e(t) \tag{4.45}$$

Equation (4.44) is a second-order differential linear equation. Its characteristic equation is written as:

$$mr^2 + cr + k = 0 \tag{4.46}$$

The roots of this equation are:

$$r_1 = \frac{-c + \sqrt{c^2 - 4mk}}{2m} \qquad r_2 = \frac{-c + \sqrt{c^2 + 4mk}}{2m}$$

Three cases should be considered:

Case 1: $\mathbf{c^2 - 4mk > 0}$ (super – damping)

In this case, both roots r_1 and r_2 are real numbers and different and

$$x_1 = c_1e^{r_1t} + c_2e^{r_2t} \tag{4.47}$$

As c, m and k are positives, it results $\sqrt{c^2 - 4mk} < c$. Therefore, both r_1 and r_2 are negatives. In this case for $t \to \infty$ results $x \to 0$. Figure 4.36 shows for the time evolution of the solution that oscillation is not admitted. This is a typical time evolution of a super-damping system where the damping force (high viscosity and others) is higher than the elastic force or the frame mass is small.

Case 2: $c^2 - 4mk = 0$ (critical – damping)

In this second case, both roots r_1 and r_2 are the same

$$r_1 = r_2 = -\frac{c}{2m}$$

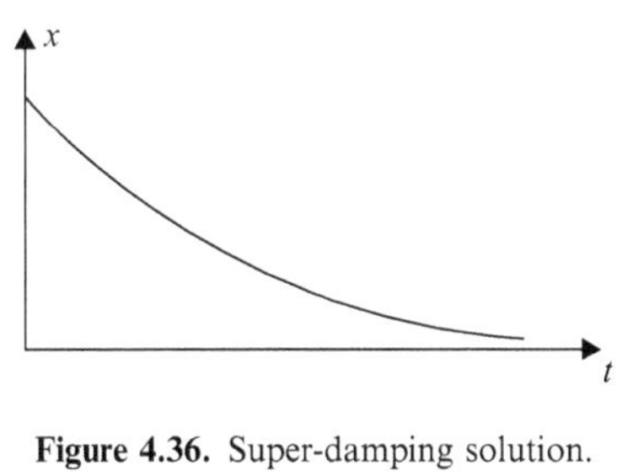

Figure 4.36. Super-damping solution.

The solution is given by:

$$x_1 = (c_1 + c_2 t)e^{-\left(\frac{c}{2m}\right)t} \tag{4.48}$$

Figure 4.37 shows a typical plot of this kind of solution. The damping is just enough to avoid any vibration. With any reduction in the viscosity, the vibration effect will appear, as we will see in the next case.

Case 3: $c^2 - 4mk < 0$ (sub – damping)

In this last case, both roots are complex:

$$r_1 = -\frac{c}{2m} + j\omega \quad r_2 = -\frac{c}{2m} - j\omega$$

where:

$$\omega = \frac{\sqrt{4mk - c^2}}{2m}$$

The solution is given by:

$$x_1 = e^{-\left(\frac{c}{2m}\right)t}(c_1 \cos\omega t + c_2 \mathrm{sen}\omega t) \tag{4.49}$$

Figure 4.38 shows a damped vibration by the factor $e^{-\left(\frac{c}{2m}\right)t}$. Since both c and m are positives the final position is that $t \to \infty$ results in $x \to 0$.

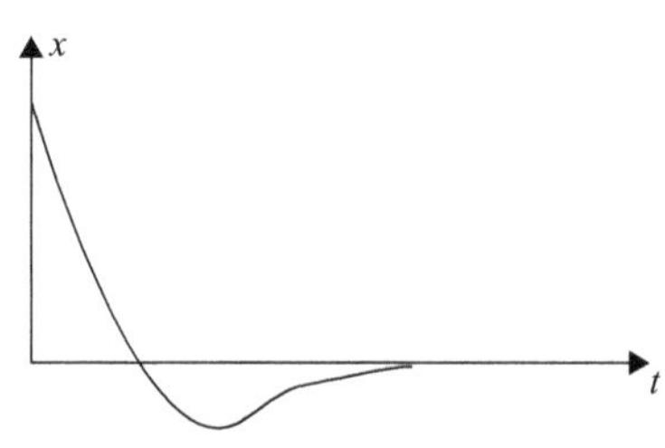

Figure 4.37. Critical damping solution.

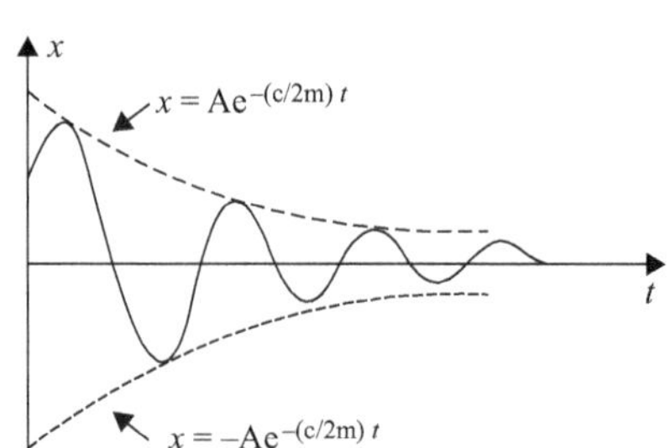

Figure 4.38. Sub-damping solution.

As the system is subjected to external forces issuing from the electromagnetic interaction, we must find a solution for:

$$m\frac{d^2x_2}{dt^2} + c\frac{dx_2}{dt} + Kx_2 = K_l e(t)$$

The typical external force exerted in this kind of device is a periodic force since the back emf is normally a sinusoidal time-dependent function. Supposing that:

$$e(t) = E_m \cos \omega_0 t \;\text{ where }\; \omega_0 \neq \omega = \sqrt{k/m}$$

we can apply the complex notation, since the steady state solution will also be a sinusoidal time-dependent function with the same frequency as the source.

As the complex representation (the phasor) of $e(t)$ is:

$$\dot{E} = E_m \angle 0$$

The most common analysis of this kind of problem issues from the lack of damping force in the system ($c = 0$).

Considering $c = 0$ we can re-write (4.45) using complex notation as:

$$(-\omega_0^2 m + k)\dot{x}_2 = K_l E_m \angle 0$$

Remarks: To change the notation from the time-dependent domain to the frequency domain we must apply the following correspondence:

$$\frac{d}{dt} \to j\omega$$

$$x(t) = x_m \cos(\omega t + \alpha) \to \dot{x} = x_m \angle \alpha$$

That results in:

$$\dot{x}_2 = \frac{K_l E_m}{-\omega_0^2 m + k} = \frac{K_l E_m}{m(\omega^2 - \omega_0^2)} \tag{4.50}$$

The time-domain representation of x_2 is:

$$x_2 = \frac{K_l E_m}{m(\omega^2 - \omega_0^2)} \cos \omega_0 t$$

Therefore, the final solution for a non-damping ($c = 0$) system is:

$$x = c_1 \cos \omega t + c_2 \,\text{sen}\, \omega t + \frac{K_l E_m}{m(\omega^2 - \omega_0^2)} \cos \omega_0 t$$

If $\omega_0 = \omega$ the applied frequency reinforces the natural frequency. As a result, big magnitudes of oscillation are observed. This kind of phenomenon is called *resonance*.

4.7 Summary

In this chapter, we presented the three main subsystems which comprise the energy conversion in an electromechanical system: the electric circuit, the mechanical subsystem and the electromechanical device. The energy conversion phenomenon is associated with both the magnetic force and the electromotive force that appears in a conductor, when it is moving in a space subjected to a magnetic field. We saw that an electromagnetic device can operate as a motor or a generator, depending on the directions of the magnetic force and its velocity. Some important basic electromechanical devices were also presented, like the elementary DC machine, the Faraday disc and a loudspeaker with axisymmetric symmetry.

Project

The magnetic system of the MagLev (high speed magnetically levitated train) is built with a slotted track which has a sequence of coils with an 80 cm pole pitch and 1.5 m useful length as shown in figure 4.39. Each coil has two turns and each set of eight coils is in series. The magnetic system located in the train is also built with a sequence of eight poles (four north and four south poles) with the same dimensions. As the train moves the sets of coils are excited in sequence to supply an electric current to the coil continuously in order to maintain energy only in the set of coils under the train. A creative sequence of switches is designed in order to keep only the set under the train energized.

The magnitude of that magnetic flux density is 0.4 Wb m^{-2}.

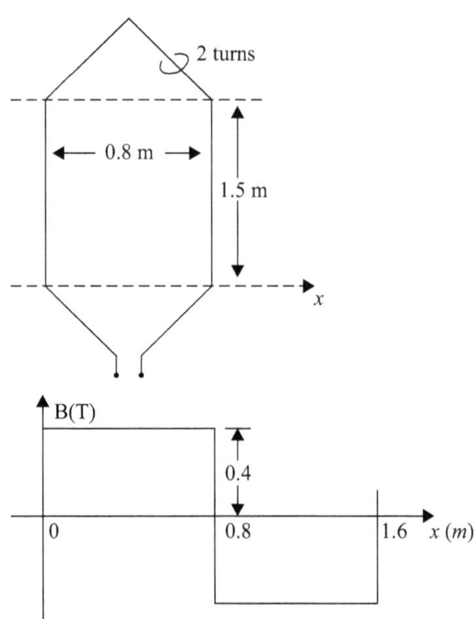

Figure 4.39. Coil geometry (1 pole representation). Magnetic flux density distribution.

The minimum speed for levitation is 180 km h^{-1} and the maximum admissible speed is 540 km h^{-1}.

The resistive force exerted by the air friction is such that:

$$F = 0.09\ v^2$$

Using copper as the material conductor of the coil, that admits a current density of 7 A m^{-2}, you have to find the cross-section of the coil wire during the maximum admissible speed operation.

The project also finds both the emf and the frequency of the electrical source for both limits of the speed.

Answer: 33mm^2; 960 V–2880 V.

Problems

4.1 The system that is shown in figure 4.40 represents the starter system of a certain type of automobile. It is started by pushing the button (1) (or turning a key) causing a magnetic contact (2) to close a switch in another circuit. This drives an electric motor, which, through appropriate shafts and gears, turns the automobile engine. Classify all components of this system based on the chart flow of figure 4.1. Try to write the basic equations involved in this operation.

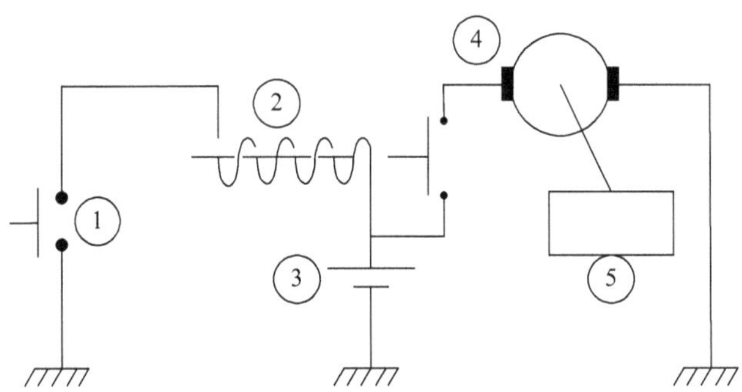

Figure 4.40. Problem 4.1: (1) manually operated switch, (2) magnetic contact,(3) battery, (4) electric motor, (5) shafts and gears to engine.

4.2 Figure 4.41 shows an infinite sequence of magnetic poles north and nouth whose dimensions are given in centimeters. Above this sequence of magnetic poles are two series connected full pitch coils, 40 cm width, 10 turns, traveling at a translational speed of 72 km h^{-1}. Considering that the maximum flux density is 1.2 T, sketch the induced emf wave indicating:

(a) the maximum value of the induced emf;

(b) the period and the frequency of the induced emf;

(c) the average emf value over a half period of the voltage wave.

Answer: 240 V; 20 ms; 50 Hz; 210 V.

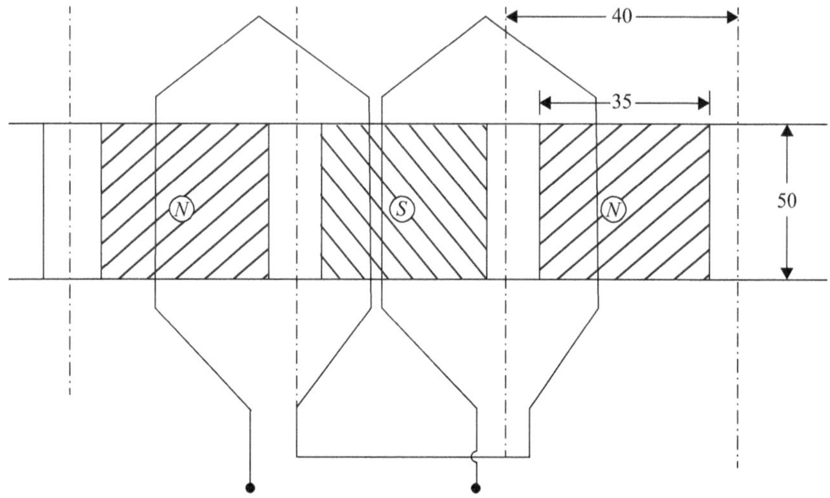

Figure 4.41. Coils traveling above magnetic flux density.

4.3 The system in problem 4.2 is used for dragging a load that requires an average mechanical power of 10.5 kW at 72 km h^{-1}.
Compute:
(a) the average force developed by the coil system;
(b) the maximum value of the developed force;
(c) sketch both the developed force and the current that should be injected in the coil as a function of time;
(d) the maximum exerted force in each side of the coil.

Answer: 525 N; 600*N*; $I_{max} = 50$ A; 150 N.

4.4 Figure 4.42 shows a rotating magnetic system composed of a surface mounted permanent magnetic in a magnetic cylinder. The permanent magnet establishes a constant magnetic flux density of 1.0 T on its surface. The PM surrounds only 165° of the total pole pitch. The diameter at the PM surface is 40 cm and its length is 30 cm. A 10 turns coil is installed longitudinally very close to the PM with its sides separated by 180°.

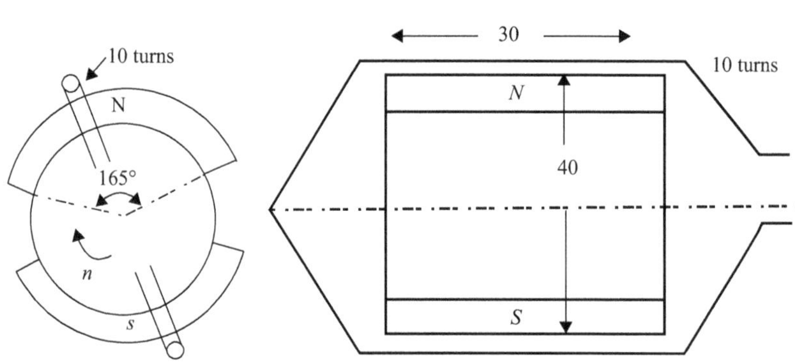

Figure 4.42. Problem 4.4. Left: cross-section. Right: longitudinal view.

The cylinder is spinning at 1800 rpm driven by a prime mover.
Compute:
(a) the maximum value of the induced emf in the coil;
(b) the average induced emf over a half period;
(c) sketch the emf wave indicating its maximum value;
(d) the period and the frequency of the induced emf.

Answers: (a) 226.2 V; (b) 207.2 V; (c) 33.3 ms; (d) 30 Hz.

4.5 At the ends of the coil from problem 4.4 a 20 Ω resistance is connected. Neglecting the inductance and all losses compute:
(a) the maximum value of the electric current carried by the coil;
(b) the maximum value of power delivered to the load;
(c) the average value of power delivered to the load;
(d) the average developed torque required from prime mover;
(e) sketch the time dependency of current, power delivered and torque developed.

Answers: (a) 11.3 A; (b) 2558 W; (c) 2345 W; (d) 12.4 Nm.

4.6 Supposing that a pulsed voltage source is connected to the ends of the coil from problem 4.4, sketch the current wave that should be maintained for an average developed torque of 20 Nm at the rotational speed of 2400 rpm.

4.7 An eddy current Faraday disc shown in figure 4.43 has the following characteristics:
conductivity $\sigma = 5.61 \times 10^7$ S m^{-1}
inside radius a = 30 mm
external radius b = 350 mm
thickness h = 12.7 mm

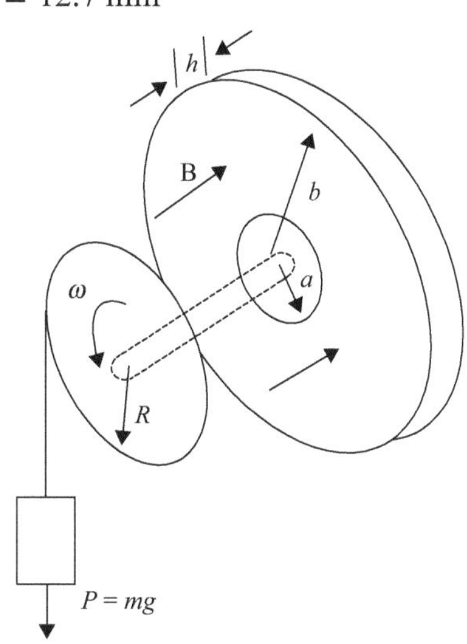

Figure 4.43. Problem 4.7.

Such a disc is excited by a normal magnetic flux density of B = 0.4 Wb m^{-2}. As the magnetic excitation covers only a small area of the surface of the disc the correcting factor f_c = 0.15 should be considered.

This disc is part of the system used to limit the angular speed of the suspending mass $m = 1000$ kg. The total moment of inertia of the moving part is $J = 5000$ kgm^2 and the friction coefficient associated with the friction losses dependent on the angular speed is 50 Nms rad^{-1}. All additional losses should be neglected. Consider $g = 9.8$ m s^{-2} and $R = 0.5$ m.

(a) write the differential equation that governs the angular speed of the system;

(b) the maximum speed of the system;

(c) plot the curve $\omega \times t$.

4.8 A phonograph pickup is shown in figure 4.44. The amplitude of vibration is 0.01 cm (each way from neutral); the frequency is 256 Hz, the coil has 10 turns, the length of each turn being 2.0 cm, and $B = 0.4$ Wb m^{-2}. What is the maximum of the induced emf wave? What is its rms value?

Answer: 0.013 sen(1608t); 0.01 V.

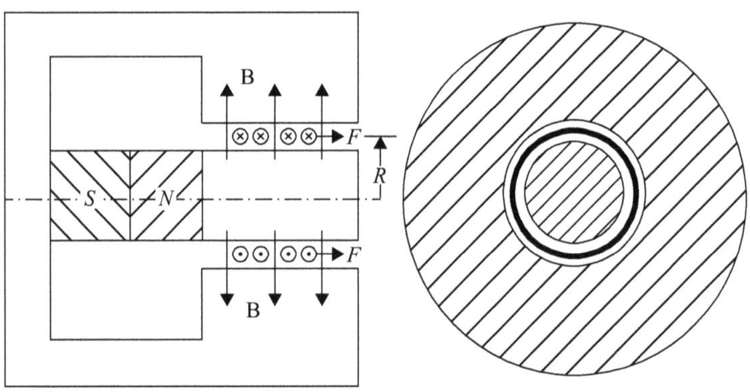

Figure 4.44. The phonograph pickup. The magnetic flux density distribution is radial and its value is 0.1 T.

4.9 The dynamic speaker in figure 4.45 has 12 turns in the voice coil and its diameter is 26 mm. If $B = 0.4$ Wb m^{-2} in the region in which the voice coil moves, and the rms current in the voice coil is 0.10 A, find the *maximum* force on the coil.

Answer: 80 mN (mean value: 40 mN).

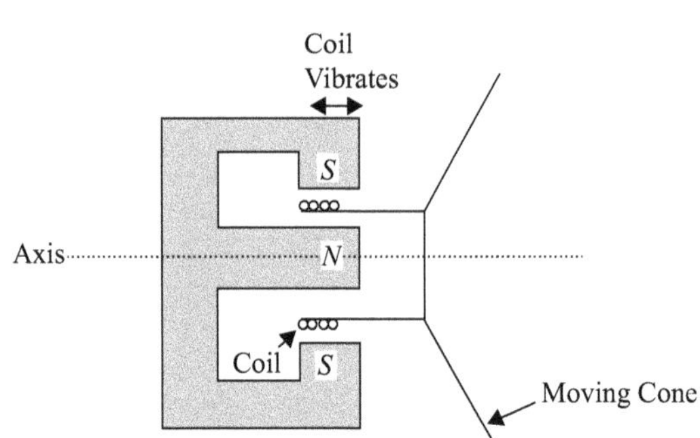

Figure 4.45. A speaker.

4.10 A cylinder (R = 20 cm and length L = 40 cm) of an elemental DC machine has four magnetic poles uniformly distributed on its surface. A set of four coils with five turns each are associated in series via its commutator as is shown in figure 4.46. A DC voltage of 450 V is feeding the winding and the spinning speed of the rotor is 1500 rpm. The resistance of the winding is 5 Ω and its inductance can be neglected. The magnitude of the magnetic flux density is 0.4 Wb m^{-2} and could be considered constant in the pole pitch.

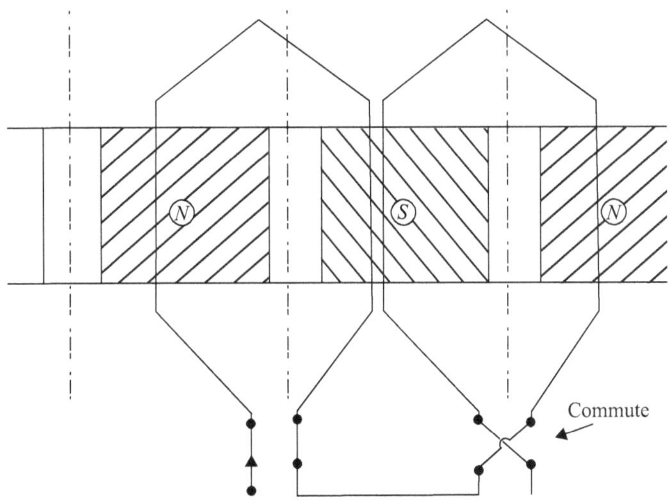

Figure 4.46. Winding of four poles—only two poles are shown.

Compute:
(a) the magnetic force developed by each coil;
(b) the magnetic torque developed by the winding.

Answer: (a) 64 N; (b) 257 N.

Further reading

[1] Gourishankar V 1966 *Electromechanical Energy Conversion* (Scranton: International Textbook Company)
[2] Skilling H H 1962 *Electromechanics: A First Course in Electromechanical Energy Conversion* (New York: Wiley)
[3] Bansal R (ed) 2004 *Handbook of Engineering Electromagnetics* (New York: Marcel Decker)
[4] Slemon G R and Straughen A 1980 *Electric Machines* (Reading, MA: Addison-Wesley)
[5] Sacarpino M 2016 *Motors for Makers: A Guide to Steppers, and Other Electrical Machines* (Indianapolis, IN: Que Publishing)
[6] Cardoso J R 2011 *Engenharia Electromagnetica* (Rio de Janeiro: Elsevier)

IOP Publishing

Electromechanical Energy Conversion Through Active Learning

J R Cardoso, M B C Salles and M C Costa

Chapter 5

The flow of electromechanical energy

5.1 Introduction

The operation of an electromechanical device involves two main kinds of energy exchange: electrical and mechanical. Although thermal energy is also involved in electromechanical operation, it is not responsible for developing forces and torques. The flow of thermal energy is due to not only Joule losses in its several windings, but also, from mechanical losses like friction and ventilation.

We can identify the following energy components involved in an electromechanical device:

E_{se}: the energy supplied by the electrical source;
E_{mag}: the stored energy in magnetic field;
E_{j}: the energy issued from Joule losses;
E_{sm}: the energy supplied by the mechanical source;
E_{kin}: the stored mechanical energy (kinetics and/or potential energy);
E_{fv}: the energy losses in friction and ventilation.

The components of both Joule and friction/ventilation losses are not reversible. The sign of the others can change, depending on the kind of electromechanical device operation. Concerning the energies issued from external sources (electrical or mechanical), the positive signal will be attributed to the one that is supplied to the device and negative to the opposite one.

If the electromechanical device is operating as a motor, the variation during a time interval of the energy supplied by the electrical source (ΔE_{se}) is positive and the variation of energy supplied by the mechanical source (ΔE_{sm}) in the same time interval is negative. Indeed, in this case the mechanical energy is delivered to the load.

Operating as a generator, the variation of energy supplied by the electrical source (ΔE_{se}) is negative because it is delivered to the electrical load and the variation of the energy supplied by the mechanical source (ΔE_{sm}) is positive.

doi:10.1088/978-0-7503-2084-9ch5

Concerning the variation of the energy stored in the magnetic field (ΔE_{mag}), during a time interval, it is positive when the magnetic flux density is growing and negative when it is decreasing. A positive variation of the stored mechanical energy (ΔE_{kin}) at the same time interval occurs when the speed of the device is growing.

Considering a time interval (Δt) the balance of energy flowing in the electromechanical device can be established as:

$$\begin{bmatrix} \Delta E_{se} \\ \text{The} \\ \text{energy} \\ \text{variation} \\ \text{supplied} \\ \text{by the} \\ \text{electrical} \\ \text{source} \end{bmatrix} + \begin{bmatrix} \Delta E_{sm} \\ \text{The} \\ \text{energy} \\ \text{variation} \\ \text{supplied} \\ \text{by the} \\ \text{mechanical} \\ \text{source} \end{bmatrix} = \begin{bmatrix} \Delta E_{mag} \\ \text{The} \\ \text{variation} \\ \text{of the} \\ \text{stored} \\ \text{energy in} \\ \text{magnetic} \\ \text{field} \end{bmatrix} + \begin{bmatrix} \Delta E_{kin} \\ \text{The} \\ \text{variation} \\ \text{of the} \\ \text{stored} \\ \text{mechanical} \\ \text{energy} \end{bmatrix} + \begin{bmatrix} \Delta E_{j} \\ \text{The} \\ \text{variation} \\ \text{of the} \\ \text{Joule} \\ \text{losses} \end{bmatrix} + \begin{bmatrix} \Delta E_{fv} \\ \text{The} \\ \text{variation} \\ \text{of the} \\ \text{friction} \\ \text{and} \\ \text{ventilation} \\ \text{losses} \end{bmatrix}$$

It is convenient to concentrate on the left-side components involving only energy variations issued from electrical phenomena and on the right-side the energy variations issued from mechanical phenomena.

$$\begin{bmatrix} \Delta E_{se} \\ \text{The} \\ \text{variation} \\ \text{of energy} \\ \text{supplied} \\ \text{by the} \\ \text{electrical} \\ \text{source} \end{bmatrix} - \begin{bmatrix} \Delta E_{mag} \\ \text{The} \\ \text{variation} \\ \text{of the} \\ \text{stored} \\ \text{energy in} \\ \text{magnetic} \\ \text{field} \end{bmatrix} - \begin{bmatrix} \Delta E_{j} \\ \text{The} \\ \text{variation} \\ \text{of the} \\ \text{Joule} \\ \text{losses} \end{bmatrix}$$

$$= -\begin{bmatrix} \Delta E_{sm} \\ \text{The} \\ \text{energy} \\ \text{variation} \\ \text{supplied} \\ \text{by the} \\ \text{mechanical} \\ \text{source} \end{bmatrix} + \begin{bmatrix} \Delta E_{kin} \\ \text{The} \\ \text{variation} \\ \text{of the} \\ \text{stored} \\ \text{mechanical} \\ \text{energy} \end{bmatrix} + \begin{bmatrix} \Delta E_{fv} \\ \text{The} \\ \text{variation} \\ \text{of the} \\ \text{friction} \\ \text{and} \\ \text{ventilation} \\ \text{losses} \end{bmatrix}$$

Considering the device as a motor—that does not impose any restriction—the energy variation supplied by the mechanical source is negative. Indeed, this energy is delivered to the load and not consumed by the motor. In this situation, we can redefine it using the concept of the electromechanical energy, that is the total amount of energy converted from electrical to mechanical by electromagnetic interaction as:

$$\begin{bmatrix} \Delta E_{\mathrm{em}} \\ \text{The} \\ \text{electro–} \\ \text{mechanical} \\ \text{energy} \\ \text{variation} \end{bmatrix} = -\begin{bmatrix} \Delta E_{\mathrm{sm}} \\ \text{The} \\ \text{energy} \\ \text{variation} \\ \text{supplied} \\ \text{by the} \\ \text{mechanical} \end{bmatrix} + \begin{bmatrix} \Delta E_{\mathrm{kin}} \\ \text{The} \\ \text{variation} \\ \text{of the} \\ \text{stored} \\ \text{mechanical} \\ \text{energy} \end{bmatrix} + \begin{bmatrix} \Delta E_{\mathrm{fv}} \\ \text{The} \\ \text{variation} \\ \text{of the} \\ \text{friction} \\ \text{and} \\ \text{ventilation} \\ \text{losses} \end{bmatrix}$$

Doing that, the variation energy balance becomes:

$$\begin{bmatrix} \Delta E_{\mathrm{se}} \\ \text{The} \\ \text{variation} \\ \text{of energy} \\ \text{supplied} \\ \text{by the} \\ \text{electrical} \\ \text{source} \end{bmatrix} - \begin{bmatrix} \Delta E_{\mathrm{mag}} \\ \text{The} \\ \text{variation} \\ \text{of the} \\ \text{stored} \\ \text{energy in} \\ \text{magnetic} \\ \text{field} \end{bmatrix} - \begin{bmatrix} \Delta E_{\mathrm{j}} \\ \text{The} \\ \text{variation} \\ \text{of the} \\ \text{Joule} \\ \text{losses} \end{bmatrix} = \begin{bmatrix} \Delta E_{\mathrm{em}} \\ \text{The} \\ \text{electro–} \\ \text{mechanical} \\ \text{energy} \\ \text{variation} \end{bmatrix}$$

Remarks: The variation of electromechanical energy (ΔE_{em}) is the summation of the variation of the energy delivered to the load ($-\Delta E_{\mathrm{sm}}$), the variation of the stored mechanical energy (ΔE_{kin}) and the variation of the mechanical losses (ΔE_{fv}).

5.1.1 The variation of the energy stored in magnetic field

From the electromagnetism—see the Poynting Theorem—the variation of magnetic energy density necessary for magnetizing a medium from B_1 to B_2(Wb m^{-2}) is:

$$\Delta w_{\mathrm{mag}} = \int_{B_1}^{B_2} H db \ (\mathrm{J\ m^{-3}}) \tag{5.1}$$

If $B_1 < B_2$ the variation of magnetic energy density is positive, representing the energy delivered by the electrical source to the medium for magnetizing it. Otherwise, if $B_1 > B_2$, it represents the medium sending back to the electrical source the energy received when it was magnetized.

Figure 5.1 shows a typical magnetization curve of a ferromagnetic material. The hatched area is the variation of magnetic energy density necessary to increase the magnetic flux density from B_1 to B_2.

This process is not reversible in non-linear magnetic material. That is, the variation of magnetic energy density necessary for restoring the magnetic flux density from B_2 to B_1 is small due to the hysteretic losses.

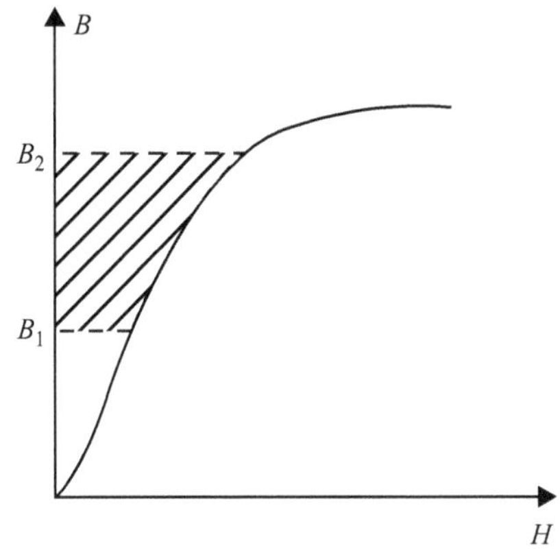

Figure 5.1. The magnetic energy density. The magnetic energy density necessary for restoring B from B_2 to B_1 is smaller than that for increasing B from B_1 to B_2 due to hysteresis.

$$\int_{B_1}^{B_2} Hdb > -\int_{B_2}^{B_1} Hdb$$

When $B_1 = 0$ and $B_2 = B$, the expression:

$$w_{\text{mag}} = \int_0^B Hdb \text{ J m}^{-3} \tag{5.2}$$

is called the magnetic energy density stored in the magnetic field. Indeed, this magnetic energy density represents the magnetic energy density (J m^{-3})necessary for magnetizing the medium from 0 to B. Therefore, the variation of magnetic energy density for magnetizing the medium from B_1 to B_2 can be written as:

$$\Delta w_{\text{mag}} = w_{\text{mag}}(B_2) - w_{\text{mag}}(B_1)$$

or:

$$\Delta w_{\text{mag}} = \int_0^{B_2} Hdb - \int_0^{B_1} Hdb \tag{5.3}$$

For the linear magnetic material whose magnetic permeability is constant, the energy density stored in the magnetic field becomes:

$$w_{\text{mag}} = \frac{1}{2}\mu H^2 = \frac{1}{2}\frac{B^2}{\mu} \text{J m}^{-3} \tag{5.4}$$

As the magnetic structure of electromechanical device is composed of a sequence of prisms as shown in figure 5.2, the total amount of magnetic energy stored in the entire volume due to the magnetic field could be written as:

$$E_{\text{mag}} = \sum_i \left[\int_0^B Hdb\right] V_i \tag{5.5}$$

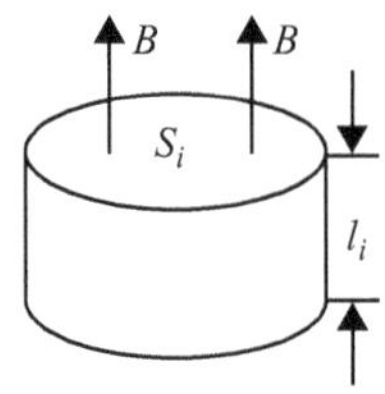

Figure 5.2. Prism of magnetic structure.

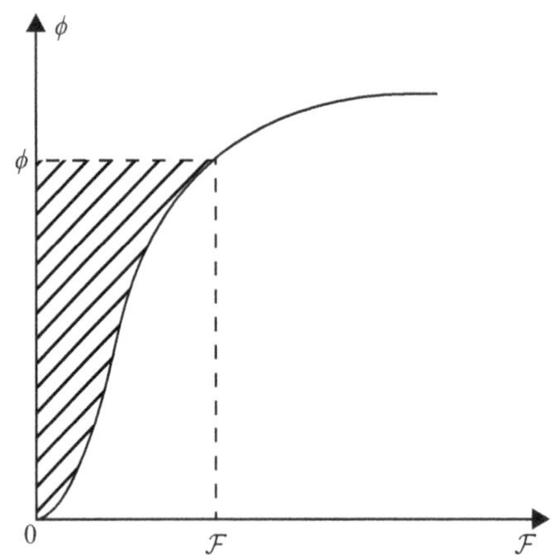

Figure 5.3. Stored energy for establishing the magnetic flux. Non-linear magnetic material.

As $V_i = S_i l_i$, the previous equation could be rewritten as:

$$E_{\text{mag}} = \int_0^B \sum_i (H_i l_i) d(bS_i) \tag{5.6}$$

As the magnetomotive force is such that $\mathcal{F} = \sum H_i l_i$ and $\phi = bS$, results in:

$$E_{\text{mag}} = \int_0^\phi \mathcal{F} d\phi \tag{5.7}$$

Using the *saturation* curve as defined in chapter 2, the hatched area of both figures 5.3 and 5.4 is the stored energy for establishing the magnetic flux in the electromechanical device.

In the linear case, the magnetomotive force (mmf) is related to the magnetic flux by $\mathcal{F} = \mathcal{R}\phi$. Therefore, the stored energy in the magnetic field in a linear medium is:

$$E_{\text{mag}} = \frac{\mathcal{R}\phi^2}{2} = \frac{\mathcal{F}\phi}{2} = \frac{\mathcal{F}^2}{2\mathcal{R}} = \frac{\lambda i}{2} \text{J} \tag{5.8}$$

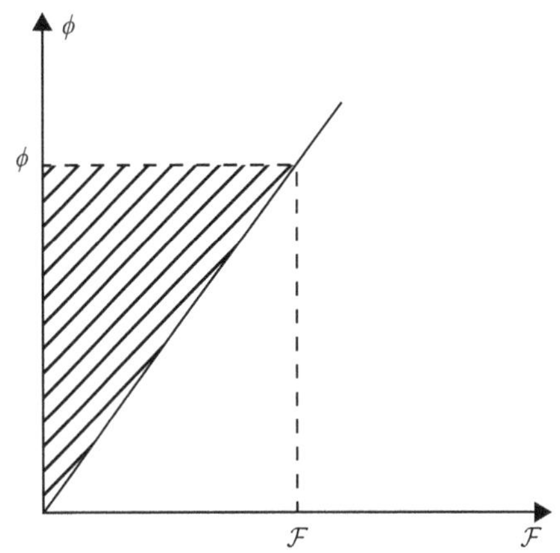

Figure 5.4. Stored energy for establishing the magnetic flux. Linear magnetic material.

We can also represent the stored energy in the magnetic field by lumped parameters like inductances. In a simply excited electromagnetic device, we can do it by observing that:

$$L = \frac{N^2}{\mathcal{R}} \text{ and } Li = N\phi$$

That results in:

$$E_{\text{mag}} = \frac{1}{2}Li^2 \tag{5.9}$$

5.1.2 The variation of the energy supplied by the electrical source

The flow of energy from the source to the device occurs during the movement of the moving parts that we can observe the variation of the magnetic flux or magnetomotive force or both.

Figure 5.5 shows a simple excited electromagnetic device. The magnetic structure presents a saturation curve dependent on the air-gap (x).

The energy delivered by the source from the position x_1 at instant t_1 to position x_2 at instant t_2 is given by:

$$\Delta E_{\text{se}} = \int_{t_1}^{t_2} v(t)i(t)dt \tag{5.10}$$

As $v(t) = e(t) + ri(t)$, we can rewrite (5.10) as:

$$\Delta E_{\text{se}} = \int_{t_1}^{t_2} e(t)i(t)dt + \int_{t_1}^{t_2} ri(t)^2dt \tag{5.11}$$

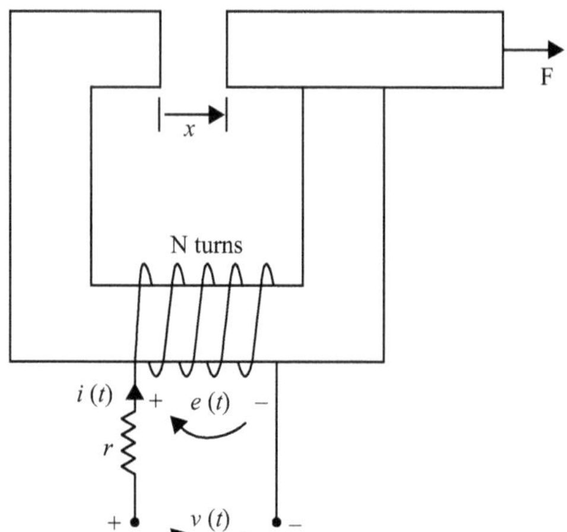

Figure 5.5. The magnetic system.

The second term of (5.11) is the variation of the energy issued for the Joule losses (ΔE_j). Therefore:

$$\Delta E_{se} - \Delta E_j = \int_{t_1}^{t_2} e(t)i(t)dt \tag{5.12}$$

This is the variation of the energy delivered by the source to the electromechanical device excluding the Joule losses. As $e(t)$ is the induced emf given by:

$$e(t) = N\frac{d\phi}{dt}$$

the previous expression could be written as:

$$\Delta E_{se} - \Delta E_j = \int_{t_1}^{t_2} N\frac{d\phi}{dt}idt = \int_{\phi_1}^{\phi_2} Nid\phi \tag{5.13}$$

or:

$$\Delta E_{se} - \Delta E_j = \int_{\phi_1}^{\phi_2} \mathcal{F}d\phi \tag{5.14}$$

5.1.3 The variation of the electromechanical energy

For evaluating the integral of equation (5.14), we must consider some common practices applied to the operation of this kind of electromagnetic device.

5.1.3.1 First case: movement under a constant current

The movement under a constant current is done by feeding the coil with a DC electrical source and considering the slow movement of the armature. Figure 5.6

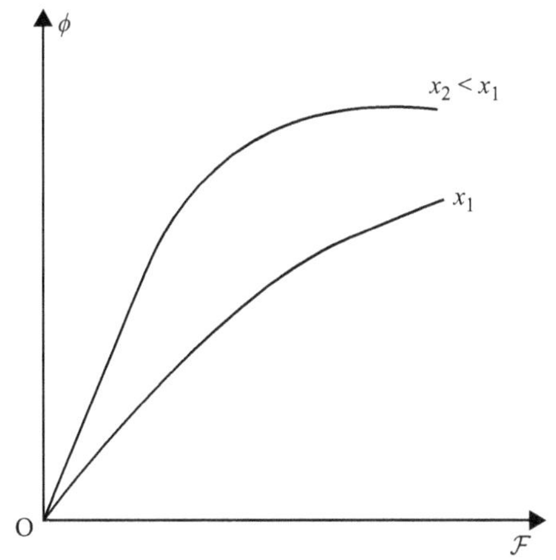

Figure 5.6. The saturation curves for different air-gaps. The air-gap x_1 is bigger than x_2.

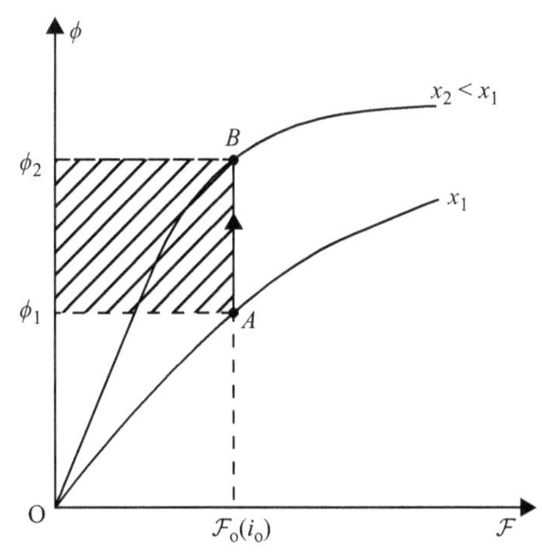

Figure 5.7. The variation of the energy supplied by the electrical source ($\Delta E_{se} - \Delta E_j$) under constant mmf.

shows two saturation curves of the same magnetic system for air-gaps $x_1 > x_2$ that are the armature positions at instants $t_2 > t_1$.

Supposing now that a DC current (i_0) is injected in the N-turns coil producing a mmf $\mathcal{F}_0 = Ni_0$. With the armature at the position x_1, a magnetic flux ϕ_1 is established in the core and point A in figure 5.7 is the 'operation point' for this condition.

Moving the armature slowly from position x_1 to position $x_2 < x_1$ without changing the excitation current (i_0), the new magnetic flux ϕ_2 in the core is the one defined by point B.

The variation trajectory from ϕ_1 to ϕ_2 is the straight segment AB, as $\mathcal{F}_0$ remains constant during the movement. Therefore, according to equation (5.14), the hatched area ($AB\phi_1\phi_2$) is the variation of the energy supplied by the electrical source excluding the Joule losses.

Concerning the variation of the stored energy in the magnetic field (ΔE_{mag}), we can write:

$$\Delta E_{\text{mag}} = E_{\text{mag}}(\phi_2) - E_{\text{mag}}(\phi_1) \tag{5.15}$$

Figure 5.8 shows the algebraic operation with those areas representing the stored magnetic energies in both positions $x_1(t_1)$ and $x_2(t_2)$. In this operation the resulting area $CB\phi_1\phi_2$ is positive and the area OAC is negative.

As the electromechanical energy variation (ΔE_{em}) is given by:

$$\Delta E_{\text{em}} = \Delta E_{\text{se}} - \Delta E_{\text{j}} - \Delta E_{\text{mag}} \tag{5.16}$$

the resulting area that represents ΔE_{em} will be:

$$\Delta E_{\text{em}} = AB\phi_1\phi_2 - CB\phi_1\phi_2 + OAC = OAB$$

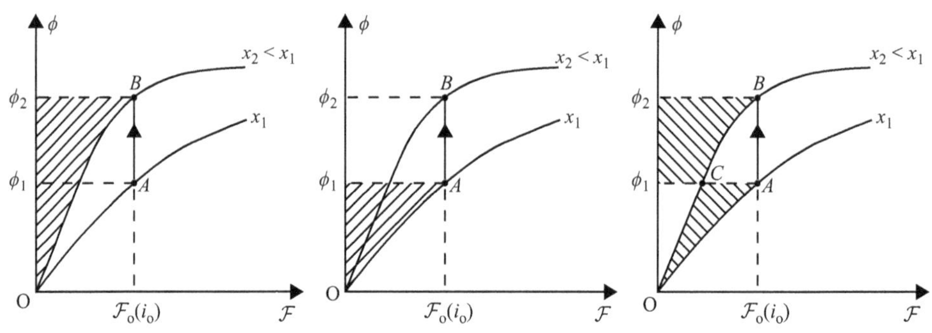

Figure 5.8. Evaluation of ΔE_{mag}. Area $CB\phi_1\phi_2 > 0$ and area $OAC < 0$.

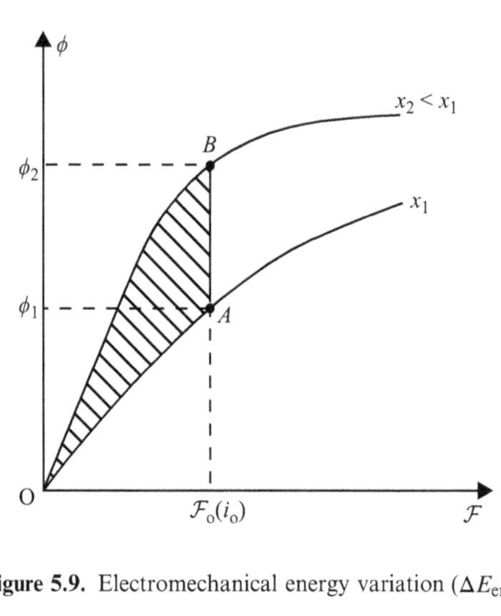

Figure 5.9. Electromechanical energy variation (ΔE_{em}).

In conclusion, during the time interval $\Delta t = t_2 - t_1$ the electromechanical energy variation (ΔE_{em}) under constant current is the area between both saturation curves limited by the straight segment of the mmf $\mathcal{F}_0 = Ni_0$ (figure 5.9).

5.1.3.2 Second case: movement under constant flux

The movement under constant flux does not involve any transfer of useful electrical energy from the electrical source to the electromechanical device because there is no resulting time variation in the magnetic flux $d\phi/dt = 0$. Therefore:

$$\Delta E_{\text{se}} - \Delta E_{\text{j}} = \int_{t_1}^{t_2} N\frac{d\phi}{dt}idt = 0$$

The only change of energy between the electromagnetic device and the electrical source is the one associated to the Joule losses. Consequently, in (5.16) the electromechanical energy variation (ΔE_{em}) is due only to the variation of the stored energy in the magnetic field, that is:

$$\Delta E_{\text{em}} = -\Delta E_{\text{mag}} \tag{5.17}$$

Figure 5.10 shows the electromechanical energy variation (ΔE_{em}) for the constant flux movement. The hatched area is the difference of stored magnetic energy in the magnetic field for both position x_1 and x_2.

In conclusion, during the time interval $\Delta t = t_2 - t_1$ the electromechanical energy variation (ΔE_{em}) under *constant flux* is the area between both saturation curves limited by the straight segment defined by the flux ϕ.

The real movement of the armature is not slow or fast enough. Therefore, the real trajectories from x_1 to x_2 and from x_2 to x_1 are shown in figure 5.11.

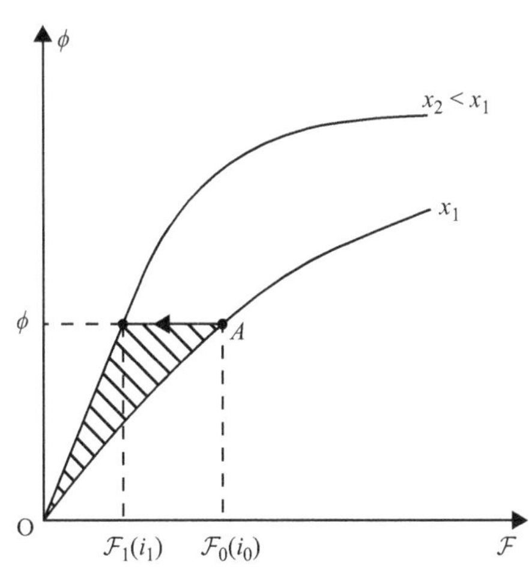

Figure 5.10. Electromechanical energy variation (ΔE_{em}).

$$\Delta E_{\text{em}} = -(\text{area OC}\phi - \text{area OA}\phi)$$

5.1.4 The co-energy

We can identify two regions in the saturation curve of a magnetic material as shown in figure 5.12.

The region I is the *magnetic energy stored in the magnetic field* that can be evaluated by the integral:

$$E_{\text{mag}} = \int_0^{\phi_0} \mathcal{F} d\phi \tag{5.18}$$

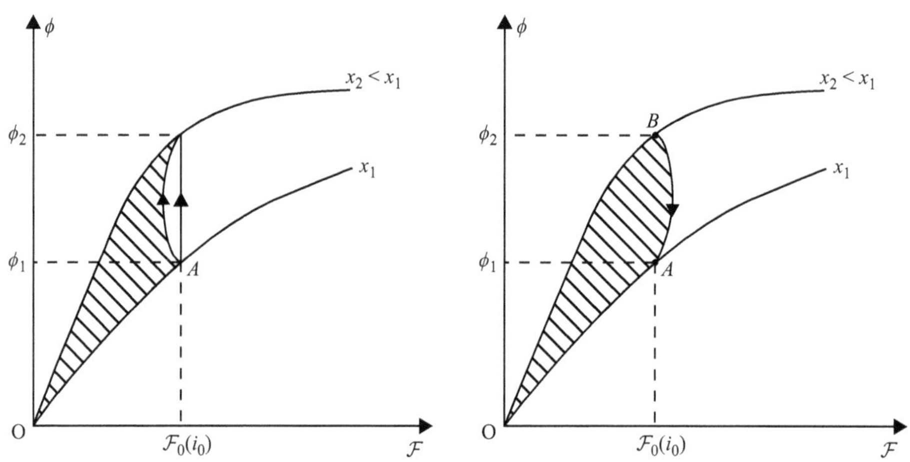

Figure 5.11. Flux versus mmf curves for motion of the armature from x_1 to x_2. Left side: movement from x_1 to x_2, Right side: movement from x_2 to x_1.

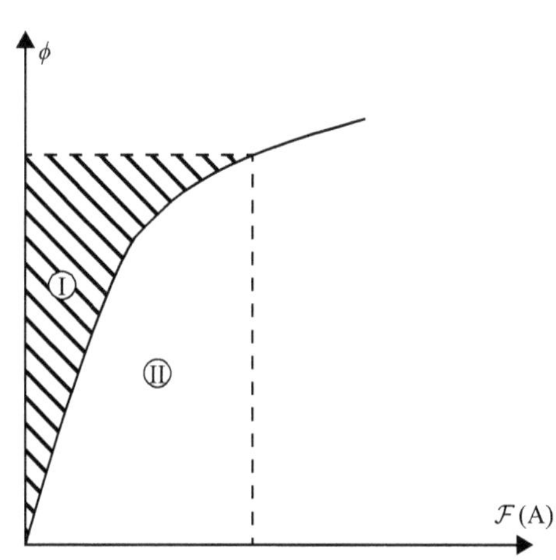

Figure 5.12. Regions of the saturation curve. I: the magnetic energy stored in the magnetic field, II: the co-energy the electromechanical energy variation.

The region II mathematically expressed by the integral:

$$E'_{\text{mag}} = \int_0^{\mathcal{F}_0} \phi d\mathcal{F} \tag{5.19}$$

is called *co-energy* that has no physical meaning. It is just an easy way of representing the region.

Based on the geometric approach of figure 5.13 and considering the co-energy concepts, we can express the electromechanical energy variation (ΔE_{em})—*under constant current*—delivered between the positions x_1 to x_2 as:

$$\Delta E_{\text{em}} = E'_{\text{mag}}(x_2) - E'_{\text{mag}}(x_1)$$

or:

$$\Delta E_{\text{em}} = \Delta E'_{\text{mag}} \tag{5.20}$$

Figure 5.13 illustrates this approach.

Considering now the case of *constant flux* from figure 5.10, we can extract that the electromechanical energy variation represented by the shaded area can be express by:

$$\Delta E_{\text{em}} = -[E_{\text{mag}}(x_2) - E_{\text{mag}}(x_1)]$$

or:

$$\Delta E_{\text{em}} = -\Delta E_{\text{mag}} \tag{5.21}$$

Figure 5.14 illustrates this approach

5.1.5 The average developed electromechanical force

Associated with the mechanical energy is the development of the force in a translational electromechanical device or the torque in a rotational device.

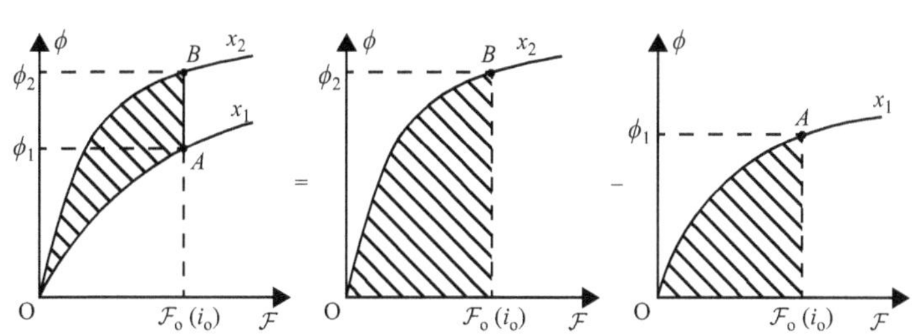

Figure 5.13. The composition of co-energies for obtaining the electromechanical energy variation under constant current.

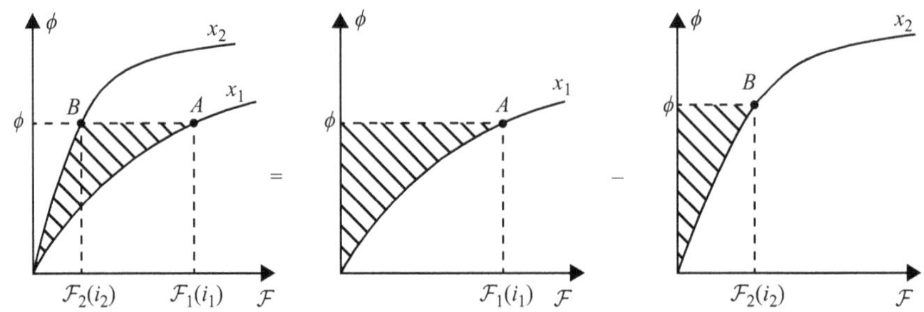

Figure 5.14. The composition of energy stored in magnetic field for obtaining the electromechanical energy variation under constant flux.

If the displacement of the armature of the magnetic system shown in figure 5.5 is small enough then we can consider the work as:

$$\tau = \Delta E_{em} = F_{dev} \Delta x \tag{5.22}$$

where:

$\Delta x = x_{final} - x_{start}$

$F_{dev:}$ the average force developed by the electromagnetic device

We obtain the average force developed in the armature of the magnetic system from:

a. For a constant current

$$F_{dev} = \frac{\Delta E'_{mag}}{\Delta x} \tag{5.23}$$

b. For a constant flux

$$F_{dev} = \frac{-\Delta E_{mag}}{\Delta x} \tag{5.24}$$

Equations (5.23) and (5.24) are important and fundamental results and are applicable, in general, to any magnetic system.

Example 5.1

Figure 5.15 shows the saturation curves of a magnetic system like the one shown in figure 5.5 for the positions $x_1 = 2$ mm and $x_2 = 3$ mm. The coil has 200 turns and its resistance is 1 mΩ.

The armature is moving from x_1 to x_2 under constant current of 2 A in 10 s. This time interval is enough large to consider the movement at constant current. The dissipated energy by friction during the movement is 50 mJ.

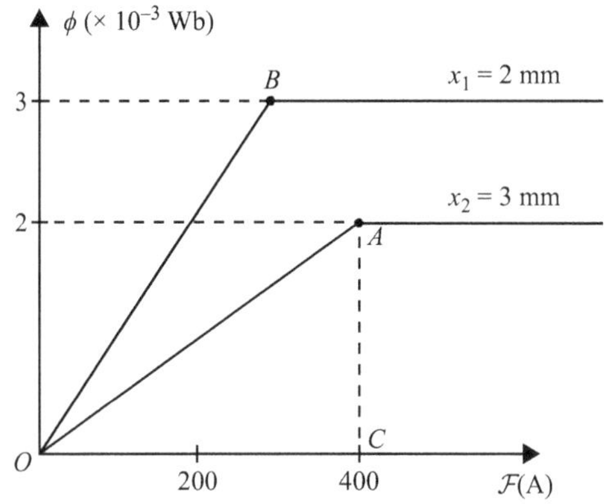

Figure 5.15. Example 5.1. Saturation curves of a magnetic system.

Determine:

(a) the energy balance in the magnetic system during this time interval;
(b) the average developed electromechanical force;
(c) the net force applied by the mechanical source;
(d) the induced emf in the coil;
(e) Sketch a flow-diagram of the balance of energy in the magnetic system.

Solution

(a) *The variation energy balance during the time interval*

The variation of the energy stored in the magnetic field

$$E_{mag}(x_1) = \frac{300 \times 3 \times 10^{-3}}{2} = 0.45 \text{ J}$$

$$E_{mag}(x_2) = \frac{400 \times 2 \times 10^{-3}}{2} = 0.40 \text{ J}$$

$$\Delta E_{mag} = E_{mag}(x_2) - E_{mag}(x_1) = -0.05 \text{ J}$$

The energy variation due to Joule losses

$$\Delta E_j = ri^2 \, \Delta t = 1 \times 10^{-3} \times 2^2 \times 10 = 0.04 \text{ J}$$

The electromechanical energy variation

Based on equation (5.20), to evaluate the electromechanical energy variation it is necessary to evaluate the co-energies first.

Co-energy at $x_1 = 2$ mm

$$E'_{mag}(x_1) = \frac{300 \times 3 \times 10^{-3}}{2} + 100 \times 3 \times 10^{-3} = 0.75 \text{ J}$$

Co-energy at $x_2 = 3$ mm

$$E'_{mag}(x_2) = \frac{400 \times 2 \times 10^{-3}}{2} = 0.40 \text{ J}$$

The co-energy variation is:

$$\Delta E'_{mag} = E'_{mag}(x_2) - E'_{mag}(x_1) = -0.35 \text{ J}$$

Finally, the electromechanical energy variation is:

$$\Delta E_{em} = \Delta E'_{mag} = -0.35 \text{ J}$$

The variation of the energy supplied by the electrical source

Based on equation (5.16), the variation of the energy supplied by the electrical source is such that:

$$\Delta E_{se} = \Delta E_{em} + \Delta E_{j} + \Delta E_{mag}$$

that results in:

$$\Delta E_{se} = -0.35 + 0.04 - 0.05 = -0.36 \text{ J}$$

The variation of the mechanical energy supplied by the source

This term is related to the electromechanical energy variation by the equation:

$$\Delta E_{em} = -\Delta E_{sm} + \Delta E_{kin} + \Delta E_{fv}$$

That results in:

$$\Delta E_{sm} = -\Delta E_{em} + \Delta E_{kin} + \Delta E_{fv}$$

or:

$$\Delta E_{sm} = -(-0.35) + 0 + 0.05 = 0.40 \text{ J}$$

(b) *The average developed electromechanical force*

This force is obtained from equation (5.23):

$$F_{dev} = \frac{\Delta E'_{mag}}{\Delta x} = \frac{-0.35}{(2-1) \times 10^{-3}} = -350 \text{ N}$$

(c) *The net force issued from the external agent*

The external agent should apply a net force given by:

$$F_{net} = \frac{\Delta E_{sm}}{\Delta x} = \frac{0.40}{10^{-3}} = 400 \text{ N}$$

Remarks: The difference between F_{net} and $| F_{dev} |$ is due to the friction effect.

(d) *The induced emf in the coil*

The induced emf is obtained by applying Faraday's law given by:

$$e = -N\frac{\Delta\phi}{\Delta t} = -200 \times \frac{(2-3)\times 10^{-3}}{10} = 20\ \text{mV}$$

(e) *The flow-diagram of the energy balance.*

The magnetic system needs an external agent for the movement because it is necessary to input mechanical energy in the system for the opening the armature. Consequently, the electric source will receive part of this energy and the term ΔE_{se} is negative. Therefore, the flow diagram in figure 5.16 should start by the input of mechanical energy represented by the term ΔE_{sm} and end by the term ΔE_{se}.

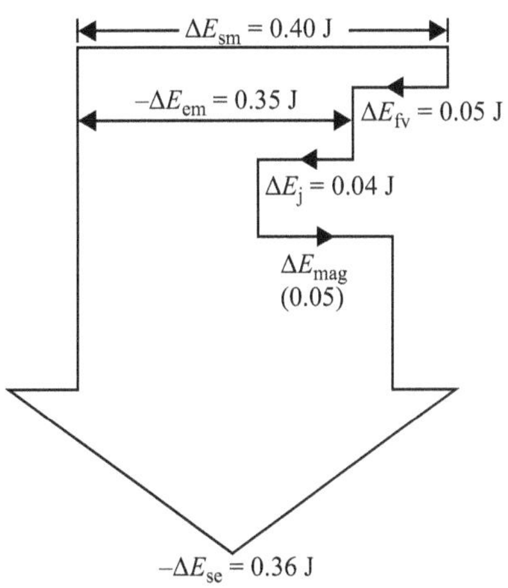

Figure 5.16. The energy flow diagram.

Example 5.2

The magnetic system in example 5.1 has its armature moving from x_1 to x_2 under *constant flux* of 2×10^{-3} Wb in 10 s.

Determine:

(a) the energy balance in the magnetic system during this time interval;
(b) the average developed electromechanical force;
(c) the net force applied by the mechanical source;
(d) the induced emf in the coil;
(e) sketch a flow-diagram of the balance of energy in the magnetic system.

Solution:

(a) *The variation energy balance during the time interval*

The variation of the energy stored in the magnetic field. As:

$$E_{mag}(x_1) = \frac{200 \times 2 \times 10^{-3}}{2} = 0.2\ \text{J}$$

$$E_{mag}(x_2) = \frac{400 \times 2 \times 10^{-3}}{2} = 0.4 \text{ J}$$

Results in:

$$\Delta E_{mag} = E_{mag}(x_2) - E_{mag}(x_1) = 0.2 \text{ J}$$

The energy variation due to Joule losses

As the mmf varies from $\mathcal{F}_1 = 200$ A, that corresponds $i_1 = 1$ A to $\mathcal{F}_2 = 400$ A,that corresponds $i_2 = 2$ A in 10 s, we can consider that the time variation of the electrical current has a linear characteristic as shown in figure 5.17.

Therefore, the energy variation due to Joule losses will be given by:

$$\Delta E_j = \int_0^{10} ri^2 dt = 10^{-3} \int_0^{10} (1 + 0.1t)^2 dt = 0.023 \text{ J}$$

The electromechanical energy variation

Based on (5.21), the electromechanical energy variation is given by:

$$\Delta E_{em} = -\Delta E_{mag} = -(0.2) = -0.2 \text{ J}$$

The variation of the energy supplied by the electrical source

Based on the same procedure applied in example 5.1, the variation of the energy supplied by the electrical source is such that:

$$\Delta E_{se} = \Delta E_{em} + \Delta E_j + \Delta E_{mag}$$

That results in:

$$\Delta E_{se} = -0.2 + 0.023 + 0.2 = 0.023 \text{ J}$$

Remarks: The change of electrical energy between the electrical source and the magnetic system is only the Joule losses, because both the magnetic flux variation and the emf are null.

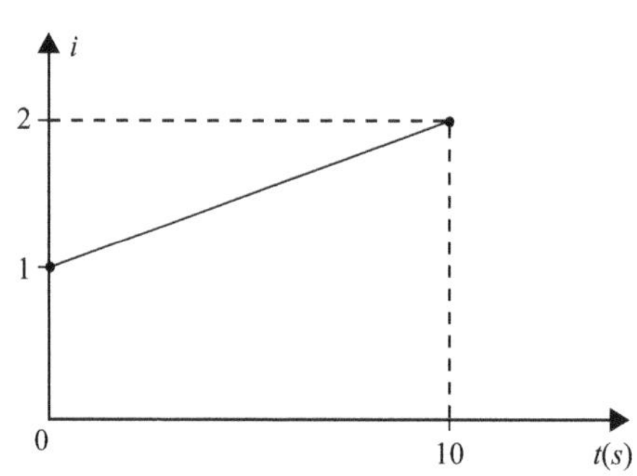

Figure 5.17. Time variation of the electrical current. This function is described by $i = 1 + 0.1\,t$.

The variation of the mechanical energy supplied by the source

This term is related to the electromechanical energy variation by the equation:

$$\Delta E_{sm} = -\Delta E_{em} + \Delta E_{kin} + \Delta E_{fv}$$

or

$$\Delta E_{sm} = -(-0.2) + 0 + 0.05 = 0.25 \text{ J}$$

b. *The average developed electromechanical force*

This force is obtained from equation (5.24):

$$F_{dev} = \frac{-\Delta E_{mag}}{\Delta x} = \frac{-(0.2)}{10^{-3}} = -200 \text{ N}$$

The additional force due to the friction losses

$$F_{fv} = \frac{-\Delta E_{fv}}{\Delta x} = \frac{-0.05}{10^{-3}} = -50 \text{ N}$$

c. The external agent should apply a net force given by:

$$F_{net} = F_{net} + F_{fv} = -250 \text{ N}$$

d. *The induced emf in the coil*

The induced emf is obtained applying Faraday's law given by:

$$e = -N\frac{\Delta\phi}{\Delta t}$$

As the magnetic flux remains constant during the movement, the emf is null.

e. *The flow-diagram of the energy balance*

As shown in previous examples, an external agent should supply mechanical energy to open the armature. As the magnetic flux remains constant during the movement, the only energy received by the electrical source is the one for Joule losses. Consequently, ΔE_{se} is negative. Therefore, the flow diagram of figure 5.18 should start by the input of mechanical energy represented by the term ΔE_{sm} and end by the term $\Delta E_{se} = -\Delta E_{j}$.

The average developed electromechanical torque

Associated with the mechanical energy there is a torque development in a rotational magnetic system.

If the angular rotor's displacement ($\Delta\theta$) of the rotational magnetic system is small enough, we can consider the work as:

$$\tau = \Delta E_{em} = T_{dev}\Delta\theta = (5.25)$$

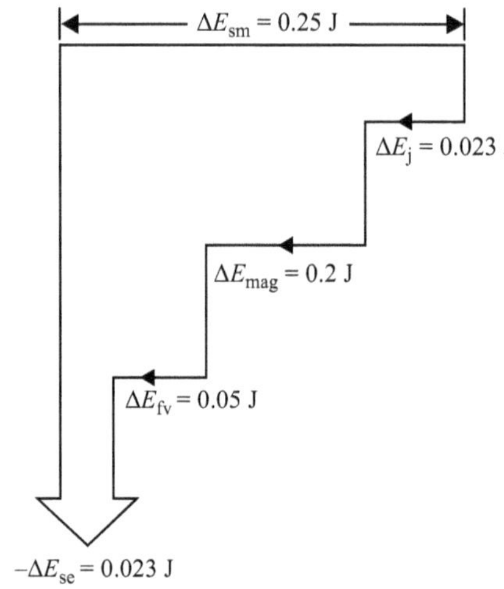

Figure 5.18. The energy flow diagram.

where:

$\Delta\theta = \theta_{\text{final}} - \theta_{\text{start}}$

T_{dev}: the average developed torque by the electromagnetic device

The average developed torque in the rotor of the rotational magnetic system is given by:

(a) For a constant current

$$T_{\text{dev}} = \frac{\Delta E'_{\text{mag}}}{\Delta\theta} \tag{5.26}$$

(b) For a constant flux

$$T_{\text{dev}} = \frac{-\Delta E_{\text{mag}}}{\Delta\theta} \tag{5.27}$$

Equations (5.26) and (5.27) are important and fundamental results and are applicable, in general, to any rotational magnetic system.

Example 5.3

The cross-section of a rotational magnetic system and its saturation curves for the positions $\theta_1 = 0°$ and $\theta_2 = 30°$ are shown in figure 5.19. The coil has 150 turns in each pole and its resistance is 0.1 Ω.

The armature is moving from θ_1 to θ_2 under a constant current of 1 A in 10 s. This time interval is large enough to consider the movement to be at constant current. All mechanical losses are neglected.

Determine:

(a) the energy balance in the magnetic system during this time interval;

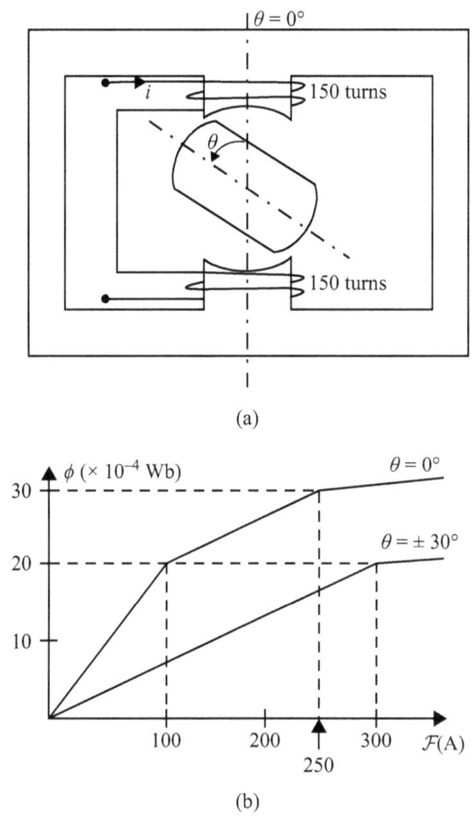

Figure 5.19. Example 5.3. (a) The rotational magnetic system and (b) saturation curves.

(b) the developed electromechanical torque;
(c) the induced emf in the coil;
(d) sketch a flow-diagram of the balance of energy in the magnetic system.

Solution

(a) *The variation energy balance during the time interval*

The variation of the energy stored in the magnetic field

$$E_{mag}(\theta_1) = \frac{100 \times 2 \times 10^{-3}}{2} + \frac{(100 + 250)}{2} \times 10^{-3} = 0.275 \text{ J}$$

$$E_{mag}(\theta_2) = \frac{300 \times 2 \times 10^{-3}}{2} = 0.30 \text{ J}$$

$$\Delta E_{mag} = E_{mag}(\theta_2) - E_{mag}(\theta_1) = 0.025 \text{ J}$$

The energy variation due to Joule losses

$$\Delta E_j = ri^2 \, \Delta t = 0.1 \times 1^2 \times 10 = 0.10 \text{ J}$$

The electromechanical energy variation

The co-energies necessary for the evaluation the electromechanical energy variation is given by:

co-energy at $\theta_1 = 0°$

$$E'_{\text{mag}}(\theta_1) = \frac{100 \times 2 \times 10^{-3}}{2} + \frac{(30 + 20)}{2} \times 15 \times 10^{-3} + 50 \times 3 \times 10^{-3} = 0.625 \text{ J}$$

co-energy at $\theta_2 = 30°$

$$E'_{\text{mag}}(\theta_2) = \frac{20 \times 30 \times 10^{-3}}{2} = 0.30 \text{ J}$$

The co-energy variation is:

$$\Delta E'_{\text{mag}} = E'_{\text{mag}}(\theta_2) - E'_{\text{mag}}(\theta_1) = -0.325 \text{ J}$$

Finally, the electromechanical energy variation is:

$$\Delta E_{\text{em}} = \Delta E'_{\text{mag}} = -0.325 \text{ J}$$

The variation of the energy supplied by the electrical source

Based on equation (5.16), the variation of the energy supplied by the electrical source is such that:

$$\Delta E_{\text{se}} = \Delta E_{\text{em}} + \Delta E_{\text{j}} + \Delta E_{\text{mag}}$$

That results in:

$$\Delta E_{\text{se}} = -0.325 + 0.1 + 0.025 = -0.20 \text{ J}$$

The variation of the mechanical energy supplied by the source

This term is related to the electromechanical energy variation by the equation:

$$\Delta E_{\text{em}} = -\Delta E_{\text{sm}} + \Delta E_{\text{kin}} + \Delta E_{\text{fv}}$$

As:

$$\Delta E_{\text{kin}} = \Delta E_{\text{fv}} = 0$$

resulting in:

$$\Delta E_{\text{sm}} = -\Delta E_{\text{em}} = 0.325 \text{ J}$$

(b) *The developed electromechanical torque*

In rotational systems, the relationship of the torque and the electromechanical energy variation is given by:

$$T_{\text{dev}} = \frac{-\Delta E'_{\text{mag}}}{\Delta \alpha} = \frac{-0.325}{\frac{\pi}{6}} = -0.62 \text{ Nm}$$

Remarks: the negative sign is an indication that the developed torque is acting to reestablish the stable condition of $\theta = 0$.

(c) *The induced emf in the coil*

As usual, the induced emf is obtained applying Faraday's law given by:

$$e = -N\frac{\Delta\phi}{\Delta t} = -300 \times \frac{(2 - 3) \times 10^{-3}}{10} = 30\ \text{mV}$$

5.2 Electromechanical devices with permanent magnet

The development of rare earth permanent magnet (PM) technology has diffused its application in several electromechanical solutions. Not only small actuators but also big electric motors are built using PMs today. From small motors for drones to big propulsion motors for ships the presence of PMs is frequent.

Figure 5.20 shows a magnetic system built with both an ideal soft magnetic material and a PM whose characteristics are shown in figure 5.21. The fringing effect and the mechanical losses can be neglected for this analysis.

Neglecting the drop of mmf in the magnetic material, the analog electric circuit of the magnetic circuit excited by a PM is given by the one shown in figure 5.22.

Solving the electric circuit of figure 5.22, we get:

$$-H_p l_p = H_g l_g \tag{5.28}$$

that result in:

$$H_p = -\frac{H_g l_g}{l_p} \tag{5.29}$$

This means that the air-gap introduces a negative field to the PM material.

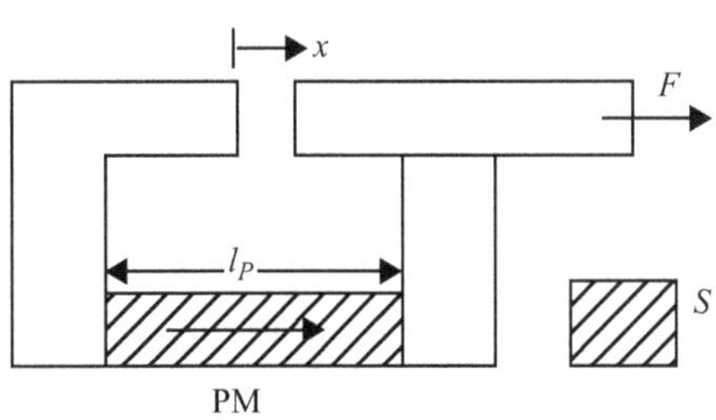

Figure 5.20. The magnetic system.

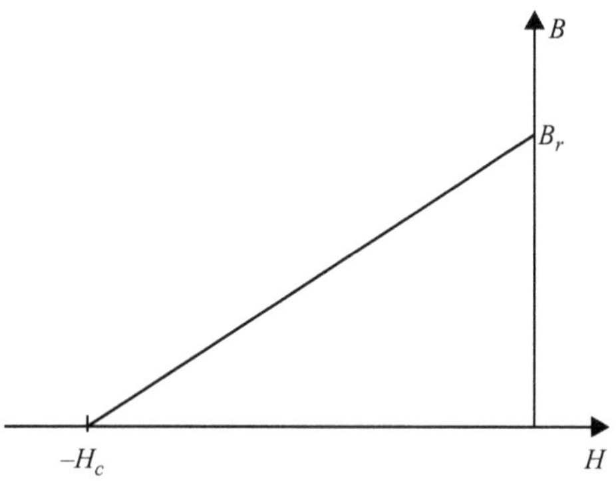

Figure 5.21. The PM characteristic.

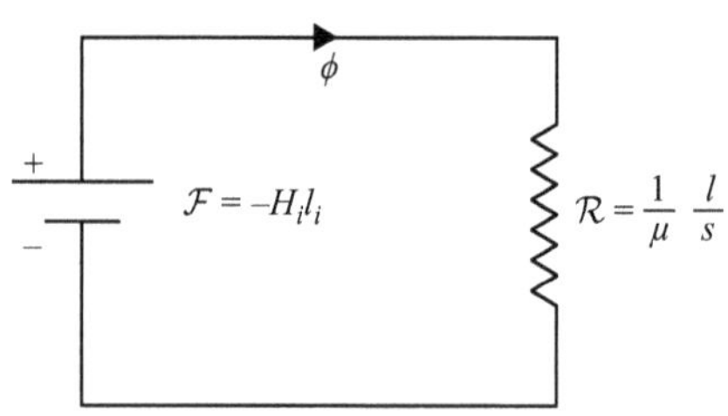

Figure 5.22. The analog electric circuit of magnetic circuit. $\mathcal{F} = -H_p l_p$: the mmf created by the PM. The magnetic flux is in the opposite sense of the magnetic field at the PM.

Neglecting all leakage flux, the magnetic flux is the same for the entire magnetic structure, so:

$$B_p S_p = B_g S_g \tag{5.30}$$

As $B_g = \mu_0 H_g$ and $H_g = -\frac{H_p l_p}{l_g}$, from equation (5.30) results in:

$$B_p = -\mu_0 \frac{S_g}{S_p} \times \frac{l_p}{l_g} H_p \tag{5.31}$$

Figure 5.23 shows the relation (B_p, H_p) for the magnetic circuit, that is a straight line of angular coefficient $-\mu_0 \frac{S_g}{S_p} \times \frac{l_p}{l_g}$. The operation point of the PM is the intersection of this straight line with the magnetization characteristics of the PM represented by the point P in figure 5.23.

So, the coordinates of P give not only the magnetic flux density of the PM, but also the magnetic flux density of the air gap.

Remarks: Aiming to avoid the PM demagnetization process, all PM devices are designed to operate in the linear part of the magnetization curve.

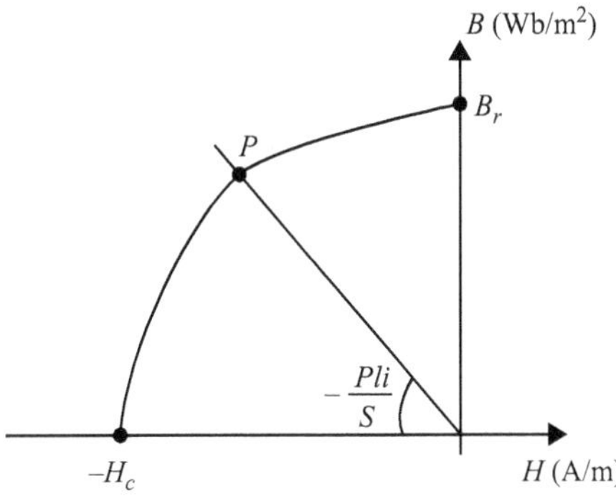

Figure 5.23. Operation point of the PM.

The energy involved in PM operation is an important parameter for a correct design of magnetic devices that contain PMs in their magnetic structure.

The product B_pH_p is called the energy product of the PM, not only because it is dimensionally identified with an associated volumetric density of energy $\left(\frac{\text{J}}{\text{m}^3}\right)$, but also because it is strongly linked to the dimensions of the PM as we will see.

Based on both equations (5.29) and (5.31), we can write that:

$$B_pH_p = B_gH_g\frac{S_gl_g}{S_pl_p} \tag{5.32}$$

as $V_g = S_gl_g$ is the air-gap volume and $V_p = S_pl_p$ is the PM volume, the equation (5.32) can be written as:

$$B_pH_p = B_gH_g\frac{V_g}{V_p} \tag{5.33}$$

Based on equation (5.8), that establishes that:

$$E_{\text{mag}} = \frac{\mathcal{F}\phi}{2} \tag{5.34}$$

applying it in the air-gap volume to evaluate the magnetic energy stored in the air, and also considering that $\mathcal{F} = H_gl_g$ and $\phi = B_gS_g$, the equation (5.34) becomes:

$$E_{\text{mag}} = \frac{H_gl_g \times B_gS_g}{2} = \frac{B_gH_gV_g}{2} \tag{5.35}$$

Substituting this result into equation (5.33), yields:

$$V_p = \frac{2E_{\text{mag}}}{B_pH_p} \tag{5.36}$$

From (5.36), we can identify that for an established magnetic energy stored in the air gap, the minimum PM volume is obtained when the energy product B_pH_p is maximum.

This is an ideal condition, but all designers avoid using this operation point due to the possibility of demagnetization of the PM during a transient condition.

It will be shown in the next chapter that the force developed by the PM device is related to the magnetic energy stored in the air-gap by:

$$\frac{F_{\text{dev}}}{S_g} = \frac{1}{2}\frac{B_g^2}{\mu_0} \tag{5.37}$$

As a result, the developed force for the magnetic system of figure 5.20 will be:

$$F_{\text{dev}} = \frac{E_{\text{mag}}}{l_g} \tag{5.38}$$

Example 5.4

The magnetic device shown in figure 5.24 is made of a soft magnetic material that can be considered ideal ($\mu \to \infty$) and a PM is rigid enough that we can consider its magnetic characteristics linear such that $B_r = 0.5$ T and $H_c = 500000$ A m^{-1}.

The length of the PM is 10 mm in the direction of magnetization and its cross-section area is 10 cm^2.

The cross-section of the air-gap area is 12 cm^2. Fringing of flux around the air gaps may be ignored. Determine for both $l_g = 2$ mm and $l_g = 5$ mm:

(a) the magnetic flux density in both the PM and the air gap;
(b) the magnetic energy stored in the air-gap magnetic field;
(c) the developed force in the movable part.

Solution

(a) The magnetic flux computation

Figure 5.25 shows the magnetic characteristics of PM.

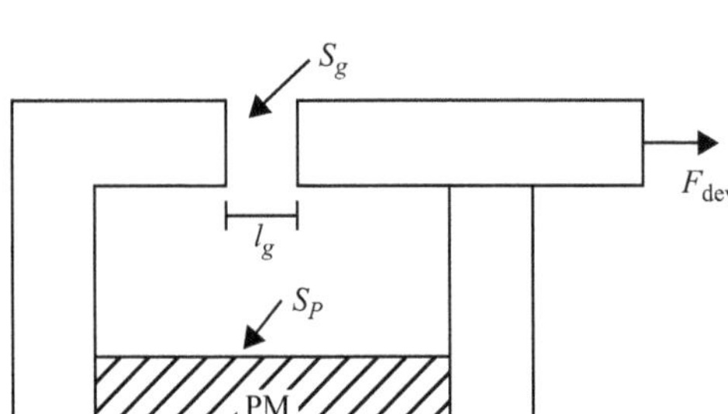

Figure 5.24. Magnetic device of example 5.4, $S_g = 12\text{ cm}^2 - S_p = 10\text{ cm}^2$, $l_g = 2 - 5$ mm; $l_p = 102$ mm, $B_r = 0.5\,T$; $H_c = -500000\left(\frac{\text{A}}{\text{m}}\right)$.

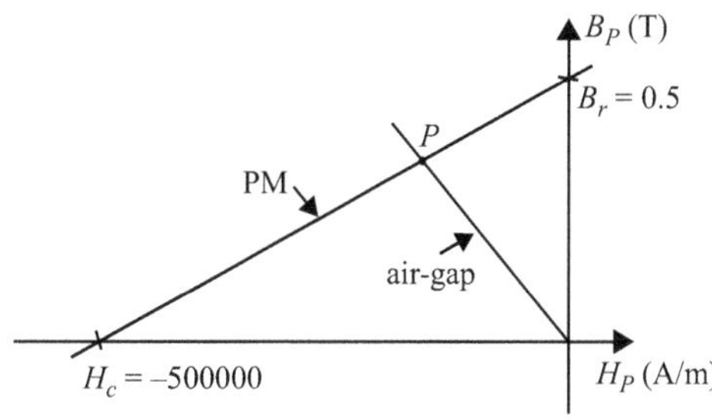

Figure 5.25. PM and air-gap characteristics.

The operation point P is the common point of two straight lines, the PM and the air-gap characteristics.

The first one is the PM characteristics given by:

$$B_p = aH_p + b$$

Considering that for $H_p = 0$ results in $B_p = B_r$ and for $B_p = 0$ results in $H_p = H_c$, the PM characteristics yields:

$$B_p = -\frac{B_r}{H_c}H_p + B_r$$

Applying the numerical values results in:

$$B_p = 10^{-6}H_p + 0.5$$

From equation (5.1), the magnetic flux density in the air-gap should satisfy:

$$B_p = -\mu_0\frac{S_g}{S_p} \times \frac{l_p}{l_g}H_p = -7.54 \times 10^{-6}H_p$$

Therefore, the magnetic field intensity H_p is such that:

$$-7.54 \times 10^{-6}H_p = 10^{-6}H_p + 0.5$$

Resulting in:

$$H_p = -58548 \text{ A m}^{-1} \text{ and } B_p = 0.44 \text{ (T)}$$

The magnetic flux density in the air-gap is evaluated from:

$$B_pS_p = B_gS_g$$

Resulting in:

$$B_g = \frac{S_p}{S_g} \times B_p = \frac{10}{12} \times 0.44 = 0.37 \text{ (T)}$$

(b) The magnetic energy stored in the air-gap

As $B_g = \mu_0 H_g$, from equation (5.35), we get:

$$E_{mag} = \frac{B_g H_g V_g}{2} = \frac{0.37 \times 294436 \times 12 \times 10^{-4} \times 2 \times 10^{-3}}{2}$$

$$E_{mag} = 0.13\ (\mathrm{J})$$

(c) The developed force is given by:

$$F_{dev} = \frac{E_{mag}}{l_g} = \frac{0.13}{2 \times 10^{-3}} = 65\ (\mathrm{N})$$

Remarks: For $l_g = 5$ mm the results are:

$H_p = -124\,500$ A m^{-1}; $B_p = 0.37$ (T); $B_g = 0.31$ (T); $H_g = 249\,000$ (A m^{-1}); $E_{mag} = 0.23$ (J) and $F_{dev} = 47$ (N)

5.3 Summary

Concerning the electrical side of an electromagnetic device, we can appoint the different kinds of energy variation:

ΔE_{se}: the variation of energy supplied by the electrical source;
ΔE_{mag}: the variation of the energy stored in the magnetic field;
ΔE_j: the variation of the energy due to the Joule losses.

As a rule, if ΔE_{se} is positive, it means that the source is injecting energy in the electromechanical device (motor operation). If it is negative, the device is giving back the energy to the source (generator operation).

The same statement is applied in ΔE_{mag}. This quantity is positive when the energy is injecting into the device to grow the magnetic field, as occurs when the electrical current is grown, and negative when the energy is giving back to the electrical source, as occurs when the electrical current is reduced. The variation of ΔE_j is always a positive quantity.

Concerning the mechanical side, we identify:

ΔE_{sm}: the variation of the mechanical energy supplied by the mechanical source, that is positive when the source introduces mechanical energy in the electromechanical device (generator operation) and negative when the device is delivering energy to the mechanical load (motor operation);

ΔE_{kin}: the variation of kinetic energy that is positive when the speed is growing and negative when it is shrinking;

ΔE_{fv}: the variation of the energy issued from both friction and ventilation, this quantity is always positive.

The quantity ($\Delta E_{em} = -\Delta E_{sm} + \Delta E_{kin} + \Delta E_{fv}$) represents the variation of the electromechanical energy that is the variation of the converted energy. If the energy is converted from electrical to mechanical ΔE_{em} (motor operation), it is positive, but it is negative if the energy is converted from mechanical to electrical (generator operation).

Any variation of electromechanical energy is associated with both force or torque development that can be evaluated considering whether they are made under current or magnetic flux constant.

For a constant current:

$$F_{dev} = \frac{\Delta E'_{mag}}{\Delta x} \text{ or } T_{dev} = \frac{\Delta E'_{mag}}{\Delta \theta}$$

For a constant flux:

$$F_{dev} = \frac{-\Delta E_{mag}}{\Delta x} \text{ or } T_{dev} = \frac{\Delta E'_{mag}}{\Delta \theta}$$

where $\Delta E'_{mag}$ is the co-energy that has no physical meaning.

Finally, the operation of a magnetic device with a PM is discussed. It is important to assure that in a PM device design the operation point is located at the linear part of the PM characteristics to avoid its demagnetization during the transient process.

5.3.1 Project of magnetic crane with a PM

The magnetic crane device shown in figure 5.26 consists of two ferrite magnets the magnetization curve of which is given by figure 5.27. To avoid irreversible demagnetization, the maximum magnetic field intensity inside the PM is 80 000 A m^{-1}. As a dimension constraint, the minimum thickness of the PM specified by the manufacturer is d_{min}.

The device is conceived to hold a mass (m) suspended in the air with an air-gap of 5 mm. The magnets are fixed to a high-permeability iron backing plate and faced by a high-permeability rigid member. Fringing of flux around the air gaps may be ignored. Determine the dimensions of the PM.

The project is conceived for five work teams with different constraints established in the following table.

d_{min} (mm)	h_{min} (mm)	mass (kg)
15	8	1
20	8	2
20	10	3
25	10	4
25	10	5

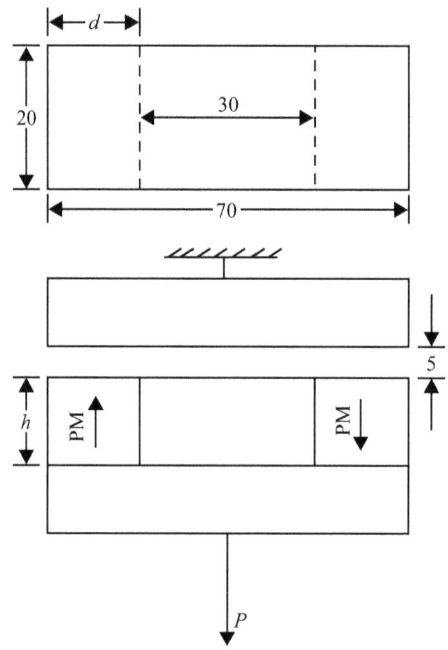

Figure 5.26. Draft of crane.

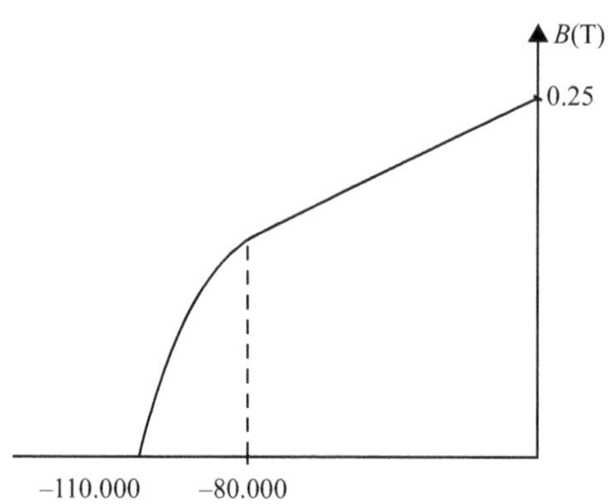

Figure 5.27. PM characteristics.

Remarks: To evaluate the developed force in a PM system at a specific air-gap, consider that the armature has moved from its original position $x_1 = 0$ to the final position x_2 corresponding to the imposed air gap.

Problems

5.1 The magnetic system of figure 5.28 is built with a magnetic material that can be considered ideal ($\mu \to \infty$). A 200 turns coil has a resistance of 1 mΩ and is fed by a DC electrical source. The system is used for a translational movement in an x direction. The armature is shifted from the position $x_1 = 1$ mm to $x_2 = 3$ mm under constant current of 5 A in 5 s.

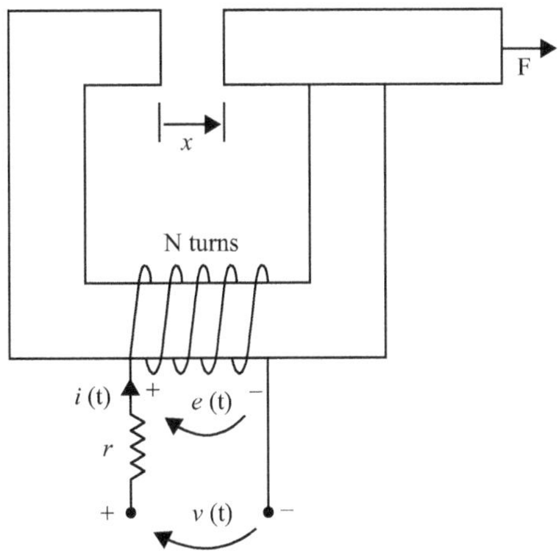

Figure 5.28. Problem 5.1. Cross-section of the core: 20 cm^2.

The losses of friction effect are 50 mJ.
The cross-section of the device is 20 cm^2.
To characterize this electromagnetic device it is necessary to obtain:

(a) the saturation curves ($\phi \times \mathcal{F}$) for both positions x_1 and x_2;
(b) the variation of the stored energy in the magnetic field;
(c) the variation of the energy due to the Joule losses;
(d) the variation of the electric energy supplied by the source;
(e) the variation of the electromechanical energy developed by the electromagnetic device;
(f) the balance of energy involved in this operation.

Answers: (a) $\phi = 0.025 \times 10^{-4}\mathcal{F}$; $\phi = 0.0084 \times 10^{-4}\mathcal{F}$; (b) −0.83; (c) 0.125 J; (d) −16.4 J; (e) −0.83 J.

5.2 The same magnetic system as shown in figure 5.28 is now built with a magnetic material that can be considered ideal ($\mu \to \infty$) until 1 Wb m^{-2} as is shown in figure 5.29. Compute again the six items for problem 5.1 considering the change of the material characteristics.

Answers: (b) −0.376 J; (c) 0.125 J; (d) −1.035 J; (e) −0.784 J.

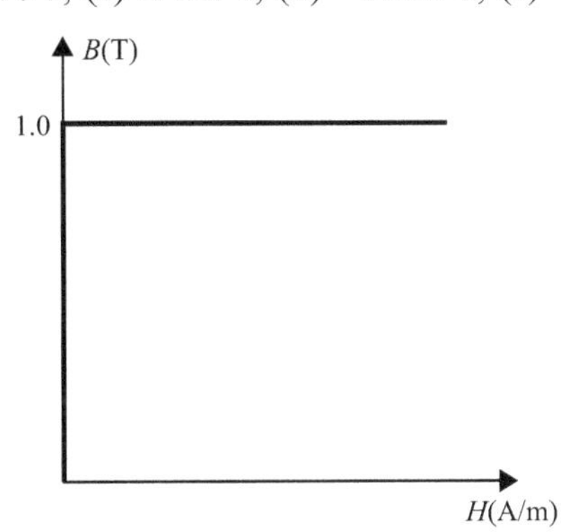

Figure 5.29. Magnetic material magnetization curve of problem 5.2.

5.3 Figure 5.30 shows the saturation curves of a magnetic system like the one shown in figure 5.27 for the positions $x_1 = 2$ mm and $x_2 = 3$ mm. The coil has 200 turns and its resistance is 1 mΩ.

The armature is moving from x_2 to x_1 under constant current of 2 A in 10 s. This time interval is enough large to consider the movement at constant current. The dissipated energy by friction during the movement is 50 mJ.

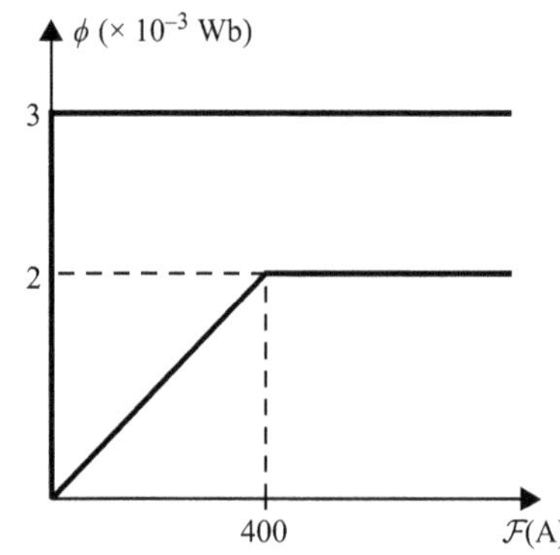

Figure 5.30. Problem 5.3. Answers: (b) 800 N; 750 N; (d) 0.02 V.

Determine:
(a) the energy balance in the magnetic system during this time interval;
(b) the average developed electromechanical force;
(c) the net force applied by the mechanical source;
(d) the induced emf in the coil;
(e) sketch a flow-diagram of the balance of energy in the magnetic system.

5.4 The magnetic system of problem 5.1 has its armature moving from x_2 to x_1 under *constant flux* of 2×10^{-3} Wb in 10 s.

Determine:
(a) the energy balance in the magnetic system during this time interval;
(b) the average developed electromechanical force;
(c) the net force applied by the mechanical source;
(d) the induced emf in the coil;
(e) sketch a flow-diagram of the balance of energy in the magnetic system.

Answers: (b) 790 N; (c) 765 N; (d) null.

5.5 The cross-section of a rotational magnetic system and its saturation curves for the positions $\theta_1 = 0°$ and $\theta_2 = 30°$ are shown in figure 5.31. The coil has 150 turns in each pole and its resistance is 1 mΩ.

The armature is moving from θ_2 to θ_1 under a constant current of 1 A in 10 s. This time interval is large enough to consider the movement at constant current. All mechanical losses are neglected.

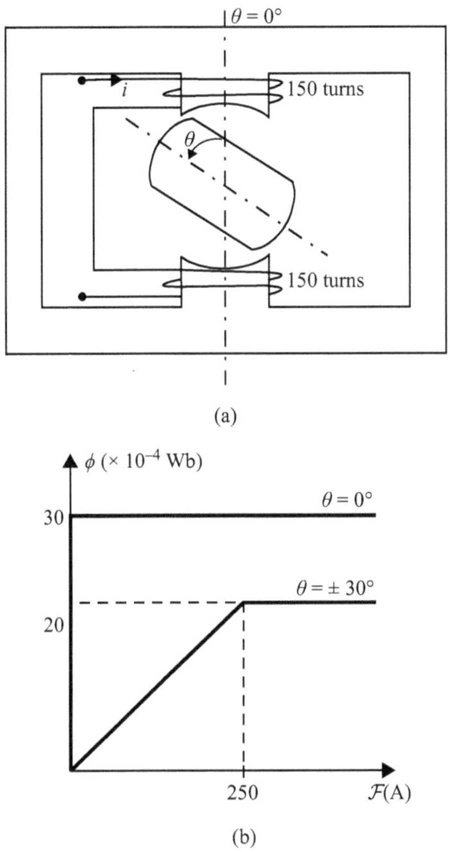

Figure 5.31. Problem 5.5. (a) The rotational magnetic system and (b) saturation curves.

Determine:
(a) the energy balance in the magnetic system during this time interval;
(b) the developed electromechanical torque;
(c) the induced emf in the coil;
(d) sketch a flow-diagram of the balance of energy in the magnetic system.

Answers: (b) 0.95 Nm; (c) 0.06 V.

5.6 A magnetic device based on a permanent magnet is made of a ferrite (a type of material suitable for making a PM), its magnetization characteristics are given on figure 5.32. The cast steel pole pieces may be assumed to have infinite permeability. Both the fringing effects and leakage flux are neglected. All dimensions are in millimeters (figure 5.33).

The flux density in the air-gap should be 0.3 Wb m^{-2} to provide the necessary force. Determine the maximum value of x to avoid the demagnetization process during transient performance.

Answer: 3.35 mm.

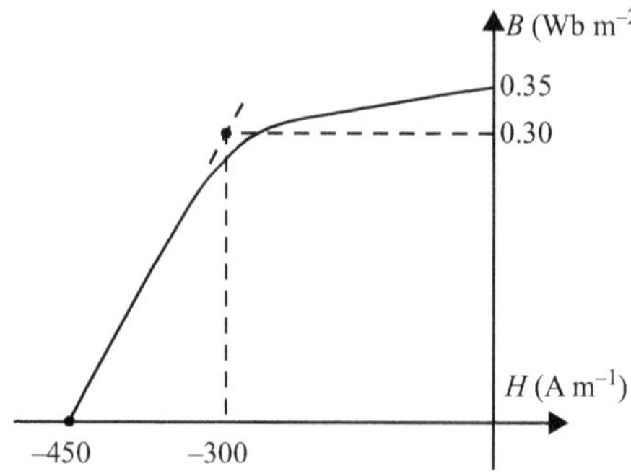

Figure 5.32. Problem 5.6. The magnetizations characteristics of the PM.

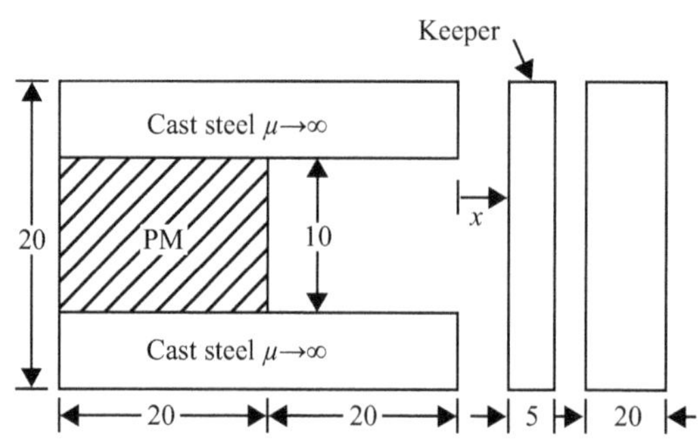

Figure 5.33. Problem 5.6. The device geometry—dimensions in millimeters.

5.7 Using the data for problem 5.6 and considering that the air-gap is 75% of the one computed in the previous problem, determine:
   (a) the reluctance of the magnetic circuit;
   (b) the magnetic field and the flux density inside PM and the air-gap;
   (c) the developed force exerted in the movable part of the device.

Answers: (a) 10^7 A Wb^{-1}; (b) 0.3 Wb m^{-2}; (c) 7 kg.

5.8 The cylindrical magnetic device in figure 5.34 is excited by two different PMs whose main characteristics are given in the table below.

Material	Residual magnetic flux density (T)	Coercive magnetic field intensity (A m^{-1})
NbFeB	1.0	795 000
Ferrite	0.38	175 000

Both characteristics can be considered linear.

The soft magnetic material is ideal and the air-gap between the plunger and the magnetic structure is negligible:
   (a) compute the ratio of the developed force using both materials separately;
   (b) compare the ratio of the energy product ($B_p \times H_p$) of both PM materials.

Answers: (a) 8; (b) 8.4.

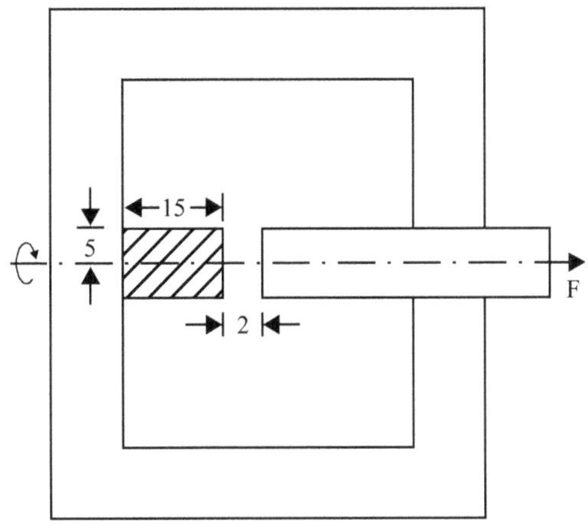

Figure 5.34. Problems 5.8 and 5.9. Dimensions in millimeters.

5.9 Using the data from problem 5.8, compare the volume of both PMs for developing the magnetic force evaluated using a ferrite PM.

Answer: 8.4.

5.10 In a permanent magnet the product ($B_p \times H_p$) is called the energy product. The *maximum* value of this product is a very important parameter for qualifying a PM. Considering that the PM characteristic is a straight line like the ones used in problem 5.8, compare the ratio of *maximum* energy product of both permanent magnets.

Answer: 9.

Further reading

[1] Gourishankar V 1966 *Electromechanical Energy Conversion* (Scanton: International Textbook Company)

[2] Skilling H H 1962 *Electromechanics: A First Course in Electromechanical Energy Conversion* (New York: Wiley)

[3] Bansal R (ed) 2004 *Handbook of Engineering Electromagnetics* (New York: Marcel Decker)

[4] Bozorth R M 1993 *Ferromagnetism* (Piscataway, NJ: IEEE Press)

[5] Fitzgerald A E, Kingsley C Jr and Umans S D 1992 *Electric Machinery* 5th edn (New York: McGraw-Hill)

IOP Publishing

Electromechanical Energy Conversion Through Active Learning

J R Cardoso, M B C Salles and M C Costa

Chapter 6

Electromechanical forces and torques

6.1 Introduction

In chapter 5 we discussed the several kinds of energies involved in the operation of an electromagnetic device. We have also discussed the development of both torques and forces issuing from the electromechanical energy conversion as an introduction to this chapter. We will learn in this chapter how to evaluate the instantaneous forces and torques in several devices and through this evaluation find the conditions that the device should obey for developing continuous work as the main characteristics of electric machines.

Our focus will be on introducing the concepts involving the operation of the most important electrical machines applied not only in industrial motion devices but also in those applied to the electrical mobility sector, such as in the electric traction of trains and electrical vehicles.

The electromechanical motion device can be fed by DC electrical current whose developed force or torque are constant or by an AC electrical current source that implies the development of a time-dependent force or torque. In such situation, the time average of those quantities should be non-null for having a permanent movement.

The rapid improvement of the power electronics industry brought together the production of high performance components and the development of advanced control technology. This combination has enable new designs of both inverters and converters to become very versatile motion drives for complex and modern applications, such as for robotics, drones, electric traction, and so on.

Therefore, this chapter is the most important in this book because all the concepts involved in the operation of electromechanical devices are presented.

As discussed earlier, during a fixed time step the energy balance in the electromechanical side of the device can be represented by the following components:

$$\Delta E_{\text{em}} = \Delta E_{\text{se}} - \Delta E_{\text{j}} - \Delta E_{\text{mag}} \tag{6.1}$$

doi:10.1088/978-0-7503-2084-9ch6

ΔE_{se} is the variation of energy supplied by the electrical source. For motor operation, this component is positive and for generator operation it is negative, corresponding to the sense of the energy being delivered to the device as positive. When the device is in generation mode, the sense of the energy is from the device to the electrical source (or the electric grid) and is considered as negative.

ΔE_{mag} is the variation of the energy stored in the magnetic field. As discussed in chapter 5, this term is positive when the magnetic field increases and negative when it decreases.

ΔE_{j} is the variation of the energy due to the Joule losses that are always positive.

Concerning the mechanical side, we have:

$$\Delta E_{em} = -\Delta E_{sm} + \Delta E_{kin} + \Delta E_{fv} \tag{6.2}$$

ΔE_{sm}: The variation of the mechanical energy supplied by the mechanical source. This term is positive when the source introduces mechanical energy to the device (generator operation mode) and negative in the motor operation, when the device is delivering mechanical energy.

ΔE_{kin}: The variation of kinetic energy that is positive when the speed is increasing and negative when it is decreasing. This term could also include the potential energy or the compliances of mechanical system.

ΔE_{fv}: The variation of the energy issuing from both friction and ventilation. This quantity is always positive.

Figure 6.1 shows the flowchart of the energy balance which is very useful in understanding the rule of the terms involved in the process of energy conversion.

6.1.1 The evaluation of energies from lumped parameters

The evaluation of energies from the saturation curves is suitable for analyzing non-linear electromagnetic devices, as we have seen in the last chapter. Nevertheless, the elegance of this formalism, which introduces the physical meaning of the operation of electromechanical devices, is not suitable for analyzing complex magnetic moving structures that host several coils.

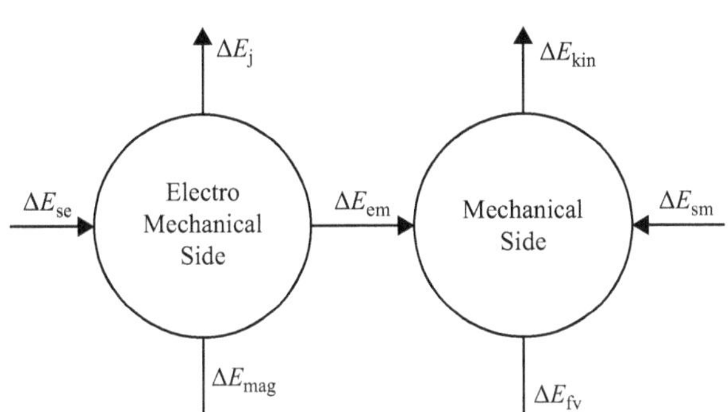

Figure 6.1. The energy balance in an electromechanical device. ΔE_{fv} and ΔE_{j} are always positive. ΔE_{kin} and ΔE_{mag} could be positive or negative. ΔE_{se} is positive in the motor operation. ΔE_{sm} is positive in the generator operation.

In such cases, it is convenient to represent the energies through the lumped parameters, like both self and mutual inductance. Using this representation, we can not only evaluate forces and torques in steady state but also in transient conditions.

All evaluations will submit to the following hypotheses:

(a) all devices are built with ideal magnetic material, therefore, no magnetic energy is stored in its magnetic structure;
(b) no fringing effect is considered, this means that no magnetic flux spreads out in the air-gap surrounding area.

Although these effects are important, neglecting them at this stage is necessary to focus on the main physical aspects of the operation of electromechanical devices.

6.2 The simply excited electromechanical device

6.2.1 First case: DC constant current operation

We start with the simpler magnetic system made with an ideal magnetic material, as presented by figure 6.2.

In this analysis, the coil is fed by a DC electrical source that injects a *constant electrical current*. From the previous chapter the average developed force for a constant current due to movement Δx is:

$$F_{\text{dev}} = \frac{\Delta E'_{\text{mag}}}{\Delta x} \tag{6.3}$$

If Δx becomes small, then equation (6.3) can be written as:

$$F_{\text{dev}} = \frac{\partial E'_{\text{mag}}}{\partial x} \tag{6.4}$$

In equation (6.4), E'_{mag} is the co-energy. As we are considering a linear magnetic system, this term also represents the magnetic energy stored in the magnetic field E_{mag} as shown in figure 6.3.

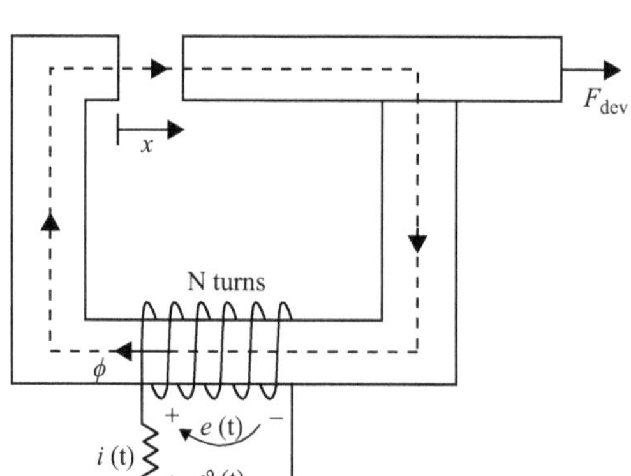

Figure 6.2. The simple excited magnetic system with ideal magnetic material.

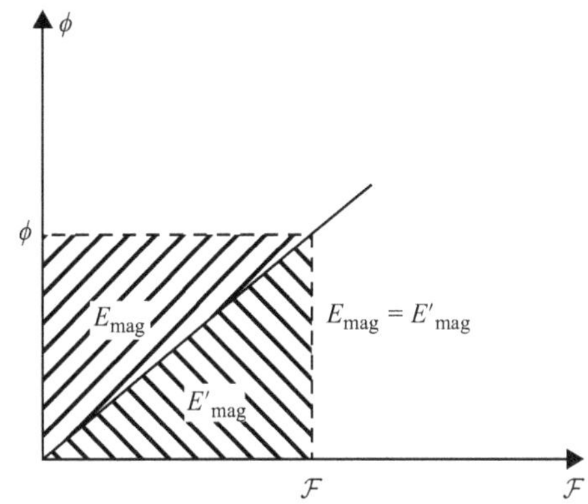

Figure 6.3. $E_{\text{mag}} = E'_{\text{mag}}$ for a linear magnetic system.

Therefore, based on equation (5.9) we can express the developed force from equation (6.4) as:

$$F_{\text{dev}} = \frac{\partial}{\partial x}\left(\frac{1}{2}Li^2\right) \tag{6.5}$$

Supposing that the current remains constant during the movement, the previous equation becomes:

$$F_{\text{dev}} = \frac{1}{2}i^2\frac{\partial L}{\partial x} \tag{6.6}$$

The relation between the number of turns of the coil and the reluctance of the magnetic circuit evaluates the self-inductance of the magnetic system:

$$L = \frac{N^2}{\mathcal{R}} \tag{6.7}$$

where:

$$\mathcal{R} = \frac{1}{\mu_0}\frac{x}{S} \tag{6.8}$$

In equation (6.8), the air-gap is represented by x, the cross-section of the device by S and the magnetic permeability of the air is given by μ_0. Consequently:

$$F_{\text{dev}} = \frac{1}{2}i^2\frac{\partial L}{\partial x} = \frac{1}{2}i^2\frac{\partial}{\partial x}\left(\frac{N^2\mu_0 S}{x}\right)$$

therefore:

$$F_{\text{dev}} = -\frac{N^2 i^2 \mu_0 S}{2x^2} \tag{6.9}$$

or,

$$F_{\text{dev}} = -\frac{ki^2}{x^2} \tag{6.10}$$

where,

$$k = \frac{N^2 \mu_0 S}{2} \tag{6.11}$$

is the magnetic system constant.

The negative signal is indicating that the sense of the developed force is opposite to the increment of x ($x > 0$). The developed force acts in the sense that enables it to reach its minimum reluctance. All magnetic systems move towards this position as a natural stable position.

Since $Hx = Ni$, equation (6.9) becomes:

$$F_{\text{dev}} = -\frac{\mu_0 H^2 S}{2} \tag{6.12}$$

An interesting outcome can be drawn from (6.12). Dividing both sides of the equation by S, the area of the surfaces bounding the air-gap, we get:

$$\frac{F_{\text{dev}}}{S} = -\frac{B^2}{2\mu_0} \tag{6.13}$$

that is, the force by unit area of the boundary is the *energy density in the magnetic field* enclosed by the boundary. This quantity is also called the *magnetic pressure* on the surface.

$$\text{Magnetic pressure} = \text{Energy density}$$

The results obtained so far, for translational magnetic systems, can be readily extended to rotational magnetic systems.

Example 6.1

The armature of the magnetic systems of figure 6.4 is restrained by a constant force f through a spring of the elastic constant K. Given that $v = 5$ V, determine the stable position of the armature using the following parameters:

$$r = 10\ \Omega,\ K = 2667\ \text{m}^{-1},\ x_0 = 3\ \text{mm},\ k = 6.283 \times 10^{-5}\ \text{H m}.$$

Solution

Concerning the electric circuit equation, we have:

$$v = ri \tag{6.14}$$

From the balance of the forces on the electromechanical side, we get:

$$f = K(x - x_0) - F_{\text{dev}} \tag{6.15}$$

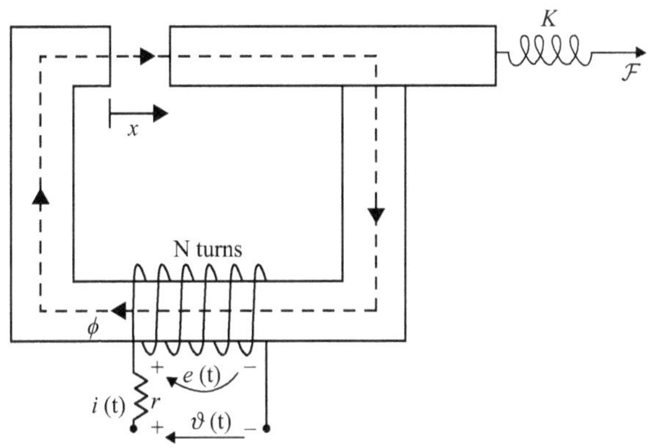

Figure 6.4. The magnetic system. K: elastic constant. k: magnetic system constant.

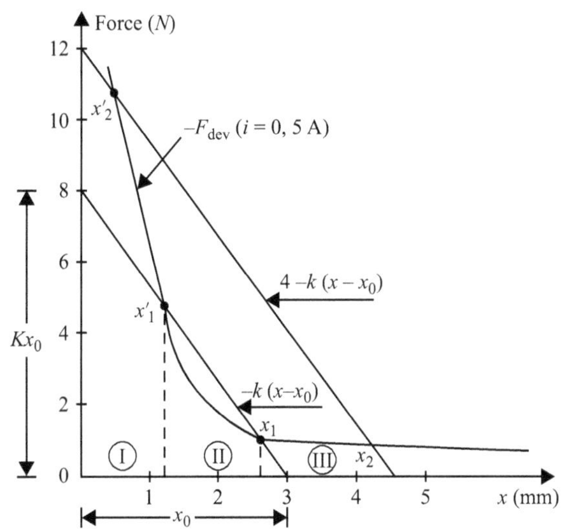

Figure 6.5. Steady-state operation of electromechanical device. x_1 and x_2: are stable solutions. x_1' and x_2': are unstable solutions.

Equation (6.15) may be written as:

$$-F_{dev} = f - K(x - x_0) \tag{6.16}$$

The visualization of the solution is shown by plotting the curves of both sides of (6.16) separately, as shown in figure 6.5.

In this figure, the second order hyperbole represents the negative of the developed force due to electromagnetic coupling. The two straight lines represent the function of the right-hand side of equation (6.16), the one on the left represents the right-hand function of (6.16) considering $f = 0$.

Two solutions, x_1 and x_1', of (6.16) are identified but only one is admissible. To analyze the admissibility we need to verify whether the proposed solution is stable.

Considering the first solution, if a small perturbation is introduced changing x_1 from x_1 to $x_1 + \Delta x$ the magnetic developed force becomes higher than the restraining force, reducing the air-gap to the original position. In contrast, if the perturbation is given in the opposite sense the new position becomes $x_1 - \Delta x$. In this new position, the restraining force is higher than the magnetic developed force increasing the air-gap to the original position. We conclude that the position x_1 is a stable solution.

Considering the second solution, if a small perturbation is introduced changing x_1' from x_1' to $x_1' + \Delta x$, the magnetic developed force becomes lower than the restraining force opening the air-gap completely. In contrast, if the perturbation is given in the opposite sense the new position becomes $x_1' - \Delta x$. In this position, the restraining force is lower than the magnetic developed force, closing the air-gap completely. We conclude that the position x_1' is an unstable solution.

Introducing an additional force in the spring, a new straight line is defined (the one on the right), and two new solutions could be found. Applying the above procedure, the new stable solution is x_2.

We can also apply this methodology to analyze a rotational actuator, like the one shown in figure 6.6.

In all rotational magnetic systems the main characteristic is the electromagnetic torque developed in its moving part. Applying the same approach used in the translational magnetic system, the electromagnetic torque developed under *constant electric current* is given by:

$$T_{\text{dev}} = \frac{1}{2} i^2 \frac{\partial L}{\partial \theta} \tag{6.17}$$

The self-inductance of the rotational magnetic system depends on the angular position of the moving part. With the rotor aligned with the stator ($\theta = 0$), the reluctance of the magnetic system is *minimum* and, consequently, that is the *maximum* self-inductance position (L_{max}). Otherwise, if the rotor is positioned orthogonally to the stator ($\theta = \pi/2$), the self-inductance is *minimum* (L_{min}).

From $\theta = 0$ to $\theta = \pi/2$, the inductance variation can be approximated by a sinusoidal function (see figure 6.7).

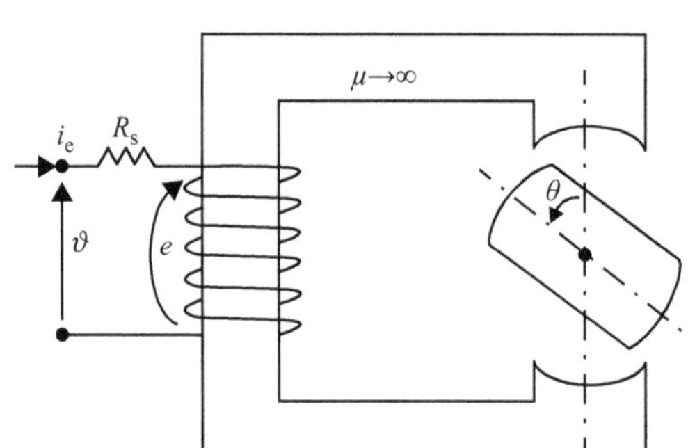

Figure 6.6. Rotational actuator. Static part: stator. Moving part: rotor.

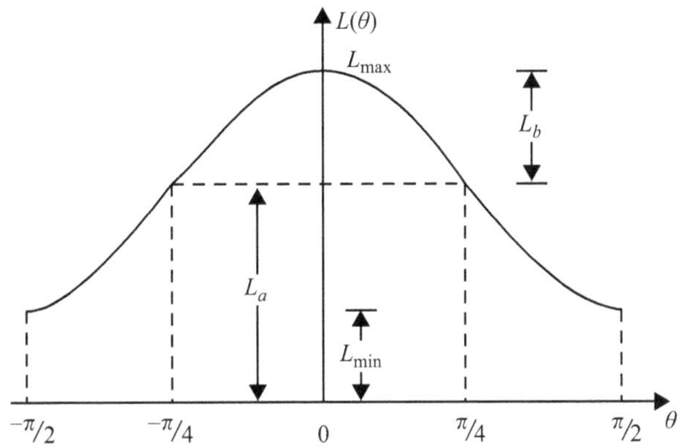

Figure 6.7. Self-inductance variation. $\theta = 0$: direct axis. $\theta = \pi/2$: quadrature axis.

The mathematical function that represents the inductance variation with the position can be written as:

$$L(\theta) = L_a + L_b \cos 2\theta \tag{6.18}$$

where:

$$L_a = \frac{L_{max} + L_{min}}{2} \tag{6.19}$$

and

$$L_b = \frac{L_{max} - L_{min}}{2} \tag{6.20}$$

Substituting this function in equation (6.17), we get:

$$T_{dev} = -i^2 L_b \sin 2\theta \tag{6.21}$$

Using (6.20), we can also write (6.21) as:

$$T_{dev} = -\frac{1}{2} i^2 L_{max} \sin 2\theta + \frac{1}{2} i^2 L_{min} \sin 2\theta \tag{6.22}$$

The developed torque depends on the difference between the maximum and minimum inductances.

Figure 6.8 shows the electromagnetic developed torque dependent on the angular position. In the position $\theta = 0$, the axes of both rotor and stator are aligned. This is a stable position because any perturbation around the origin develops torque that re-establishes the original position in $\theta = 0$. Otherwise, if the axes of both the rotor and the stator are orthogonal (or in quadrature), an unstable position is characterized. Now, any perturbation around $\theta = \pi/2$ will never re-establish the original position.

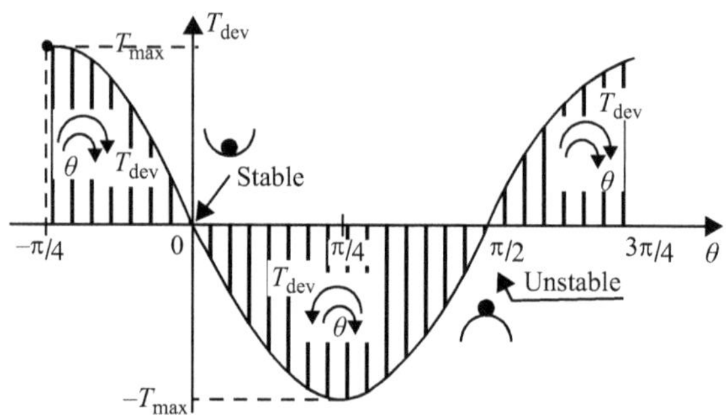

Figure 6.8. Steady state electromagnetic developed torque. $\theta = 0$: stable position. $\theta = \pi/2$: unstable position. The maximum developed torque is: $T_{max} = i^2 L_b$.

Remarks: An easy way to determine if the position is stable or not is to evaluate the first derivative of the torque. When $\frac{dT_{dev}}{d\theta} < 0$, as is happening in position $\theta = 0$, we are over a stable position.

This kind of magnetic system has no continuous movement when fed by a DC current, because for that it is necessary to have a continuous non-null average torque. It works just as an actuator for detecting the angular position.

Example 6.2

Figure 6.9 shows a cross-section representation of a 2-poles rotational actuator. The angular pole pitch is $\alpha = \pi/2$. Each pole of the stator is wound with an $N = 200$ turns coil that is connected in series with concordant polarities, producing magnetic flux in the same direction. The magnetic material is ideal. The radius of the rotor is $R = 10$ cm, and the length of air-gap between the rotor and the stator is $g = 2$ mm on either side (Note that there are two air-gaps). The axial length perpendicular to the plane of the paper is $h = 30$ cm. The ratio of both minimum and maximum inductance is 0.6. Determine for an exciting current of 5 A:

(a) the maximum inductance of the coil;
(b) the inductance as function of the position;
(c) the developed torque as a function of the position.

Solution

(a) The maximum inductance is observed when the axes of both the rotor and the stator are aligned in $\theta = 0$. To determine this we must evaluate the reluctance first. The reluctance in $\theta = 0$ is given by:

$$\mathcal{R}_{min} = \frac{1}{\mu_0} \times \frac{2g}{\alpha R h} \tag{6.23}$$

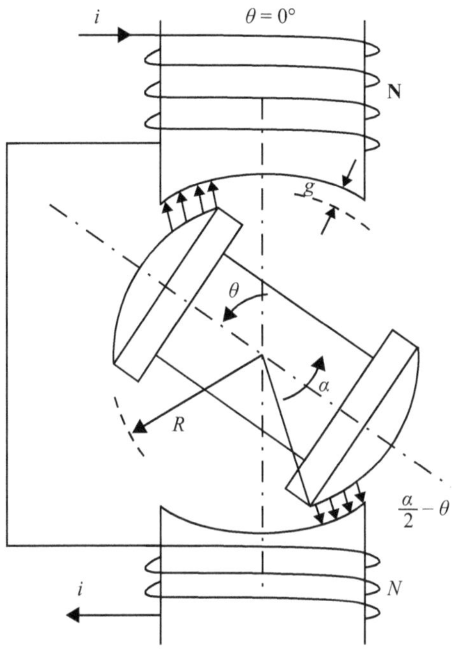

Figure 6.9. Rotational actuator. Winding $N = 200$ turns/pole; $R = 10$ cm. $h = 30$ cm; $g = 2$ mm; $\frac{L_{min}}{L_{max}} = 0.6$; $\alpha = \pi/2$; $i = 5$A.

where:

$l = 2g$: is the total air-gap of the device and $S = \alpha Rh$ the surface crossed by the magnetic flux

$$\mathcal{R}_{min} = \frac{1}{4\pi \times 10^{-7}} \times \frac{2 \times 2 \times 10^{-3}}{\frac{\pi}{2} \times 30 \times 10^{-3}} = 0.675 \times 10^{5}\text{A Wb}^{-1}$$

Therefore, the maximum inductance will be:

$$L_{max} = \frac{(2N)^2}{\mathcal{R}_{min}}$$

that results in:

$$L_{max} = \frac{(2 \times 200)^2}{0.675 \times 10^5} = 2.37 \text{ H}$$

(b) For evaluating $L(\theta)$, we need L_{min} first. As $\frac{L_{min}}{L_{max}} = 0.6$, results in:

$$L_{min} = 0.6 \times 2.37 = 1.42 \text{ H}$$

Therefore:

$$L_a = \frac{L_{max} + L_{min}}{2} = \frac{2.37 + 1.42}{2} = 1.895 \text{ H}$$

and

$$L_b = \frac{L_{\max} - L_{\min}}{2} = \frac{2.37 - 1.42}{2} = 0.475 \text{ H}$$

Consequently,

$$L(\theta) = 1.895 + 0.475 \times \cos 2\theta \text{ H}$$

(c) The developed torque is obtained from (6.21), that results in:

$$T_{\text{dev}} = -i^2 L_b \sin 2\theta = -5^2 \times 0.475 \times \sin 2\theta = -11.875 \times \sin 2\theta \text{ Nm}$$

The maximum developed torque is obtained for $\theta = \pm\frac{\pi}{2}$ and its value is 11.875 Nm.

6.2.2 A little bit of the transient analysis

In this chapter, we developed a methodology for analyzing the performance of simply excited electromagnetic devices fed by DC electric current.

We have also considered that the movement of the moving part did not affect the magnitude of the DC electric current. In fact, this statement is true only if the speed of the moving part is very slow. Otherwise, this situation does not occur in real life.

Figure 6.10 shows the translational magnetic systems in two situations. Figure 6.10(a) shows the static of the magnetic system, in such a situation that the DC electric current is constant, and its value is ($i = v/r$). Figure 6.10(b) shows the same device where its armature has moved to a new position $x + \Delta x$ with a non-negligible speed. What would happen with the electric current during this transition?

When we solved this problem, we considered that the movement of the armature was performed at a very slow speed. For each new position, a new value of magnetic flux is observed, but this change occurs during a large time interval with no expressive rate of change of the magnetic flux ($\Delta\phi/\Delta t \approx 0$). Therefore, no induced emf appears and the electric current remains constant. Otherwise, when the armature movement is fast, this effect cannot be neglected.

Kirchhoff's law applied to the coil circuit gives:

$$v = e + ri \tag{6.24}$$

Then,

$$i = \frac{v - e}{r} \tag{6.25}$$

The first case is when the armature moves fast to the right side (opening). In this situation, the magnetic flux is tending toward decreasing and the induction of emf due to the flux variation is such that:

$$e = N\frac{d\varnothing}{dt} < 0 \tag{6.26}$$

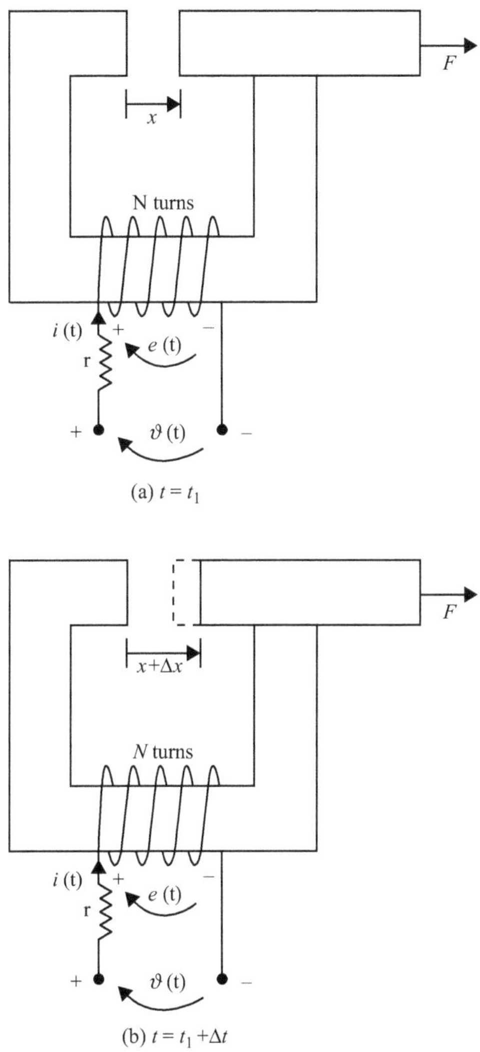

Figure 6.10. Armature movement. (a): armature position x and magnetic flux $\varnothing$. (b): armature position = $x + \Delta x$, magnetic flux $\varnothing + \Delta\varnothing$. emf: $e = N\frac{\Delta\varnothing}{\Delta t}$.

Therefore, based on equation (6.25), the electrical current increases due to the increment of the numerator. Since the armature movement stops, the induced emf becomes null and the electrical current goes back to the original value.

In contrast, if we close the armature fast the magnetic flux tend toward increasing and the induced emf is such that:

$$e = N\frac{d\varnothing}{dt} > 0 \tag{6.27}$$

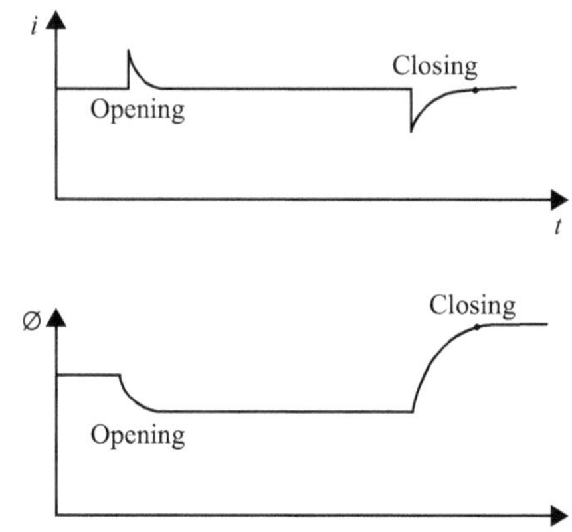

Figure 6.11. Time variation of both electrical current and magnetic flux. Opening: emf < 0 the electrical current increases and magnetic flux decreases. Closing: emf > 0 the electrical current decreases and magnetic flux increases.

Based on equation (6.25), the electrical current decreases because the numerator also decreases. Since the armature movement stops, the electrical current goes back to the original value because the magnetic flux remains constant.

Figure 6.11 shows the time variation of both the electrical current and the magnetic flux in the coil of the translational magnetic device.

6.2.3 Second case: the constant flux operation

There are several situations where the magnetic system works under a time-dependent magnetic flux variation with a constant pattern. Indeed, this is what occurs when the coil is fed by an AC voltage source. In a steady state, the magnetic flux established in the core has a sinusoidal time variation pattern with constant magnitude that can be mathematically expressed by:

$$\varnothing(t) = \varnothing_{\max} \sin \omega t \tag{6.28}$$

where:

$\varnothing_{\max}$: the maximum value of $\varnothing(t)$;
$\omega = 2\pi f$: the angular frequency of the function.

As the magnetic flux is a time-dependent function, an induced emf appears in the coil, such that:

$$e = N\frac{d\varnothing}{dt} = \omega N \varnothing_{\max} \cos \omega t \tag{6.29}$$

As expressed by equation (6.29), the induced emf is also a time-dependent sinusoidal function:

$$E_{\max} = \omega N \varnothing_{\max} \tag{6.30}$$

The rms value of the induced emf is given by:

$$E = \frac{E_{\max}}{\sqrt{2}} = \frac{\omega N \varnothing_{\max}}{\sqrt{2}} \tag{6.31}$$

As $\omega = 2\pi f$ the final version of the rms value of the induced emf is:

$$E = 4.44\, f N \varnothing_{\max} \tag{6.32}$$

Kirchhoff's law in the coil circuit gives:

$$v = e + ri \tag{6.33}$$

We can neglect the resistance of the coil first, because the conductor is made with very good electrical material, like copper. Therefore, we can also write that $v = e$ and:

$$V = E = 4.44\, f N \varnothing_{\max} \tag{6.34}$$

where V is the rms value of the voltage source.

If the coil is fed by an AC voltage source with a constant rms module, it is implied that the magnitude of the magnetic flux ($\varnothing_{\max}$) in rms value remains constant. This is the reason we entitle this topic *the constant flux operation.*

6.3 The translational magnetic system under constant magnetic flux operation

Returning to the translational magnetic system, the stored energy in the magnetic field is given by:

$$E_{\text{mag}} = \int_0^{\phi} \mathcal{F} d\phi \tag{6.35}$$

In the linear magnetic system, the relation of the magnetomotive force (mmf) and the magnetic flux is given by the equation $\mathcal{F} = \mathcal{R}\phi$. Therefore, the stored energy in the magnetic field in linear medium is:

$$E_{\text{mag}} = \frac{\mathcal{R}\phi^2}{2} \tag{6.36}$$

As we discussed earlier, the developed force under constant flux for $\Delta x \to 0$ is given by:

$$F_{\text{dev}} = \frac{-\partial E_{\text{mag}}}{\partial x} \tag{6.37}$$

Applying (6.36) in equation (6.37), we can write:

$$F_{\text{dev}} = \frac{-\partial}{\partial x}\left(\frac{\mathcal{R}\phi^2}{2}\right) \tag{6.38}$$

As the magnetic flux $\varnothing$ is not *x-dependent*, it results in:

$$F_{\text{dev}} = -\frac{\phi^2}{2}\frac{\partial \mathcal{R}}{\partial x} \tag{6.39}$$

As the magnetic flux in the AC current excitation is given by:

$$\varnothing = \varnothing_{\max} \sin \omega t$$

and,

$$\mathcal{R} = \frac{1}{\mu_0} \times \frac{x}{S}$$

From equation (6.39), the developed force results in:

$$F_{\text{dev}} = -\frac{(\varnothing_{\max} \sin \omega t)^2}{2} \times \frac{1}{\mu_0 S} \tag{6.40}$$

Applying the trigonometric identity:

$$\cos^2 \alpha - \sin^2 \alpha = \cos 2\alpha \tag{6.41}$$

That results in:

$$\sin^2 \alpha = \frac{1}{2} + \frac{1}{2}\cos 2\alpha \tag{6.42}$$

We can rewrite equation (6.40) as:

$$F_{\text{dev}} = -\frac{\varnothing_{\max}{}^2}{4\mu_0 S} - \frac{\varnothing_{\max}{}^2}{4\mu_0 S}\cos 2\omega t \tag{6.43}$$

Figure 6.12 shows the time variation of the developed force. We can identify two terms.

The first is a time-independent term or a constant term of (6.43):

$$F_{\text{avg}} = -\frac{\varnothing_{\max}{}^2}{4\mu_0 S} \tag{6.44}$$

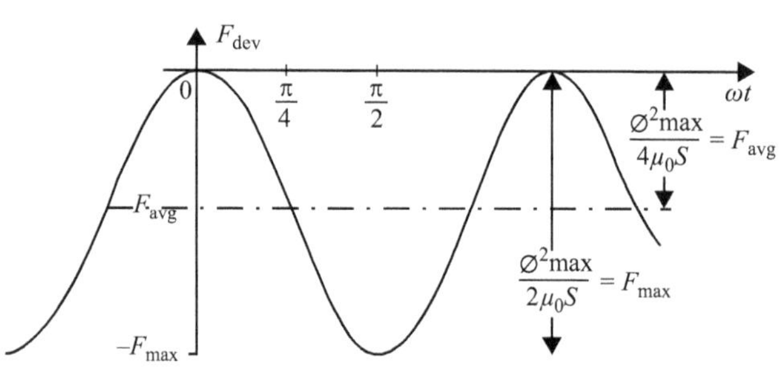

Figure 6.12. The time-dependent developed force. F_{avg}: the time average of the developed force. $F_{\max}$: the maximum value of the developed force. $F_{\min}$: the minimum value of the developed force is null.

The second is the time-dependent term:

$$f(t) = -\frac{\varnothing_{\max}{}^2}{4\mu_0 S}\cos 2\omega t \tag{6.45}$$

This term is a time variation function that varies sinusoidally with the doubling of the source frequency. As all sinusoidal function, its time average value is null.

Therefore,

$$\text{Med}[F_{\text{dev}}] = F_{\text{avg}} = -\frac{\varnothing_{\max}{}^2}{4\mu_0 S} \tag{6.46}$$

The negative signal indicates that the sense of the average force is that of closing the armature.

Remarks: Note that $F_{\min} = 0$. This device with this magnetic structure is not suitable for lifting a mass because the developed force becomes null twice in each cycle of the voltage source. During its operation, we can observe a characteristic vibration with the doubling of the source frequency.

The electric current in the coil is such that:

$$Ni(t) = \mathcal{R}\varnothing(t) = \frac{1}{\mu_0} \times \frac{x}{S}\varnothing(t) \tag{6.47}$$

Using equation (6.28), results in:

$$i(t) = \frac{x\varnothing_{\max}\sin\omega t}{\mu_0 NS} \tag{6.48}$$

The rms value of $i(t)$ is given by:

$$I = \frac{x\varnothing_{\max}}{\sqrt{2}\,\mu_0 NS} \tag{6.49}$$

In the first case, we discuss the performance of the magnetic system during a DC constant current operation (figure 6.13). In that case, the electric current remains constant independent of the moving air-gap. The same phenomenon does not occur when the magnetic system is fed by a constant AC voltage source because the current depends on the moving air-gap, as we can see in equation (6.49).

6.3.1 The shading coil

It is possible to avoid both the vibration and *null points* in the developed force curve using a conductor ring called the *shading coil* that can be installed in the armature, as shown in figure 6.14.

With the installation of the shading coil, the total magnetic flux is split into two components ($\phi = \phi_1 + \phi_2$). The component ϕ_1 is associated with the main current

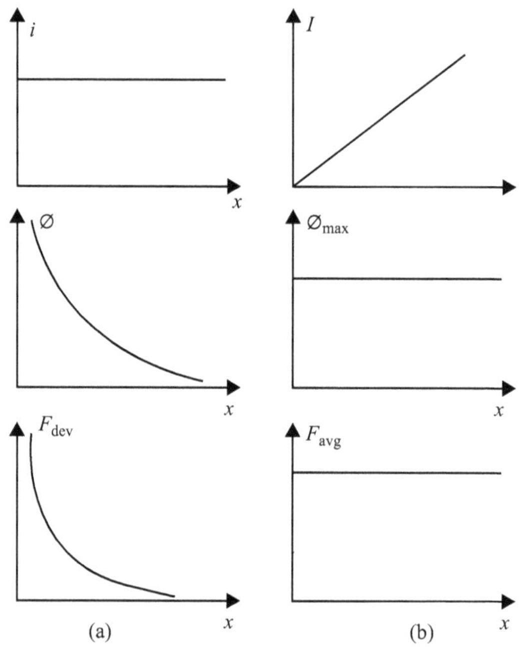

Figure 6.13. Performance comparison as a function of air-gap depending on the electric source. (a) DC constant current (i = constant). (b) AC constant voltage ($\varnothing_{max}$ = constant). The shapes of the curves are valuable for small variations of air-gap.

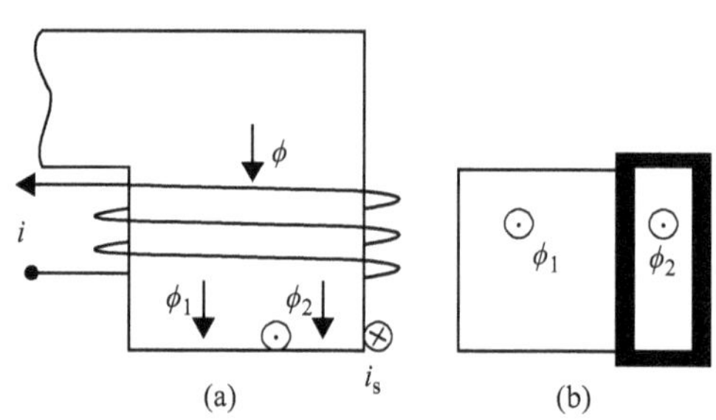

Figure 6.14. The shading coil. (a) front view and (b) bottom view.

and the component ϕ_2 is associated to the shading coil current. As both alternating current are not in phase, the magnetic fluxes are also not in phase.

Therefore, the developed force issuing from each coil is displaced as shown in figure 6.15. The final developed force is the summation of them, which gives no null values and no vibration, as it appears when the shading coil is not used.

6.4 The rotational magnetic system under constant magnetic flux operation

The performance of the rotational magnetic system under DC constant current was discussed earlier in this chapter. It was observed that under DC constant current

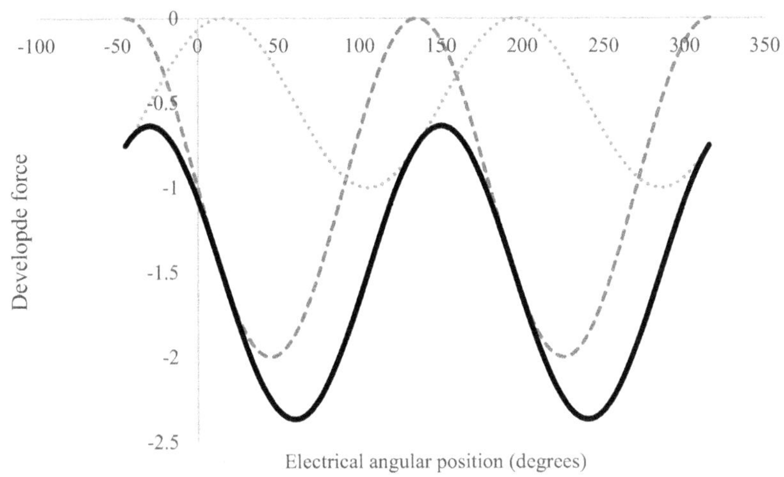

Figure 6.15. The effect of the shading coil in the developed force. Dashed line: developed force due to ϕ_1. Dotted line: developed force due to ϕ_2. Full line: developed force due to $\phi = \phi_1 + \phi_2$.

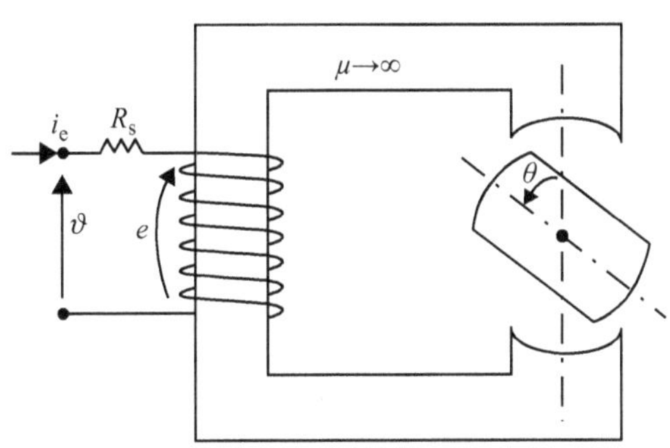

Figure 6.16. Rotational magnetic system under constant AC voltage. Magnetic material with $\mu \to \infty$. Coil resistance $r \approx 0$.

conditions the simply excited rotational magnetic system works as an actuator that can detect position. The limitation occurs because there is no condition to develop continuously an average torque that is different from null (zero).

With those aspects in mind, we can discuss the same electromagnetic device fed by an AC current source that supplies electric energy to the coil under constant voltage and constant frequency. The same hypothesis adopted in the first analysis will be considered, i.e., the magnetic material is ideal and the resistance of the coil is negligible.

As the module of the AC voltage source is constant, the maximum value of the magnetic flux is also constant (figure 6.16).

Applying the previous procedure, the instantaneous developed electromagnetic torque is given by:

$$T_{\text{dev}} = \frac{-\partial E_{\text{mag}}}{\partial \theta} \tag{6.50}$$

as,

$$E_{\text{mag}} = \frac{\mathcal{R}\phi^2}{2} \tag{6.51}$$

The developed torque for a constant flux becomes:

$$T_{\text{dev}} = -\frac{\phi^2}{2}\frac{\partial \mathcal{R}}{\partial \theta} \tag{6.52}$$

Concerning the reluctance, we can consider a sinusoidal position variation for representing it. The minimum value of the reluctance ($\mathcal{R}_{\text{min}}$) is observed when the rotor is aligned with the stator ($\theta = 0$) and its maximum value ($\mathcal{R}_{\text{max}}$) occurs when the rotor is in quadrature with the stator ($\theta = \pi/2$), as shown in figure 6.17.

This kind of variation can be mathematically expressed by:

$$\mathcal{R}(\theta) = \mathcal{R}_a + \mathcal{R}_b \cos 2\theta \tag{6.53}$$

with:

$$\mathcal{R}_a = \frac{\mathcal{R}_{\text{max}} + \mathcal{R}_{\text{min}}}{2} \tag{6.54}$$

and

$$\mathcal{R}_b = \frac{\mathcal{R}_{\text{max}} - \mathcal{R}_{\text{min}}}{2} \tag{6.55}$$

Substituting equation (6.53) in (6.52), results in:

$$T_{\text{dev}} = \phi^2 \mathcal{R}_b \sin 2\theta \tag{6.56}$$

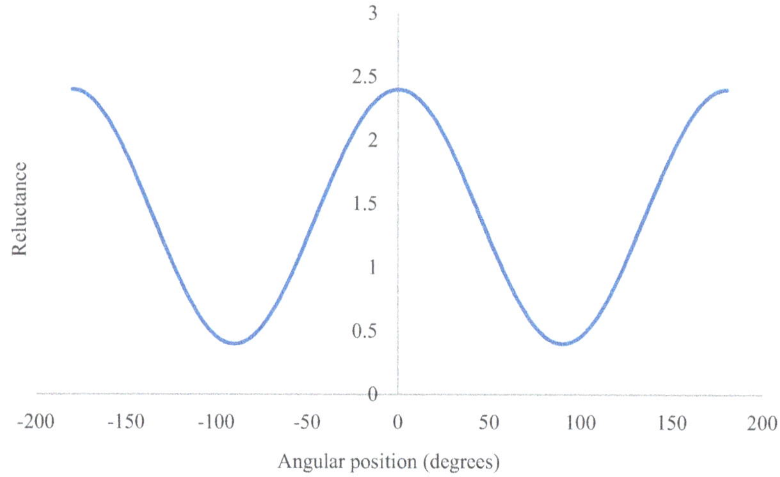

Figure 6.17. Reluctance variation with the mechanical position. For $\theta = 0$, $\mathcal{R}(\theta) = \mathcal{R}_{\text{min}}$ and for $\theta = \pi/2$, $\mathcal{R}(\theta) = \mathcal{R}_{\text{max}}$.

6.4.1 Imposing rotor angular speed

We are now supposing that the rotor is running at an angular speed that is different from the angular frequency of the AC voltage source ($\omega_r \neq \omega$) and is non-null ($\omega_r \neq 0$). The angular frequency of the AC voltage source is $\omega = 2\pi f$. For convenience, the origin of the time scale will be chosen such that at $t = 0, \phi = 0$ and:

$$\varnothing(t) = \varnothing_{\max} \sin \omega t \text{ Wb} \tag{6.57}$$

Also, at $t = 0$, let the rotor have some angular initial position δ, so that:

$$\theta = \omega_r t + \delta \tag{6.58}$$

Substituting equations (6.57) and (6.58) in (6.56):

$$T_{\text{dev}} = (\varnothing_{\max} \sin \omega t)^2 \mathcal{R}_b \sin(2\omega_r t + 2\delta)$$

Applying the trigonometrical identity

$$\sin^2 \alpha = \frac{1}{2} + \frac{1}{2}\cos 2\alpha \tag{6.59}$$

in the previous expression, it results in:

$$T_{\text{dev}} = \frac{\varnothing_{\max}{}^2 \mathcal{R}_b}{2} \sin(2\omega_r t + 2\delta) + \frac{\varnothing_{\max}{}^2 \mathcal{R}_b}{2} \cos 2\omega t \times \sin(2\omega_r t + 2\delta)$$

Applying the following identity

$$\sin \alpha \times \sin \beta = \frac{1}{2}[\sin(\alpha - \beta) + \sin(\alpha + \beta)]$$

in the previous equation, yields:

$$T_{\text{dev}} = T_1 + T_2 + T_3 \tag{6.60}$$

where:

$$T_1 = \frac{\varnothing_{\max}{}^2 \mathcal{R}_b}{2} \sin(2\omega_r t + 2\delta) \tag{6.61}$$

$$T_2 = \frac{\varnothing_{\max}{}^2 \mathcal{R}_b}{4} \sin[2(\omega_r + \omega)t + 2\delta] \tag{6.62}$$

$$T_3 = \frac{\varnothing_{\max}{}^2 \mathcal{R}_b}{4} \sin[2(\omega_r - \omega)t + 2\delta] \tag{6.63}$$

The condition for having an effective developed torque, as we previously discussed, is that the time-dependent function of the developed torque has a non-null average value. Analyzing each term of the developed torque expressed by (6.60), we can conclude that:

(1) the average value of the first term is null, because T_1 is a time-dependent sinusoidal function independent on the value of ω_r;

(2) the average value of T_2 is also null, except for $\omega_r = -\omega$, when:

$$T_{2\text{avg}} = \frac{\varnothing_{\max}{}^2 \mathcal{R}_b}{4} \sin 2\delta \tag{6.64}$$

(3) the average value of T_3 is also null, except for $\omega_r = \omega$, when:

$$T_{3\text{avg}} = \frac{\varnothing_{\max}{}^2 \mathcal{R}_b}{4} \sin 2\delta \tag{6.65}$$

As,

$$\text{Med}[T_{\text{dev}}] = \text{Med}[T_1 + T_2 + T_3] \tag{6.66}$$

we can conclude that the only condition for having a non-null average developed torque of equation (6.60) is:

$$\omega_r = \pm\omega \tag{6.67}$$

In those conditions, the average developed torque is:

$$T_{\text{avg}} = \frac{\varnothing_{\max}{}^2 \mathcal{R}_b}{4} \sin 2\delta \tag{6.68}$$

Final remarks: A rotational magnetic system is able to develop a non-null average torque since its angular speed is the same as the angular frequency of the AC voltage source, or its rotation is the same as the frequency of the electrical source $n = f$. It could be expressed, for example, both in revolutions per second (rps) or in revolutions per minute (rpm).

The electrical current in coil can be evaluated considering that:

$$Ni(t) = \mathcal{R}(\theta)\varnothing(t) \tag{6.69}$$

and imposing $\omega_r = \omega$, yields:

$$i(t) = \frac{[\mathcal{R}_a + \mathcal{R}_b \cos 2(\omega t + \delta)][\varnothing_{\max} \sin \omega t]}{N}$$

Then, it can be expressed by:

$$i(t) = i_\omega(t) + i_{3\omega}(t) \tag{6.70}$$

where,

$$i_\omega(t) = \frac{\mathcal{R}_a \varnothing_{\max}}{N} \sin \omega t - \frac{\mathcal{R}_b \varnothing_{\max}}{2N} \sin(\omega t + \delta) \tag{6.71}$$

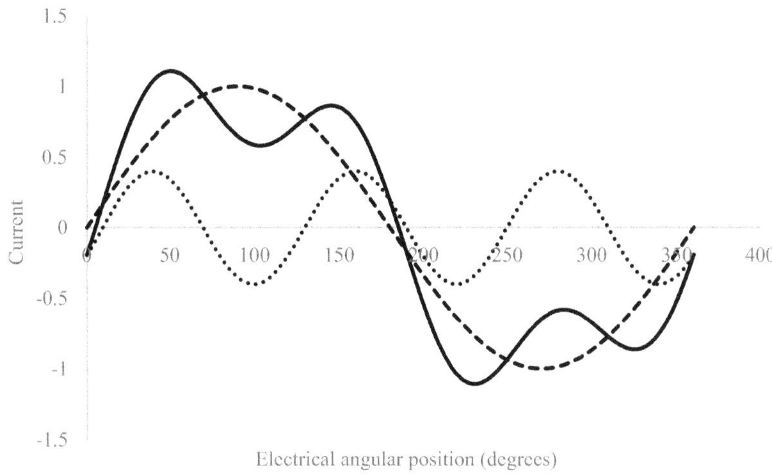

Figure 6.18. The coil's current $i(t) = i_{\omega}(t) + i_{3\omega}(t)$. $i_{\omega}(t)$. Dashed line: component with the same frequency as the AC voltage source. $i_{3\omega}(t)$. Dotted line: component with a frequency three times the frequency of the source. $i(t)$. Full line: resultant and non-sinusoidal time varying electrical current.

$$i_{3\omega}(t) = \frac{\mathcal{R}_b \varnothing_{\max}}{2N} \sin(3\omega t + \delta) \tag{6.72}$$

The first term $[i_{\omega}(t)]$ is a sinusoidal time varying current with the same frequency of the AC voltage source and the second one $[i_{3\omega}(t)]$ is also a sinusoidal time varying current but with the frequency three times the frequency of the AC voltage source (figure 6.18).

The electrical current in the coil contains a 'third harmonic' component. This is an undesirable feature that limits the use of this kind of machine, it is usable only for small machines that require accurate and constant speed. The other undesirable feature is that this machine has no starting torque, therefore, it requires an external agent to bring it to 'synchronous' speed.

6.5 Summary

The representation of the stored energy as a function of lumped parameters is very useful for analyzing the performance of an electromagnetic device. Although this feature is very useful in this task, it is important to understand that to do it we need to consider the linearity of the magnetic circuit. The stored magnetic energy and the co-energy are expressed using the same equation in linear magnetic circuit. Consequently, the same fundamental expression can be used for evaluating the force in a translational magnetic system and also the torque in a rotational magnetic system.

No electromagnetic devices that operate under constant current are able to develop a continuous effort that results in motion. Therefore, it is not possible to build either an electric motor or generator using this approach. The electromagnetic device under constant current acts as an actuator which is useful for controlling another mechanism not for converting energy in motion.

Otherwise, if the electromagnetic device is fed by an AC voltage source that imposes a constant flux, a suitable combination between the rotational speed and the frequency of the source can develop a continuous torque that enables the device to work as an electric motor or generator.

The simply excited rotational electromagnetic device, even fed by an AC voltage source, results in a time-dependent electric current with a non-sinusoidal function. The shape of this electric current is the summation of the fundamental and the third harmonic components of the frequency of the source (fundamental component is the same frequency of the source and the third harmonic component is three times the frequency of the source).

6.5.1 Project: design of a rotational magnetic system

This activity is intended to give an introduction to the design concept of a 2-pole rotational magnetic system (see figure 6.16). The data-sheet for this electromagnetic device is:

Output power: 100 W at 3600 rpm
Electrical source: 200 V–60 Hz
Manufacturer's constraints:
Minimum air-gap admissible: $g = 1$ mm
Length (h) = pole pitch (τ_p) where: $\tau_p = \pi R/p$ and p is the number or pole pairs
Current density in the coil conductor $J = 3$ A mm^{-2}
The relation of the cross-section of the conductor with an outer isolation layer (s_i) and the conductor with no isolation layer (s_{ni}) is such that: $s_i/s_{ni} = 1.4$
Designer's constraints:
The maximum value of the magnetic flux density is $B_{\max} = 0.4$ Wb m^{-2}
The cross-section of the coil is rectangular with relation $b/a = 1.62$
Magnetic material is ideal ($\mu \to \infty$)
No fringing effects are considered
Only the fundamental component of the electric current (I_ω) should be considered

In this first approach, the designer should suggest the following dimensions:

(a) the radius of the rotor;
(b) the length of the electromagnetic device;
(c) the number of turns of the coil;
(d) the cross-section of the conductor;
(e) the dimension of the rectangular coil;
(f) the mass of the copper;
(g) the resistance of the coil $r = \rho N \frac{l_{avg}}{s_c}$ where l_{avg} is the average length of the spiral turn.

Problems

6.1 Figure 6.19 shows the magnetic structure of an electromagnetic contactor. An operating current of 10 A in a coil of 300 turns on an iron core of 4 cm^2 cross-section area would be reasonable for a device of a medium size. Let the pivoted armature be 1 cm from the pole face. If we consider the pivoted construction, this can only be an average value, but it will serve for this evaluation. Putting the numbers into equation (6.9), the force turns out to be 22.6 N.

(a) Will it operate on alternating current? If so, what rms current will give the same average force? Why? Neglecting resistance, what rms voltage will be required at the terminals (consider a sinusoidal function with 60 Hz)?

The armature is now allowed to move until the average distance from the pole face is reduced from 1.0 cm to 0.5 cm.

(b) What is the new value of the force, if the same direct voltage and current are used as with the larger air-gap? (Motion of the armature is slow).

(c) Now what is the force if alternating current is used and the terminal voltage is kept the same?

(d) Returning to part (b) and considering that the armature is allowed to close quite rapidly, what will be the effect on the flux and the force while the armature is closing? Explain in terms of the energy-balance equation, but qualitatively; numbers are not required.

Answers: (a) 10 A; 4 V; (b) 90.4 N; (c) 22.6 N.

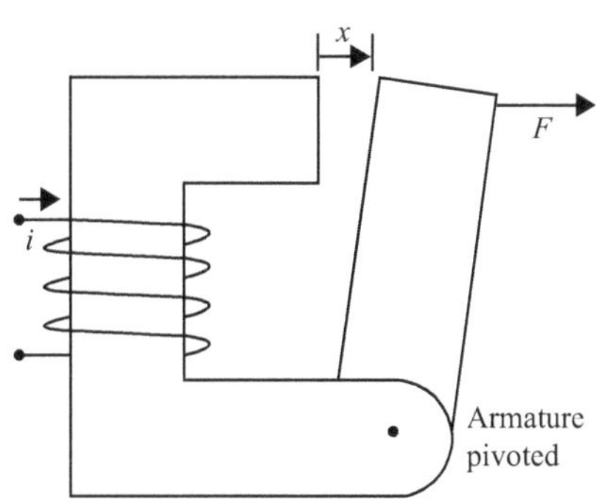

Figure 6.19. Contactor with armature pivoted.

6.2 The magnetic structure of figure 6.20 is built with a very high permeability magnetic material. Compute the mmf required in the coil to produce a magnetic flux density of 1.2 Wb m^{-2} in the central leg in which the coil is wound. Neglect both the leakage of magnetic field and the fringing effect at the air-gap.

(a) find the magnetic pressure at each surface of the air-gap;

(b) find the mechanical force of attraction;

(c) solve the problem again, assuming average permeability of 3000 times the permeability of air for all iron;

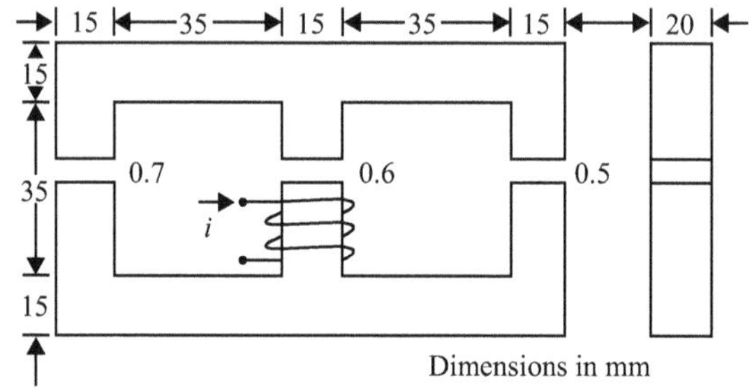

Figure 6.20. The magnetic structure.

(d) find the mechanical force of attraction with the relative permeability of the iron equal to 3000 as in part c.

Answers: mmf = 8520 A (a) 99.471 N m^{-2}; 572.957 N m^{-2}; 194.964 N m^{-2}; (b) 26 000 N.

6.3 The magnetic structure of figure 6.21 is built with a magnetic material with very high permeability, it can be considered an ideal material. The cross-section of the structure is uniform in all parts and the air-gap y is constant.

A DC voltage source injects a DC current in the coil.

Consider the following statements:

I: The developed force increases as the y air-gap decreases.

II: The developed force remains constant as y air-gap varies.

III: The developed force is dependent on the square of y.

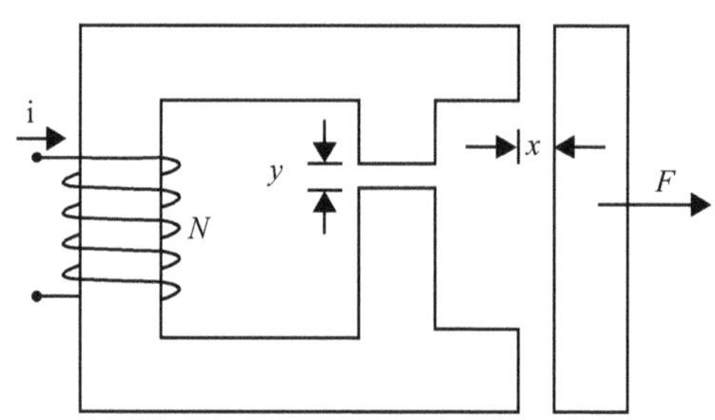

Figure 6.21. Problems 6.3 and 6.4.

Based on those statements we can conclude:

(a) statement I is false;

(b) statement II is correct;

(c) statement III can be correct if x remains constant and y can be varied freely.

Justify your answer based on the final equation that expresses the developed force as a function of both x and y.

6.4 Consider now that the magnetic device of figure 6.21 is fed by an AC voltage source that supplies a constant (rms) voltage to the coil. Answer questions (a), (b) and (c) (from the previous problem) based on this operation condition.

6.5 The electromagnetic device shown in the figure 6.22 is a practical form of magnet. It is cylindrical about a vertical axis. When the coil current is zero the plunger drops until the gap $l_g = 1.5$ cm. When the coil is energized by a 2.0 A direct current the plunger rises, stopping when $l_g = 0.5$ cm. The air-gap between the shell and the plunger is uniformly 0.02 cm. The coil has 1500 turns. Assume infinite permeability of the iron and make other suitable assumptions. When the coil is energized, compute:

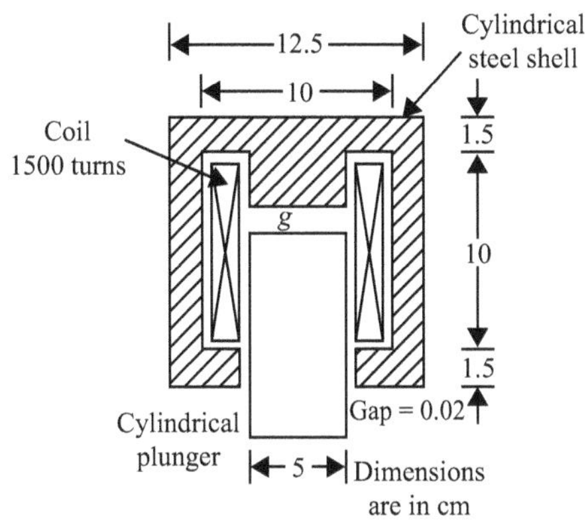

Figure 6.22. The magnetic actuator.

(a) the flux density between the flat faces of the center core and the plunger when the gap is $l_g = 1.5$ cm;
(b) the magnetic pressure and the force on the plunger;
(c) repeat, computing flux density, magnetic pressure, and forces when $l_g = 0.5$ cm.

Answers: (a) 4.9×10^{-4} Wb; (b) 2.5×10^4 Nm^{-2}; (c) 14.6×10^{-4} Wb; 21.8×10^4 Nm^{-2}.

6.6 A reluctance motor of the type illustrated in figure 6.23 has a stator and rotor cross-section 25 mm × 25 mm and a rotor 50 mm in length. The stator coil has 3000 turns and is excited from 115 V, 25 Hz supply. The air-gap is 0.5 mm and the ratio between the maximum and minimum reluctance is 1.8. The coil resistance is negligibly small.
(a) Explain how this machine can establish a continuous rotational movement.
(b) What rotational speed is the developed torque non-null?
(c) What is the maximum value of average mechanical power that this machine can develop?
(d) What is the rms value of the coil current?

Hint:

$$I = \sqrt{I_\omega^2 + I_{3\omega}^2}$$

Answers: (b) 25 rev s^{-1}; (c) 2.4 W; (d) 145 mA.

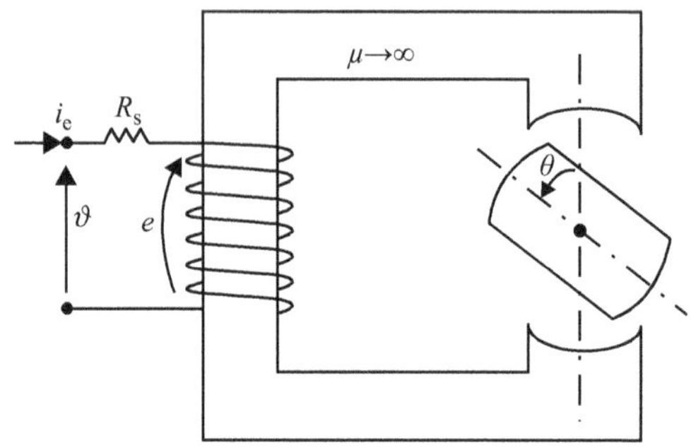

Figure 6.23. The rotational actuator.

6.7 Assume that the magnetic system of figure 6.24 is made with an ideal magnetic material. The coil has an inductance of 2 H when the length of the air-gap is 8 mm. If a constant current of 2 A flows in the coil, compute:

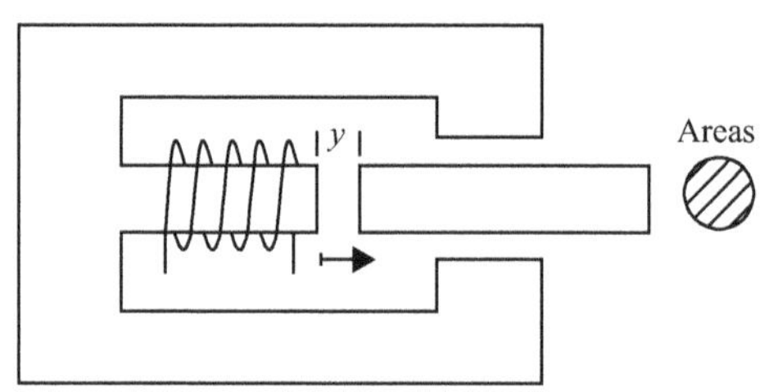

Figure 6.24. Problems 6.7 and 6.8.

(a) the energy stored in the magnetic field when the air-gap length is 8 mm;
(b) the mechanical force on the armature when the air-gap is 8 mm;
(c) the mechanical work done by the armature when it is moved from an air-gap of 8 mm to an air-gap of 14 mm.

Neglect the virtual air-gap between the plunger and the magnetic structure.

Answers: (a) 4 J; (b) 500 N; (c) 1.71 J.

6.8 Supposing now that the magnetic device of figure 6.24 is fed by an 60 Hz-AC voltage source and considering that the coil has 1130 turns and the cross-section of the plunger 25 mm^2, compute:

(a) the rms value of the voltage source that should be connected to the coil to have the same force evaluated for (b) in the previous problem;
(b) sketch the wave that represents the time-dependent force and highlight the maximum and minimum values, the frequency and its average value.

Answer: (a) 238 V.

6.9 The magnetic structure of figure 6.25(a) is made of a non-linear magnetic material, its characteristics $B - H$ are given in figure 6.25(b). These characteristics are segmented to make it easier to evaluate the area above it. The stacking factor is 0.96 and the coil has 2000 turns. The armature is fixed at an air-gap of $x = 10$ mm and a DC electric current is flowing through the coil that is large enough to produce a magnetic flux density of 1.2 T in the air-gap. Leakage flux and fringing at the air-gap may be neglected. Compute:

(a) the required DC current in the coil;
(b) the energy stored in the air-gap;
(c) the energy stored in the magnetic material;
(d) the total field energy.

Answers: (a) 9.6 A; (b) 10.52 J; (c) 0.083 J; (d) ~10.6 J.

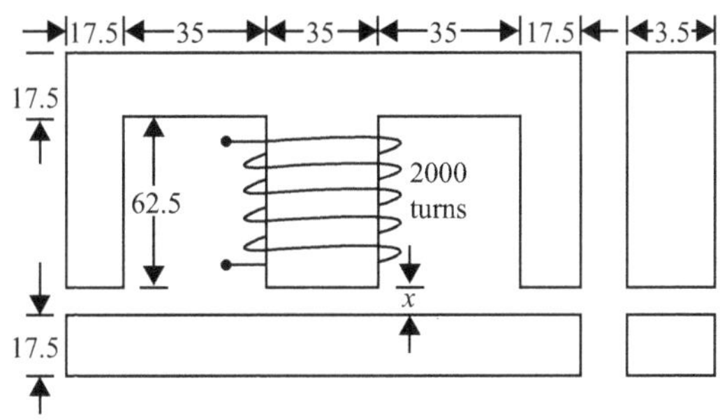

(a) Magnetic structure

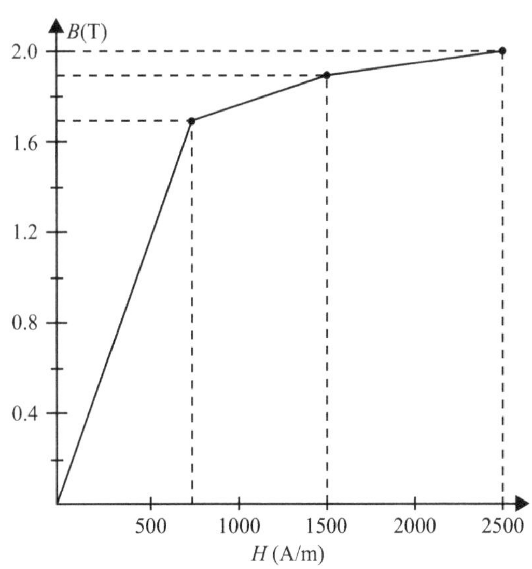

(b) Magnetization characteristics

Figure 6.25. Problems 6.9 and 6.10. Dimensions in mm.

6.10 Supposing now that the air-gap is slowly increased from 10 to 12 mm under constant DC current. Determine:
   (a) the variation of the energy stored in the air-gap;
   (b) the variation of the energy stored in the magnetic material;
   (c) the variation of the total field energy stored;
   (d) the average developed force exerting in the armature during this movement.

Answers: (a) 30.6%; (b) 31.3%; (c) 30.7%; (d) 2100 N.

Further reading

[1] Fitzgerald A E, Kingsley C Jr. and Umans S D 1992 *Electric Machinery* 5th edn (New York: McGraw-Hill)
[2] Falcone A G 1979 *Eletromecanica* (São Paulo: Editora Edgard Blucher Ltda) (in Portuguese)
[3] Meisel J 1984 *Principles of Electromechanical-Energy Conversion* (Malabar, FL: R.E. Krieger)
[4] Slemon G R and Straughen A 1980 *Electric Machines* (Reading, MA: Addison-Wesley)
[5] Skilling H H 1962 *Electromechanics: A First Course in Electromechanical Energy Conversion* (New York: Wiley)
[6] Krause P, Wasynczuk O and Pekarek S 2012 *Electromechanical Motion Devices* 2nd edn (New York: Wiley)
[7] Gourishankar V 1966 *Electromechanical Energy Conversion* (Scranton: International Textbook Company)
[8] Bansal R (ed) 2004 *Handbook of Engineering Electromagnetics* (New York: Marcel Decker)
[9] Inan U S and Inan A S 1999 *Engineering Electromagnetics* (Menlo Park, CA: Addison-Wesley)
[10] Hendershot J R and Miller T J E 1994 *Design of Brushless Permanent-Magnet Motors* (Hillsboro, OH: Magna Physics Pub.)

IOP Publishing

Electromechanical Energy Conversion Through Active Learning

J R Cardoso, M B C Salles and M C Costa

Chapter 7

Multiply excited electromechanical systems

7.1 Introduction

All electromagnetic devices used for powering industrial processes and all other electrical appliances are projected with several coils magnetically coupled. The stored magnetic energy of the electromagnetic systems has several components. First, components of each coil are represented by its self-inductances. Second, components of the coupled field are represented by its mutual inductance with other coils of the device.

In general, these lumped parameters could vary with both their relative position and the saturation state of the magnetic material. As a result, the analysis of the performance of the electromagnetic device becomes very complex if both effects are considered.

As was adopted in the previous chapters, we are considering all magnetic material with magnetic permeability high enough to be considered as an ideal material.

7.2 Flux linkages in multiple excitation

Figure 7.1 shows two coils wound in the same magnetic structure. In such an assembly, part of the magnetic flux produced by coil (1) while carrying an electrical current, links with the other (linkage magnetic flux). The other part of the magnetic flux produced by coil (1) links only with itself. The same thing happens with coil (2), while carrying an electrical current.

As we can see in figure 7.1, the flux linkages with the coil (1) are composed of two components such that:

$$\lambda_1 = \lambda_{11} + \lambda_{12} \tag{7.1}$$

where:

λ_{11}: flux linkage with coil (1) produced by its own current i_1;
λ_{12}: flux linkage with coil (1) produced by the current i_2 carried by coil (2);
λ_1: total flux linkage with coil (1).

doi:10.1088/978-0-7503-2084-9ch7

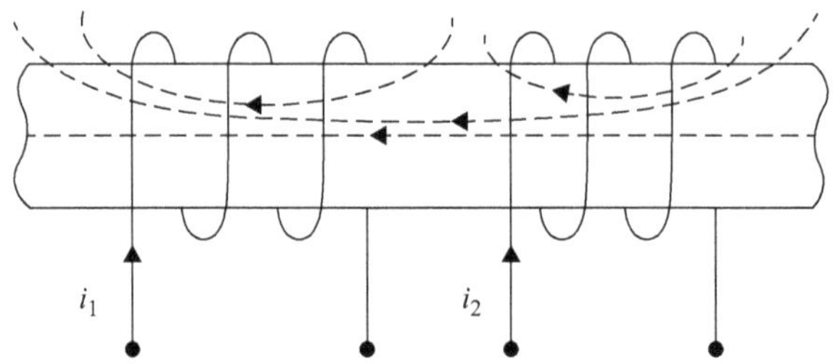

Figure 7.1. Two coils magnetically coupled.

Such components of the flux linkages can be expressed as a function of the coil self-inductance and the mutual inductance between the coils as:

$$\lambda_1 = L_{11}i_1 + M_{12}i_2 \tag{7.2}$$

where:
L_{11}: self-inductance of coil (1);
M_{12}: mutual inductance between coil (1) and coil (2).

Using the same procedure, the total flux linkage with coil (2) is given by:

$$\lambda_2 = L_{22}i_2 + M_{21}i_1 \tag{7.3}$$

where:
L_{22}: self-inductance of coil (2);
M_{21}: mutual inductance between coil (2) and coil (1).

From the electromagnetism, it is possible to demonstrate that:

$$M_{21} = M_{12} = M$$

Using this statement, equations (7.2) and (7.3) can be rewritten generically as:

$$\lambda_i = L_{ii}i_i + M_{ij}i_j \tag{7.4}$$

where i is the coil of interest and j is the other coil which the produced magnetic flux links to the coil of interest. Equation (7.4) can be expanded to include many other coils with linkage magnetic flux.

Returning to the energy balance equation developed in chapter 5, the variation of the electromechanical energy considering an incremental displacement is such that:

$$dE_{\text{em}} = dE_{\text{se}} - dE_{\text{mag}} - dE_{\text{j}} \tag{7.5}$$

The variation of the electrical energy supplied by the source can be evaluate according to the following procedure.

The electric power input in the electromagnetic device is:

$$p_{\text{se}} = \frac{dE_{\text{se}}}{dt} = v_1i_1 + v_2i_2 \tag{7.6}$$

So that,

$$dE_{se} = v_1 i_1 dt + v_2 i_2 dt \tag{7.7}$$

as

$$v_1 = r_1 i_1 + \frac{d(L_{11} i_1)}{dt} + \frac{d(M i_2)}{dt} \tag{7.8}$$

Expanding equation (7.8) we get:

$$v_1 = r_1 i_1 + L_{11}\frac{di_1}{dt} + i_1\frac{dL_{11}}{dt} + M\frac{di_2}{dt} + i_2\frac{dM}{dt} \tag{7.9}$$

Following the same procedure, we can write:

$$v_2 = r_2 i_2 + L_{22}\frac{di_2}{dt} + i_2\frac{dL_{22}}{dt} + M\frac{di_1}{dt} + i_1\frac{dM}{dt} \tag{7.10}$$

Substituting equations (7.9) and (7.10) in (7.7), results in:

$$\begin{aligned} dE_{se} = {} & r_1 i_1^2 dt + r_2 i_2^2 dt + L_{11} i_1 di_1 + L_{22} i_2 di_2 + M i_1 di_2 \\ & + M i_2 di_1 + i_1^2 dL_{11} + i_2^2 dL_{22} + 2 i_1 i_2 dM \end{aligned} \tag{7.11}$$

Concerning the variation of the stored magnetic energy, we obtain from:

$$E_{mag} = \frac{\lambda i}{2} \tag{7.12}$$

that:

$$E_{mag} = \frac{\lambda_1 i_1}{2} + \frac{\lambda_2 i_2}{2} \tag{7.13}$$

Substituting (7.2) and (7.3) in (7.13) results in:

$$E_{mag} = \frac{L_{11} i_1^2}{2} + \frac{L_{22} i_2^2}{2} + M i_1 i_2 \tag{7.14}$$

As the incremental variation of the stored magnetic energy is given by:

$$\begin{aligned} dE_{mag} = {} & L_{11} i_1 di_1 + L_{22} i_2 di_2 + M i_1 di_2 + M i_2 di_1 \\ & + i_1 i_2 dM + \frac{i_1^2}{2} dL_{11} + \frac{i_2^2}{2} dL_{22} \end{aligned} \tag{7.15}$$

Concerning the Joule losses, an incremental variation of them is given by:

$$dE_j = r_1 i_1^2 dt + r_2 i_2^2 dt \tag{7.16}$$

Substituting equations (7.11), (7.15) and (7.16) in (7.5), results in:

$$dE_{\text{em}} = \frac{i_1^2}{2}dL_{11} + \frac{i_2^2}{2}dL_{22} + i_1 i_2 dM \tag{7.17}$$

For a translational magnetic system, we have;

$$dE_{\text{em}} = F_{\text{dev}}\, dx \tag{7.18}$$

so that:

$$F_{\text{dev}} = \frac{dE_{\text{em}}}{dx}$$

or:

$$F_{\text{dev}} = \frac{i_1^2}{2}\frac{dL_{11}}{dx} + \frac{i_2^2}{2}\frac{dL_{22}}{dx} + i_1 i_2 \frac{dM}{dx} \tag{7.19}$$

The developed torque is the main characteristic for a rotational electromagnetic device. In this situation,

$$dE_{\text{em}} = T_{\text{dev}}\, d\theta \tag{7.20}$$

so that:

$$T_{\text{dev}} = \frac{dE_{\text{em}}}{d\theta}$$

or:

$$T_{\text{dev}} = \frac{i_1^2}{2}\frac{dL_{11}}{d\theta} + \frac{i_2^2}{2}\frac{dL_{22}}{d\theta} + i_1 i_2 \frac{dM}{d\theta} \tag{7.21}$$

For convenience, we deduced the expression for a double-excited magnetic system; however, these results could be expanding for a multiply excited magnetic system. The expression for the developed torque in a triple excited rotational magnetic system is given by:

$$\begin{aligned} T_{\text{dev}} = {} & \frac{i_1^2}{2}\frac{dL_{11}}{d\theta} + \frac{i_2^2}{2}\frac{dL_{22}}{d\theta} + \frac{i_3^2}{2}\frac{dL_{33}}{d\theta} \\ & + i_1 i_2 \frac{dM_{12}}{d\theta} + i_1 i_3 \frac{dM_{13}}{d\theta} + i_2 i_3 \frac{dM_{23}}{d\theta} \end{aligned} \tag{7.22}$$

with:

$M_{ij} = M_{ji}$ is the mutual inductance between coil (i) and coil (j).

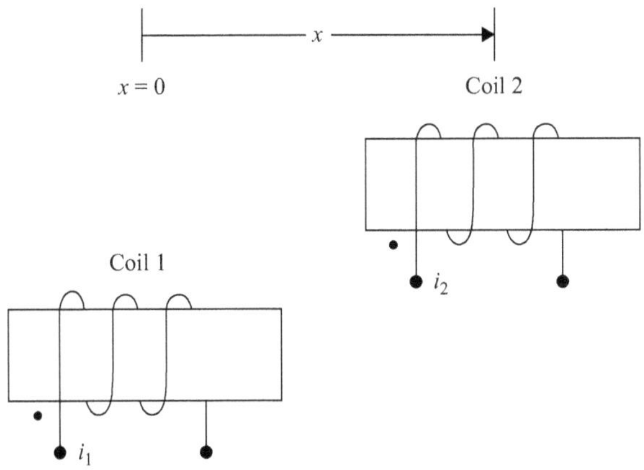

Figure 7.2. Double-excited magnetic system. Translational movement in the x-direction.

Example 7.1

In the double-excited translational magnetic system of figure 7.2, the self-inductances of coils are such that $L_{11} = L_{22} = 15\,(H)$.The mutual inductance between the two coils is $M = 5 + 10x\,(H)$. The resistances of both coils can be neglected.

(a) if coil (1) carries a constant current of 12 A and coil (2) carries a constant current of −6 A, find the mechanical work done when x is increased from 0 to 2 m;

(b) during the motion of the coils in part (a), compute the energy supplied by the sources connected to the two coils;

(c) if coil (1) carries a sinusoidal current of $i_1 = 12\sqrt{2} \times \sin \omega t$ and coil (2) carries a sinusoidal current of $i_2 = 6\sqrt{2} \times \sin(\omega_x t + \delta)$, find the relationship of ω and ω_x for having a non-null developed force;

(d) with the value of ω evaluated in part (c), compute the average developed force.

Solutions

(a) Mechanical work done when x is increased from 0 to 2 m.

From equation (7.16) we have:

$$dE_{\text{em}} = \frac{i_1^2}{2} dL_{11} + \frac{i_2^2}{2} dL_{22} + i_1 i_2 dM$$

As the self-inductances of coil (1) and coil (2) are constants, the increment of the electromechanical energy becomes:

$$dE_{\text{em}} = i_1 i_2 dM$$

so that,

$$E_{\text{em}} = \int i_1 i_2 dM \tag{7.23}$$

The work done when x is increased from 0 to 2 m is given by:

$$\tau = \int_0^2 i_1 i_2 \left[\int_{M(x=0)}^{M(x=2)} dM \right] dx \tag{7.24}$$

or,

$$\tau = i_1 i_2 [M(x = 2) - M(x = 0)]$$

in numbers:

$$\tau = -12 \times 6 \times [25 - 5] = -1440 \text{ J}$$

The negative sign indicates that the work is performed in the opposite direction to x.

(b) The energy supplied by the sources connected to the two coils:
as the resistance is negligible and both the self-inductance of the coil and the current are constants, we obtain:
For source (1)

$$dE_{se1} = v_1 i_1 dt$$

that results in:

$$dE_{se1} = i_1 i_2 dM \tag{7.25}$$

integrating (7.25), yields:

$$E_{se1} = \int_{M(x=0)}^{M(x=2)} i_1 i_2 dM \tag{7.26}$$

or:

$$E_{se1} = i_1 i_2 [M(x = 2) - M(x = 0)] = -6 \times 12 \times [25 - 5]$$

that results in:

$$E_{se1} = -1440 \text{ J}$$

Using the same approach, it is easy to demonstrate that:

$$E_{se2} = E_{se1} = -1440 \text{ J}$$

Therefore, the total energy supplied by the sources is:

$$E_{se} = E_{se1} + E_{se2} = -2880 \text{ J}$$

In conclusion, the total energy supplied by the electrical sources is shared half by the variation of stored electromagnetic energy and half by the work done during the movement, that is,

$$\tau = E_{mag} = \frac{E_{se}}{2} \tag{7.27}$$

(c) The relation of ω and ω_x for having a non-null developed force:
From equation (7.19) we have,

$$F_{dev} = i_1 i_2 \frac{dM}{dx}$$

as $i_1 = 12\sqrt{2} \times \sin \omega t$, $i_2 = 6\sqrt{2} \times \sin(\omega_x t + \delta)$ and $\frac{dM}{dx} = 10$, the previous equation becomes:

$$F_{dev} = 1440 \times \sin(\omega_x t + \delta) \times \sin \omega t$$

applying the identity:

$$\sin \alpha \times \sin \beta = \frac{1}{2}[\sin(\alpha - \beta) + \sin(\alpha + \beta)]$$

the previous equation becomes:

$$F_{dev} = 720 \times \{\sin[(\omega_x - \omega)t + 2\delta] + \sin[(\omega_x + \omega)t + 2\delta]\} \qquad (7.28)$$

Equation (7.28) can be expressed as:

$$F_{dev} = 720 \times \sin[(\omega_x - \omega)t + 2\delta] + 720 \times \sin[(\omega_x + \omega)t + 2\delta] \qquad (7.29)$$

The first term of equation (7.29) is a time sinusoidal function whose time average value equals null except when $\omega_x = \omega$.

(d) The average developed force.
In such a situation ($\omega_x = \omega$), the average value of the function is:

$$F_{dev} = 720 \times \sin 2\delta N$$

because the time average value of the second term is null.

Remarks: The angle δ is out-of-phase for both currents with the same frequency and is commonly called the *power angle*.

Figure 7.3 shows the time-dependent variation of the developed force and both their components.

7.3 The double-excited rotational magnetic system

Figure 7.4 shows a kind of rotational magnetic system that is composed of two parts. The rotor is the part that moves and the stator is the part that stays in place. The space separating the rotor and the stator is called the air gap, which makes the angular movement possible. This kind of configuration is far more efficient than the one presented in figure 6.6 of the previous chapter and its design is closer to a rotating electrical machine.

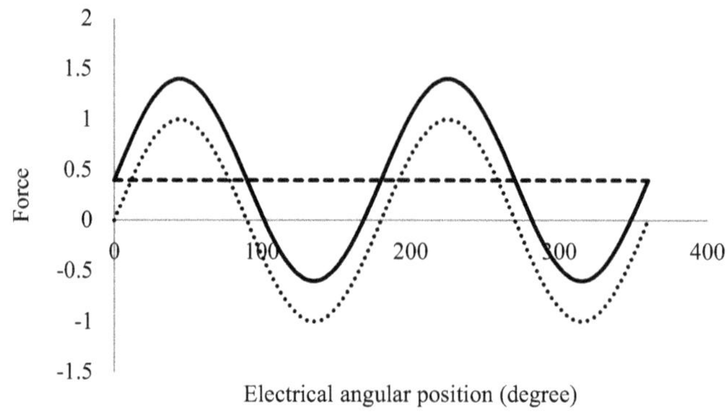

Figure 7.3. The developed force. Dashed line: the average value of the developed force—(7.28) first term ($\omega_x = \omega$). Dotted line: the time sinusoidal component—(7.28) second term ($\omega_x = \omega$). Full line: the time variation of the developed force—(7.28) both terms ($\omega_x = \omega$).

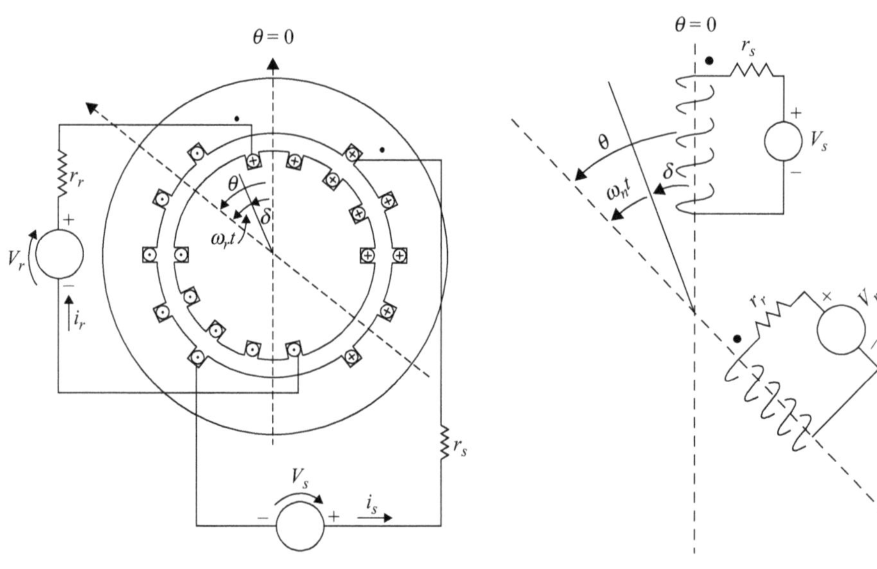

Figure 7.4. (a) Cross-section of the rotational magnetic system. The outer part is the stator; the inner part is the rotor and (b) electrical representation of the coils.

Both the stator and the rotor are longitudinally slotted to host the conductors of a distributed winding. The vertical line ($\theta = 0$) is the axis of the stator winding that is displaced by θ of the axis of the rotor winding.

Despite the slots, both stator and rotor inductances can be considered constants because they do not change enough with the relative position of coils. So that:

$$dL_s = dL_r = 0$$

Based on equation (7.21), the developed torque is given by:

$$T_{\text{dev}} = i_1 i_2 \frac{dM}{d\theta} \tag{7.30}$$

As we will see in chapter 8, the distributed winding produces a sinusoidal distribution of magnetic flux density that enables us to consider that the mutual inductance also varies sinusoidally with the relative position of the coils.

With the convention adopted for the currents (see figure 7.4(a)) when $\theta = 0$ we have $M(0) = M_{\text{max}}$ and $\theta = \pi$ results in $(\pi) = -M_{\text{max}}$. Therefore, the mutual inductance between the two windings may be expressed by the relation:

$$M(\theta) = M_{\text{max}} \cos\theta \tag{7.31}$$

Substituting equation (7.31) in (7.30) results in:

$$T_{\text{dev}} = -i_s i_r M_{\text{max}} \sin\theta \tag{7.32}$$

First case

If the device is stationary ($\omega_n = 0$) and fed by DC electrical sources, the electrical currents in the coils are $I_s = \frac{V_s}{r_s}$ and $I_r = \frac{V_r}{r_r}$. Therefore, from equation (7.32), the developed torque results in:

$$T_{\text{dev}} = -I_s I_r M_{\text{max}} \sin\theta \tag{7.33}$$

Under this kind of operation, the rotational magnetic system acts as an actuator that develops a constant magnetic torque dependent on the relative position of both coils. This is a very useful angular position detector.

Figure 7.5(a) shows the developed torque under DC excitation for different angular positions. We can identify that when the rotor position is in the range $-\pi < \theta < 0$, the developed torque is positive (in the sense that it increases θ) towards the equilibrium point at $\theta = 0$, it is shown in the right side of figure 7.5(b). When the rotor position is in the range $0 < \theta < \pi$, the developed torque is negative, in the sense that it decreases θ towards the equilibrium point at $\theta = 0$, it is shown in the left side of figure 7.5(b). However, in both conditions the sense of the developed torque is that where the mutual inductance is *maximum*.

Remarks*:* Unlike the simply excited magnetic system, the sense of the current in one coil can change the sense of the developed torque.

Second case

The stator winding carried by an AC current of frequency f_s such that:

$$i_s = \sqrt{2} I_s \cos\omega_s t \tag{7.34}$$

where $\omega_s = 2\pi f_s$ and f_s is the source frequency.

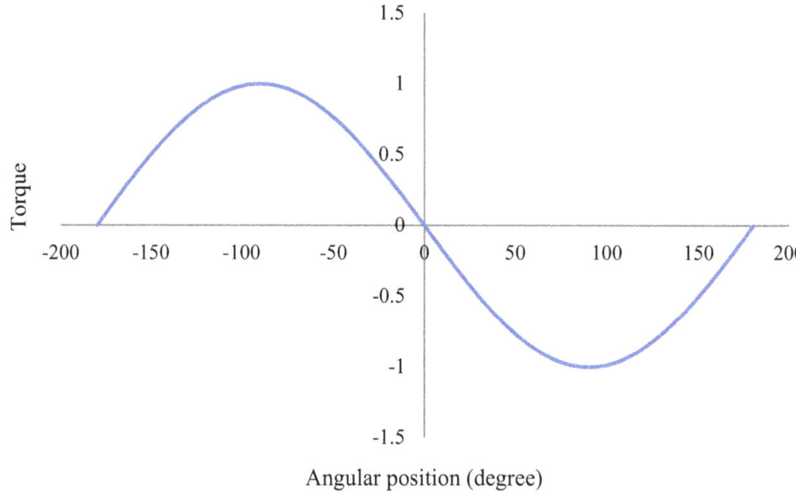

(a)

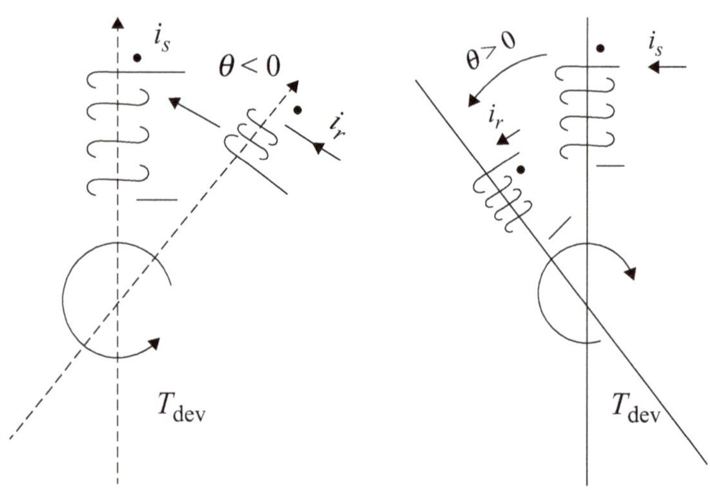

(b)

Figure 7.5. (a) The developed torque as a function of the relative position of the coils. (b) The sense of the developed torque (the circular indication) depends on the relative position of the coils.

Let the rotor winding excited by another independent DC source that injects an electrical constant current be given by $i_r = I_r$.

Also, let the rotor rotate at angular velocity

$$\omega_n = \frac{d\theta}{dt} \text{rad s}^{-1} \tag{7.35}$$

Therefore, the position of the rotor at instant t is

$$\theta = \omega_n t + \delta \text{ rad} \tag{7.36}$$

The expression of the instantaneous developed torque is obtained substituting equations (7.34) and (7.35) in (7.33) results in:

$$T_{\text{dev}} = -\sqrt{2}\, I_s I_r M_{\text{max}} \cos \omega_s t \times \sin(\omega_n t + \delta) \tag{7.37}$$

Applying the identity:

$$\cos \alpha \times \sin \beta = \frac{1}{2}[\sin(\alpha + \beta) - \sin(\alpha - \beta)]$$

equation (7.37) yields:

$$T_{\text{dev}} = T_{\text{dev1}} + T_{\text{dev2}}$$

where:

$$T_{\text{dev1}} = -\frac{\sqrt{2}\, I_s I_r M_{\text{max}}}{2} \sin[(\omega_s + \omega_n)t + \delta] \tag{7.38}$$

and

$$T_{\text{dev2}} = \frac{\sqrt{2}\, I_s I_r M_{\text{max}}}{2} \sin[(\omega_s - \omega_n)t - \delta] \tag{7.39}$$

The only possibility for having a non-null time average developed torque occurs when $\omega_s = |\omega_n|$, which means that the rotor should be rotating at the same angular speed as the angular frequency of the source. The sense of the rotation is not important in this simple double excitation magnetic system. As an example, if the frequency of the source is 60 Hz, the rotor should run at 60 cycles s^{-1} (or 3600 r min^{-1}).

This kind of rotational magnetic system is an elementary *synchronous machine*. The average developed torque under this condition is given by:

$$T_{\text{avg}} = -\frac{\sqrt{2}\, I_s I_r M_{\text{max}}}{2} \sin \delta \tag{7.40}$$

Reflecting on equation (7.40), we conclude that if $\delta < 0$ the average torque T_{avg} is positive. Therefore, the sense of the torque and the angular velocity are the same, so the electromagnetic device works as a motor. Otherwise, if $\delta > 0$ the average torque is negative, and the electromagnetic device works as generator, because the torque and the angular velocity are in the opposite sense.

Remarks: The speed such that $\omega_s = |\omega_n|$ is called *synchronous speed*. No other speed is suitable for developing a useful torque in the electromagnetic device.

Third case

The general application is the one where the current carried by the stator coil is a sinusoidal time-dependent function expressed by:

$$i_s = \sqrt{2}\,I_s \cos \omega_s t \tag{7.41}$$

and the current carried by the rotor coil is also a sinusoidal time-dependent function, but with the different frequency of the stator and expressed by:

$$i_r = \sqrt{2}\,I_r \cos(\omega_r t + \alpha) \tag{7.42}$$

and finally, the angular position of the rotor can be expressed usually in radians by:

$$\theta = \omega_n t + \delta \tag{7.43}$$

Equation (7.32) is presented here for convenience:

$$T_{\text{dev}} = -i_s i_r M_{\max} \sin \theta$$

resulting in:

$$T_{\text{dev}} = -2I_s I_r M_{\max} \cos \omega_s t \times \cos(\omega_r t + \alpha)\sin(\omega_n t + \delta) \tag{7.44}$$

After some mathematical manipulation the expression of the developed torque may be expressed as a summation of four components

$$T_{\text{dev}} = T_{\text{dev1}} + T_{\text{dev2}} + T_{\text{dev3}} + T_{\text{dev4}} \tag{7.45}$$

where,

$$T_{\text{dev1}} = -\frac{1}{2} I_s I_r M_{\max} \sin[(\omega_s - (\omega_r - \omega_n))t - \alpha + \delta]$$
$$T_{\text{dev2}} = -\frac{1}{2} I_s I_r M_{\max} \sin[(\omega_s + (\omega_r - \omega_n))t + \alpha - \delta]$$
$$T_{\text{dev3}} = -\frac{1}{2} I_s I_r M_{\max} \sin[(\omega_s - (\omega_r - \omega_n))t - \alpha - \delta]$$
$$T_{\text{dev4}} = -\frac{1}{2} I_s I_r M_{\max} \sin[(\omega_s + (\omega_r + \omega_n))t + \alpha + \delta]$$

The four components of the developed torque are composed of a time varying function that has null time average value. However, these functions could have non-null average developed torque if some combination of the velocities ω_s, ω_r and ω_n are done.

Let the ω_s and the ω_r be two different angular frequencies, considering the condition:

$$\omega_s = \omega_r - \omega_n \text{ rad s}^{-1}$$

Imposing this condition in each four components yields:

$$T_{dev1} = -\frac{1}{2} I_s I_r M_{max} \sin[-\alpha + \delta]$$
$$T_{dev2} = -\frac{1}{2} I_s I_r M_{max} \sin[2\omega_s t + \alpha - \delta]$$
$$T_{dev3} = -\frac{1}{2} I_s I_r M_{max} \sin[-\alpha - \delta]$$
$$T_{dev4} = -\frac{1}{2} I_s I_r M_{max} \sin[2\omega_s t + \alpha + \delta]$$

For this brief analysis of each component, we can see that only T_{dev1} and T_{dev3} has a non-null time average value. Both T_{dev2} and T_{dev4} are time sinusoidal functions with a null time average value.

We conclude, the time developed torque is an oscillating function with a non-null average value given by:

$$T_{avg} = T_{dev1} + T_{dev3}$$

or:

$$T_{avg} = -I_s I_r M_{max} \cos\alpha \sin\delta \tag{7.46}$$

and an oscillating torque such that:

$$T_{osc} = T_{dev2} + T_{dev4}$$

or:

$$T_{osc} = -I_s I_r M_{max} \cos\delta \cos[2\omega_s t + \varphi] \tag{7.47}$$

where $\varphi = \arctan[\text{cotan}(\alpha)]$.

From equation (7.46), it is possible to infer the necessary condition for having energy conversion in a rotational magnetic system. As we have seen in chapter 6, the mutual inductance is dependent on the product of the number of turns of the coils, such that:

$$M = kN_s N_r \tag{7.48}$$

Therefore, we can rewrite equation (7.46) as:

$$T_{avg} = -kN_s I_s N_r I_r M_{max} \cos\alpha \sin\delta$$

or,

$$T_{avg} = -k\mathcal{F}_s\mathcal{F}_r \cos\alpha \sin\delta \tag{7.49}$$

We conclude, for having an effective average torque, both mmfs ($\mathcal{F}_s$ and $\mathcal{F}_r$, stator and rotor, respectively) should not spatially aligned ($\delta \neq 0$) and not orthogonally *out-of-phase* ($\alpha \neq \pi/2$).

As $-\pi/2 < \alpha < \pi/2$, it results in $\cos\alpha > 0$. Therefore, the operation mode of the electromagnetic device is identified by the signal of $\sin\delta$. If $0 < \delta < \pi/2$, it results in $\sin\delta > 0$ with a negative torque that characterizes the generator operation mode where the torque is in the opposite sense of the movement. For $-\pi/2 < \delta < 0$, the magnetic system operates as a motor with the torque in the same sense of the movement.

Example 7.2

The electromagnetic device of figure 7.4 was built with two similar coils in the stator and rotor. The electrical parameters associated to those coils are:

Resistance of stator: 0.4 Ω	Resistance of rotor: 0.4 Ω
Inductance of stator: 45 mH	Inductance of rotor: 45 mH
Maximum mutual inductance: 40 mH	

(a) Compute the developed torque delivered by the rotor when the angle between both coils' axis is 45° and the stator coil is fed by a 4 V DC electrical source and the rotor coil by a 8 V DC source.
(b) Consider that the stator coil is fed by an AC source that injects an electrical current given by $i_s = \sqrt{2} \times 10 \times \cos 377t$ (A) and the rotor is fed by an 8 V DC electrical source.
   (i) At what speed in rps should the rotor run to develop a useful torque?
   (ii) What is the average developed torque when the rotor is running at the speed evaluated in part (i).
   (iii) Write the equation issued from the electrical side of the electromagnetic device. Is it possible to propose an equivalent electric circuit for this operation?
(c) Consider that the stator coil is fed by an AC source that injects an electrical current given by $i_s = \sqrt{2} \times 10 \times \cos 377t$ (A) and the rotor coil carries an electrical current described by $i_r = \sqrt{2} \times 20 \times \cos[18.85t + 45°]$ (A).
   (i) At what speed in rps should the rotor run to develop a useful torque?
   (ii) What is the average developed torque when the rotor is running at the speed evaluated in the previous part?

Solution:

(a) The electrical actuator.

The stator coil is fed by a 4 V DC electrical source and the rotor coil by a 8 V DC source resulting in:

$$i_s = \frac{4}{0.4} = 10\ (\text{A}) \text{ and } i_r = \frac{8}{0.4} = 20\ (\text{A})$$

Applying equation (7.33) yields:

$$T_{\text{dev}} = -10 \times 20 \times 40 \times 10^{-3} \sin 45° = -5.6\ \text{Nm}$$

(b) $i_s = \sqrt{2} \times 10 \times \cos 377t$ (A) and $i_r = 20$ (A) resulting in:
the angular velocity of the rotor should be

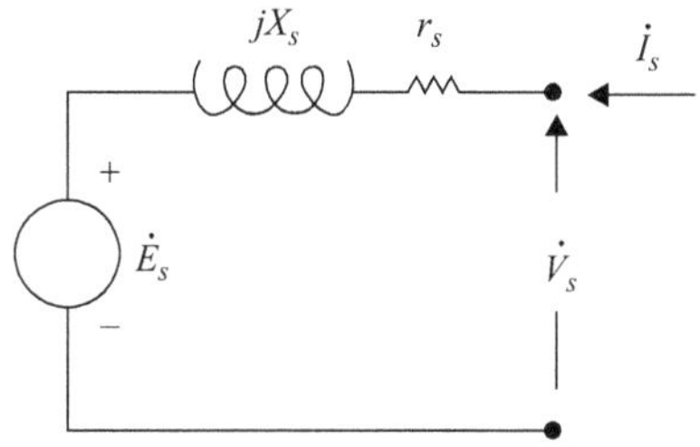

Figure 7.6. Equivalent electric circuit of synchronous machine.

$$\omega_r = \omega = 377 \text{ rad s}^{-1}$$

as $\omega_r = 2\pi n_r$ we get:

$$n_r = \frac{377}{2\pi} = 60 \text{ rps(or 3600 rpm)}$$

The average developed torque for 60 rps is given by (7.40):

$$T_{\text{avg}} = -\frac{\sqrt{2} \times 10 \times 20 \times 40 \times 10^{-3}}{2} \sin \delta = 5.6 \sin \delta \text{ Nm}$$

The equations from the electric sides of the device are:

The general equation from the stator side is:

$$v_s = r_s i_s + L_s \frac{di_s}{dt} + i_s \frac{dL_s}{dt} + M \frac{di_r}{dt} + i_r \frac{dM}{dt}$$

as both L_s and i_r are constants results:

$$v_s = r_s i_s + L_s \frac{di_s}{dt} + i_r \frac{dM}{dt}$$

As all terms in the previous equation are sinusoidal time-dependent functions, we can represent them using a complex notation like:

$$\dot{V}_s = r_s \dot{I}_s + jX_s \dot{I}_s + \dot{E}_s \tag{7.50}$$

where:

$X_s = \omega L_s = 17\ \Omega$: synchronous reactance of the machine

$\dot{E}_s = \frac{1}{\sqrt{2}} j\omega M_{\max} i_r = 213$ (V): intern induced emf

Based on (7.50), the equivalent electric circuit that represents the electric side of the stator is given by figure 7.6.

(c) As $i_s = \sqrt{2} \times 10 \times \cos 377t$ (A) and $i_r = \sqrt{2} \times 20 \times \cos[18.85t + 10°]$ (A), we obtain:

For developing a non-null useful torque, the angular velocity should be

$$\omega_s = \omega_r \pm \omega_n$$

as $\omega_s = 377$ rad s^{-1} and $\omega_r = 18.85$ rad s^{-1} yields:

$$\omega_n = |377 - 18.85| = 358.15 \text{ rad s}^{-1}$$

or, $n_n = 57$ rps (or 3420 rpm)

From (7.46), the average developed torque when the rotor is running at $n_n = 57$ rps is given by:

$$T_{avg} = -I_s I_r M_{max} \cos\alpha \sin\delta$$

Substituting the quantities by its values yields:

$$T_{avg} = -10 \times 20 \times 40 \times 10^{-3} \times \cos 45° \sin\delta$$
$$T_{avg} = -5.6 \sin\delta \text{ Nm}$$

7.4 Rotor with a salient pole

Some devices have no cylindrical rotor as figure 7.4 shows. A kind of rotor called a salient pole is shown in figure 7.7.

Not only the reluctance torque due to the rotor asymmetry but also the mutual inductance torque due to the magnetic coupling between coils is present is this magnetic system. The mutual inductance torque is the one that varies with $\sin\delta$ and the reluctance torque varies with $\sin 2\delta$.

Both types of torque comprise the total average developed torque in a salient pole machine, as shown in figure 7.8.

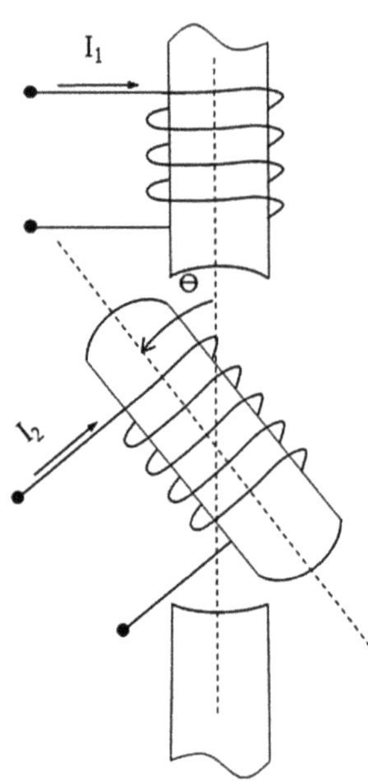

Figure 7.7. Salient pole machine.

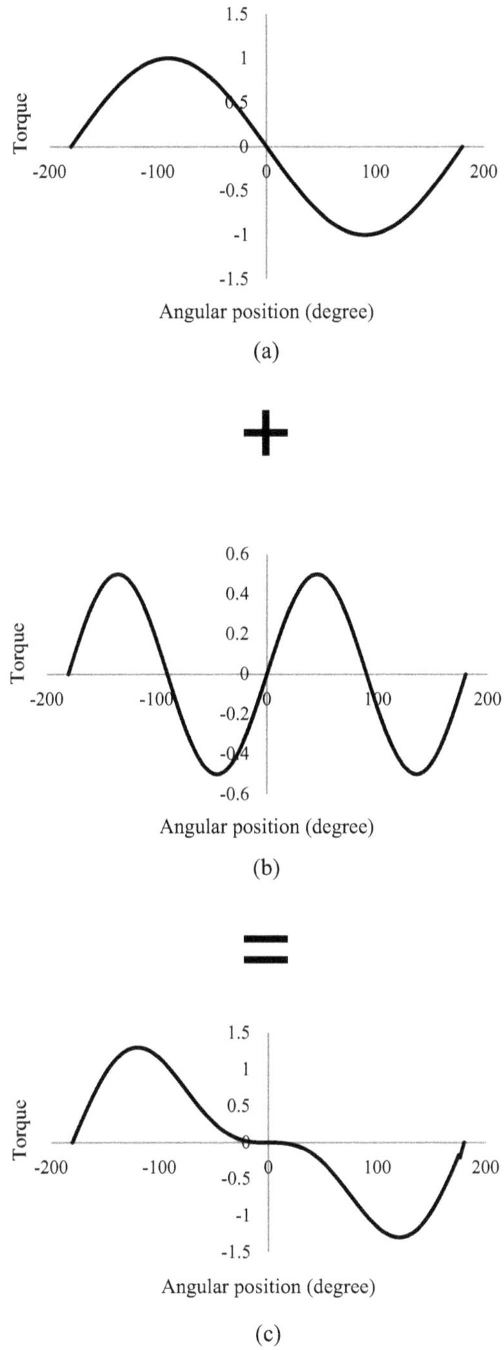

Figure 7.8. Average developed torque in a salient pole machine. (a) The mutual inductance torque contribution. (b) The reluctance torque contribution. (c) The total average torque.

7.5 Summary

Electromagnetic devices with multiple coils are the most common equipment used in the industry. In our homes we have several such devices, for example our refrigerators, washing machines, small toys and many others. That is why knowledge of the working principles of this kind of equipment is very important for an electrical engineer.

We introduced first the flux linkages in a linear multiple excited magnetic circuit to give the main concepts of both self-inductance and mutual inductance.

From the energy balance, we deduced both the force in a translational magnetic system and the torque in a rotational one. With those expressions, it is possible to evaluate the performance of the magnetic system using the electric equation issued from the electric side of the device.

For a rotational magnetic system, the main equations involved in the operation of the device are as follows.

Electric equation for the stator coil:

$$v_s = r_s i_s + L_s \frac{di_s}{dt} + i_s \frac{dL_s}{dt} + M \frac{di_r}{dt} + i_r \frac{dM}{dt} \tag{7.51}$$

Electric equation for the rotor coil:

$$v_r = r_r i_r + L_r \frac{di_r}{dt} + i_r \frac{dL_r}{dt} + M \frac{di_s}{dt} + i_s \frac{dM}{dt} \tag{7.52}$$

The developed torque:

$$T_{\text{dev}} = \frac{i_1^2}{2} \frac{dL_{11}}{d\theta} + \frac{i_2^2}{2} \frac{dL_{22}}{d\theta} + i_1 i_2 \frac{dM}{d\theta} \tag{7.53}$$

Although those equations were deduced for a magnetic system with two coils, it can be extended for any number of coils that the magnetic system could have.

As the developed torque (or force) is a time-dependent function, the requirement for developing a continuous torque (or force) is that the average value of the instantaneous developed torque should be non-null.

Finally, we conclude that it is possible to identify a suitable equivalent electric circuit for modeling the magnetic system operating in any situation.

Problems

7.1 Two coils are magnetically coupled through a linear magnetic circuit, as shown in figure 7.9. Both the inductance and the mutual inductance are known. Applying the definition of flux linkage:

(a) Identify in figure 7.9 whether the mutual inductance is in concordance or not.

(b) Determine the equivalent inductance if both coils are associated in series with the mutual inductance in concordance. In this case, the magnetic flux produced by both are in the same sense.

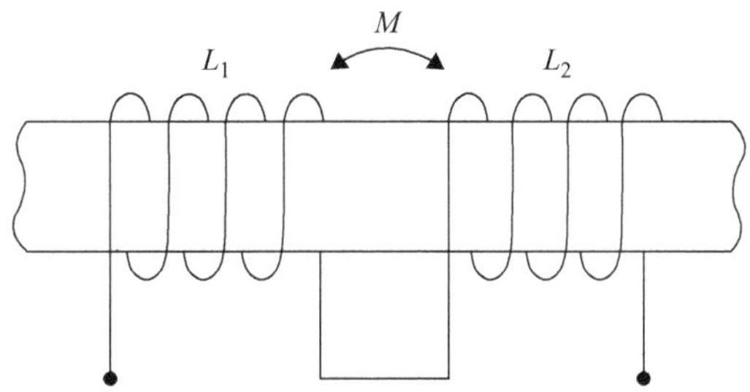

Figure 7.9. Two coils magnetically coupled.

(c) Determine the equivalent inductance if the coils are associated in series, but with non-concordant mutual inductance. In this case, the magnetic flux produced by both are in opposite sense.

7.2 Figure 7.10 shows a cross-section of an elementary electromechanical device. The coil 1–1' is located at the stator and the coil 2–2' is located at the rotor, which is able to vary the angular position.

The maximum mutual inductance between both coils is $M = 0.1\ H$.

Suppose that the DC currents $i_1 = 2$ (A) and $i_2 = 10$ (A) are flowing in the coils and a torque of 1 Nm is applied to the shaft in the clockwise direction. Compute the values and comment on whether the resultant position is stable or not (see the remarks in figure 6.8 of chapter 6):

(a) The steady state value of the angle between the coil's axis.

(b) Repeat item (a) for the torque equal to 2 Nm.

Answer: 30°, unstable.

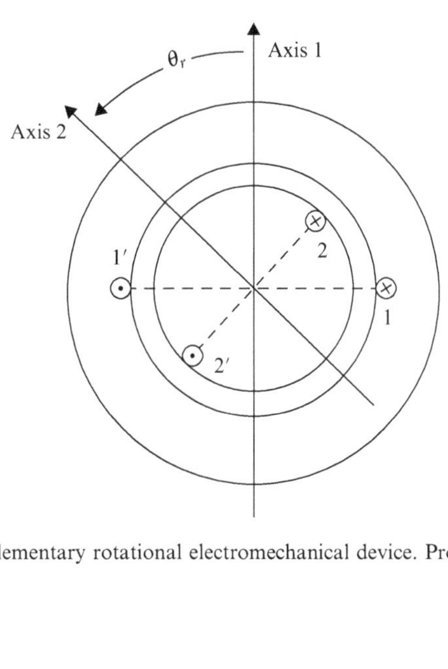

Figure 7.10. Elementary rotational electromechanical device. Problems 7.2–7.5.

7.3 Consider now reversing the direction of i_1 in the coil 1–1' of figure 7.10. Analyze the range of stable operation in this condition.

Answer: $\frac{\pi}{2} < \theta_r < \frac{3\pi}{2}$.

7.4 Supposing that both currents that flow in the coils of the device shown in figure 7.10 are positive and constant. Analyze the following situations:

(a) An external torque is applied to increase θ_r from 0° to 45° and then released. Assume damping exists. What is the final position of the rotor?

(b) Repeat for θ_r increased to (i) 90°, (ii) 120°, (iii) 180°, (iv) 210°.

Answers: (a) $\theta_r = 0$; (i) 0, (ii) 0, (iii) 0 or 2π, (iv) 2π.

7.5 In the system shown in figure 7.10, the maximum value of the mutual inductance is $M = 0.1$ H.

Considering that $i_1 = 20 \cos \omega_e t \ (A)$, $i_2 = 2 \ (A)$ and the rotational speed $\omega_r = \omega_e$, compute the power angle δ if an external torque of 1 Nm is applied in the clockwise direction.

Answer: $\theta_r = -30°$.

7.6 The rotational magnetic system of example 7.2 has its coil associated in series, as figure 7.11 shows.

Compute the average developed torque delivered by the rotor when the angle between the coil's axis is 45°. Use the following details to solve this problem:

(a) the association is fed by 10 V DC;

(b) the association is fed by 115 V–60 Hz.

Answers: (a) 4.4 Nm; (b) 0.1 Nm.

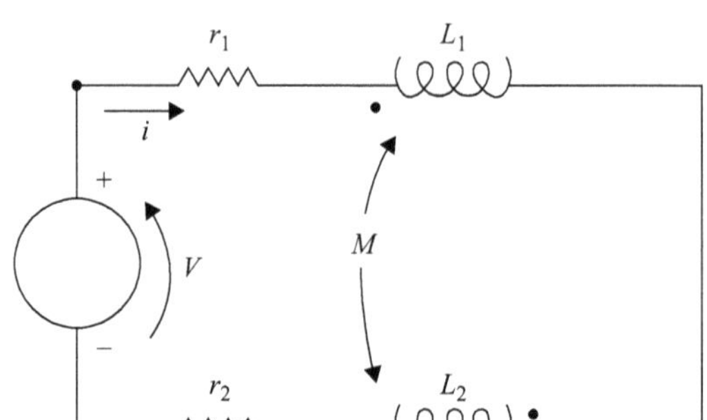

Figure 7.11. Problems 7.6 and 7.7.

7.7 Consider that the stator coil is fed by independent 60 Hz AC electric source and the rotor excited by a 25-Hz AC electric source. Explain at what speed will the device convert electrical energy to mechanical energy and vice versa.

Answer: 35 rev s^{-1} and 85 rev s^{-1}.

7.8 In a double-excited translational magnetic system similar to example 7.1 the self-inductances of coils are such that (figure 7.12) $L_{11} = L_{22} = 10$ (mH).

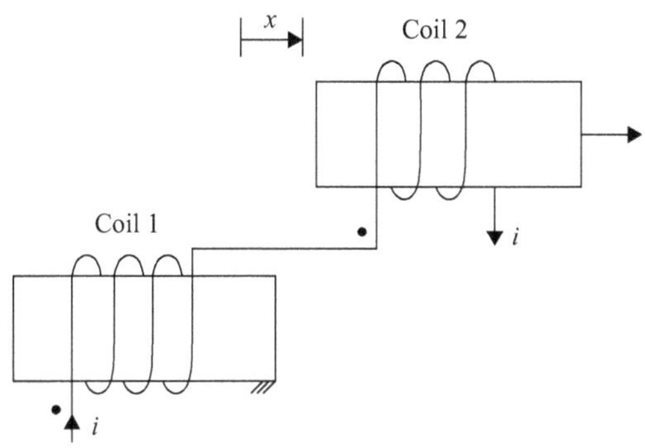

Figure 7.12. Double-excited magnetic system. Translational movement in the *x*-direction. Problems 7.8 and 7.9.

The mutual inductance between the two coils is $M = 2 + 5x$ (mH). The resistances of both coils are 0.1 Ω.

Compute the average developed force delivered by the rotor, when the position of the moving coil is $x = 20$ cm and the association is fed by a DC voltage source of 10 V. The coils are associated in series.

Answer: 12.5 Nm.

7.9 The association of coils in problem 7.7 is now fed by an AC voltage source. Compute the time-dependent developed force indicating the average value and the frequency of the wave force. Compare the frequencies of the wave force and the source. Are you able to explain it?

7.10 Consider that the coils are fed by an independent AC electric source in which the stator coil is excited by a 60 Hz AC source and the moving coil is excited by a DC current source. Explain at which speed the device will convert electrical energy to mechanical energy and vice versa.

Further reading

[1] Kuhlmann J H 1940 *Design of Electrical Apparatus* 2nd edn (New York: Wiley)

[2] Krause P, Wasynczuk O and Pekarek S 2012 *Electromechanical Motion Devices* 2nd edn (Hoboken, NJ: Wiley)

[3] Boldea I and Nasar S A 2010 *The Induction Machines Design Handbook* 2nd edn (Boca Raton, FL: CRC press)

[4] Meisel J 1984 *Principles of Electromechanical-Energy Conversion* (Malabar, FL: R.E. Krieger)

[5] Gieras J F 2016 *Electrical Machines: Fundamentals of Electromechanical Energy Conversion* 1st edn (Boca Raton, FL: CRC Press)

[6] Gonen T 2011 *Electrical Machines with MATLAB®* 2nd edn (Boca Raton, FL: CRC Press)

IOP Publishing

Electromechanical Energy Conversion Through Active Learning

J R Cardoso, M B C Salles and M C Costa

Chapter 8

Synchronous machine: the windings

8.1 Design aspects of a synchronous machine

The most common application of a synchronous machine (SM) is as an electrical energy generator. We find SMs not only in large hydroelectric but also in thermoelectric generation plants like nuclear, coal and gas power plants. In all those cases, the large synchronous generators are responsible for converting mechanical energy produced from the prime mover into electrical energy and delivering it to the transmission system to be taken to the loads on the consumer side.

SMs are also used in wind turbines, in microgrids and in small power plants like in industries, buildings and *standby* installations to serve as backup power generators. In this last case, the electrical energy supply should operate with high reliability. Although the use of a SM as a motor is not that common in industry, the features of this kind of motor are important in processes where an accurate constant speed is required. Another import feature of the SM is the possibility of injecting or consuming reactive power in or from the grid, simply by modifying the excitation field. In the last two decades, the interest in operating the SM with variable speed has grown—microturbines, wind turbines and motors for electric vehicles are good examples of that.

The stator of a SM is made of sheet steel of Fe–Si with uniformly distributed slots that host the conductors of polyphase winding. Normally, the triphasic construction is preferred as all electrical distribution systems work using this kind of system.

Figure 8.1 shows a cross section of the sheet steel of a small SM stator. The assemblage of the stator is made of a stack of sheet steel sufficiently compacted to ensure a good magnetic performance of the magnetic circuit.

Figure 8.2 shows the stator of a SM filled with two coils inserted in the electrical isolated slots. No electric contact between the conductor and the sheet steel is allowed. During normal operation, the temperature rises in the conductors making the construction even more complex. To solve this problem, a family of high performance isolated and flexible materials was developed to withstand not only the

doi:10.1088/978-0-7503-2084-9ch8

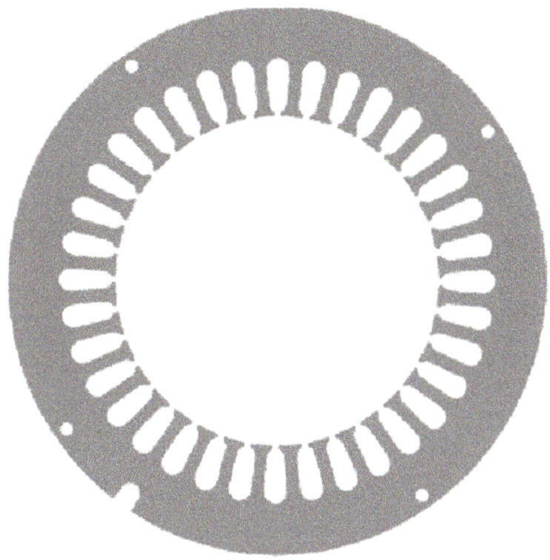

Figure 8.1. Cross section of a small SM. The 4-small holes on the sheet are used to guide the assemblage.

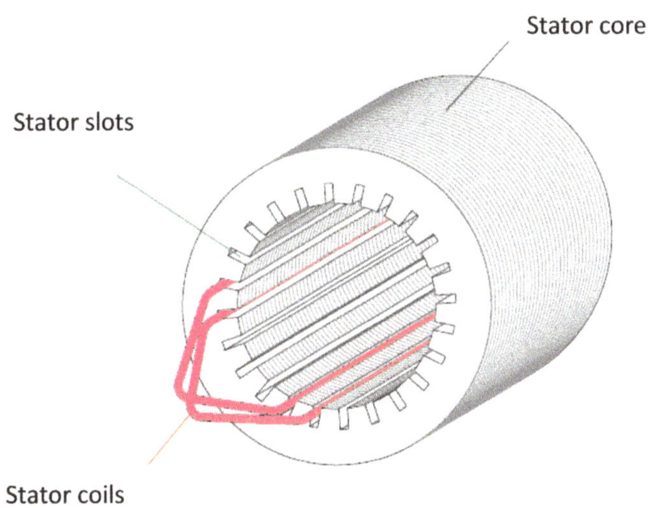

Figure 8.2. The assemblage of a SM stator. The conductors are inserted inside the isolated slots. Source: Courtesy of EQUACIONAL LTDA Brazil.

high electric field that exists between them, but also to support the high temperatures.

Two different rotor assemblages are used in a SM. They depend on the required running speed by the prime mover. A non-salient pole construction, shown in figure 8.3, is suitable for high-speed (normally from 1800 to 3600 rev min^{-1}) applications. This is very reliable to withstand the centrifugal forces issuing from the rotation.

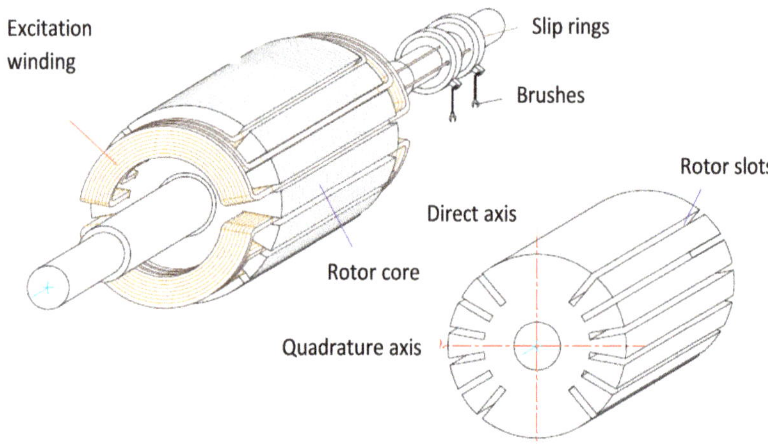

Figure 8.3. Non-salient pole rotor of a SM. Left: non-salient pole rotor built with of a stack of sheet steel. Right: solid body of non-salient pole rotor. Source: courtesy of EQUACIONAL LTDA Brazil.

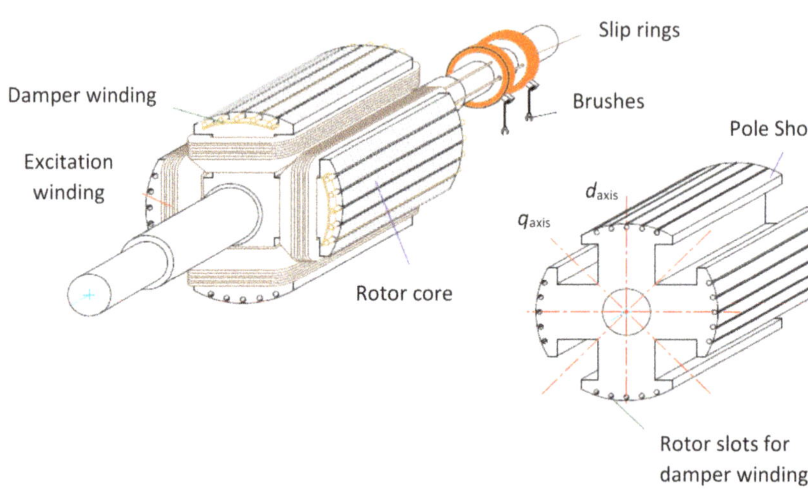

Figure 8.4. Salient pole rotor. Right: 4 pole rotor made of a stack of sheet steel. Left: 4 poles hosting a concentrated winding series connected. Source: courtesy of EQUACIONAL LTDA Brazil.

The rotor core can be made of a stack of sheet steel as we can see on the right of the figure 8.3. The rotor core is slotted for hosting the conductors of one-phase winding, which carries a DC electric current. For injecting the DC electric current in the rotor winding, it is necessary to introduce two slip rings isolated from the shaft to collect the electric current from the brushes and apply them to the winding.

For application at low speed, as observed in large hydroelectric plants, the salient pole alternative is commonly adopted due to its simple construction.

Figure 8.4 shows a salient four-pole rotor suitable for a small SM, like the one used in small generators for stand-by application. The concentrated winding on the pole carries the DC current that produces alternative polarity (N–S–N–S) in the polar pieces. As for the non-salient SM, a set of slip rings and brushes should be installed in the shaft for injecting DC current in the rotor pole winding.

8.2 How a synchronous machines works

A synchronous machine is a double excited electromechanical system where each coil carries a different kind of electric current. The stator coil, also called the armature winding, carries AC electric current. The rotor coil, also called the field winding (or excitation winding), is fed by DC electric current.

If the SM is operating as a generator, a prime mover drives the rotor by introducing mechanical energy to its shaft. The DC electrical current injected in the field winding produces a constant magnetic flux that crosses the air-gap and establishes a magnetic field density distribution in the air-gap. The interaction of both, the magnetic flux density distribution and the movement of the rotor, induces an electromotive force (emf) in the stationary armature winding. Due to a suitable construction of both, armature winding and rotor geometry, the induced emf is (*nearly*) a time dependent sinusoidal function whose frequency depends on the rotor speed.

Due to this emf, a voltage will appear at the ends of the armature winding. If a load is connected to these terminals, electrical energy is delivered to the load.

The interaction of the AC current of the armature winding and the magnetic flux density field distribution in the air-gap produces a non-null average electromagnetic torque in the opposite sense of the movement. Figure 8.5 shows a schematic representation of a SM operating as a generator.

If a SM is operating as a motor, the armature winding is connected to the AC electrical source that injects an AC electric current. The interaction of the armature AC current and the magnetic flux density distribution in the air-gap develops a non-null average electromagnetic torque in the same sense of the movement, only if the rotor speed is compatible with the frequency of the source. Figure 8.6 shows a schematic representation of a SM operating as a motor.

8.2.1 The magnetic flux density distribution by the armature winding

Figure 8.7 shows the most elementary armature winding of a SM. The stator is longitudinally slotted with two slots 180° apart that host conductors of an N turns

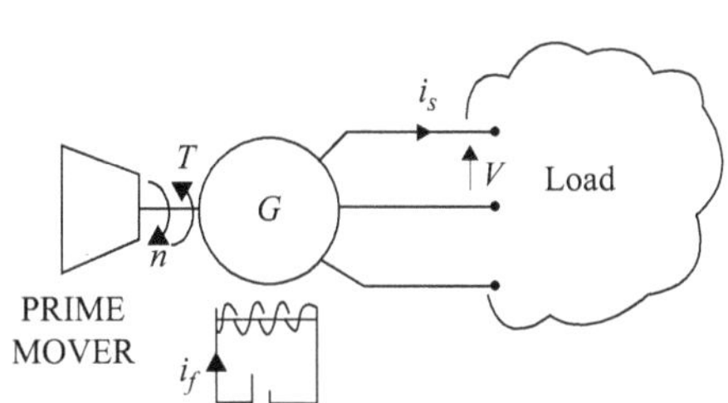

Figure 8.5. Electrical scheme of a synchronous generator.

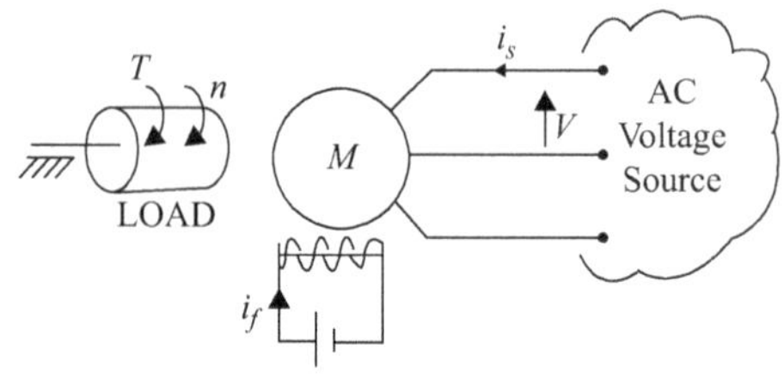

Figure 8.6. Electrical scheme of a synchronous motor.

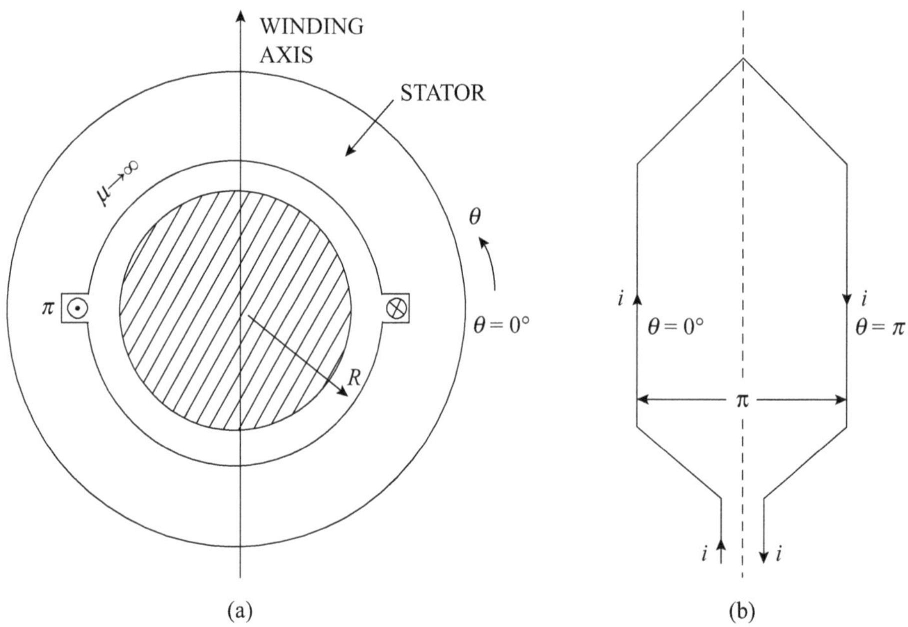

Figure 8.7. The elementary winding of an AC machine. (a) Cross section of the stator (b) Developed winding. ⊗: the positive current goes down into the plane of the figure. ⊙: the negative current goes up out of the plane of the figure.

winding. The vertical line is the axis of the stator winding and is oriented by the right-hand rules accompanying the arbitrary positive sense of the electric current.

Using the same approach of chapter 7, we are considering that the magnetic material of both stator and rotor are ideal ($\mu \to \infty$). If an electrical current is injected at the end indicated by ⊗ and comes out through the end indicated by ⊙, the lines of the magnetic flux density have the geometry shown in figure 8.8.

Applying Ampère's law to the closed line C, yields:

$$H_g l_g + H_g l_g = Ni \tag{8.1}$$

because no mmf drop is observed inside the magnetic material.

As a result, we can conclude that the total mmf Ni is consumed by the two air-gaps that are part of the closed line C.

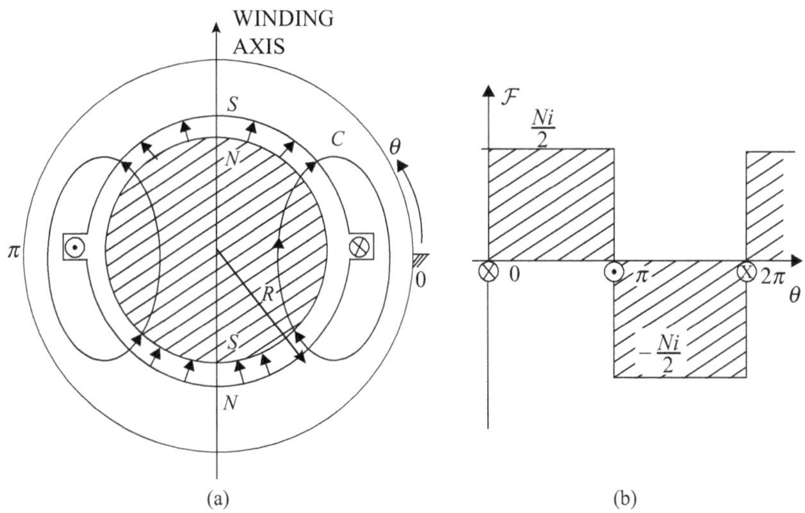

Figure 8.8. Flux lines produced by positive current. (a) Closed line C. (b) Magnetomotive force (mmf) distribution.

We are considering that the mmf drop where magnetic flux density is directed from rotor to stator as positive and negative otherwise. The plot of figure 8.8 represents the mmf distribution as a function of the position θ in the air-gap.

From equation (8.1), the magnetic field intensity H_g can be calculated as:

$$H_g = \frac{Ni}{2l_g} \text{ (A m}^{-1}\text{)} \tag{8.2}$$

As $B_g = \mu_0 H_g$ the magnetic flux density will be:

$$B_g = \frac{\mu_0 Ni}{2l_g} \text{ (Wb m}^{-2}\text{)} \tag{8.3}$$

Figure 8.9 shows the magnetic flux density distribution as a function of the position θ in the air-gap.

We conclude, for any rotational magnetic system with constant air-gap, we can write:

$$B_g = \mu_0 H_g = \frac{\mu_0}{2l_g}\mathcal{F} \tag{8.4}$$

where $\mathcal{F}$ is the mmf distribution.

The magnetic flux through the surface of magnetic pole is known as the *flux per pole* and can be evaluated by:

$$\phi_p = B_g \times S_p \text{ (Wb)} \tag{8.5}$$

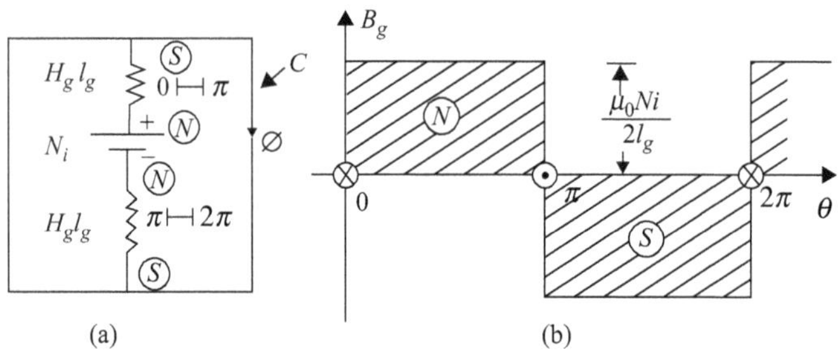

Figure 8.9. The magnetic flux density distribution: (a) magnetic circuit (b) magnetic flux density distribution. N: north pole of the rotor. S: south pole of the rotor.

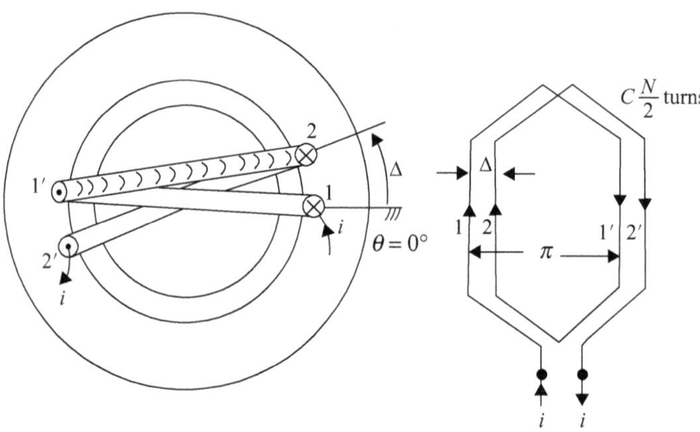

Figure 8.10. Distributed AC winding.

where $S_p = \pi R h$ and h is the length of stator.

We can write:

$$\phi_p = B_g \pi R h \text{ (Wb)} \tag{8.6}$$

If the winding is fed by an AC electric current, the square wave of magnetic flux density distribution changes maintaining its shape but its magnitude now varies according to the current variation.

The ideal operation condition of an AC winding is to produce a spatial sinusoidal magnetic flux density distribution in the air-gap. Therefore, the elementary winding of figure 8.8 is not suitable for achieving this goal.

8.3 The AC distributed winding

We can improve the quality of the shape of the magnetic field distribution, distributing the coil along the surface of the stator. This distribution makes its shape closer to a spatial sinusoidal, as is shown in figure 8.10.

In this figure, the original winding composed of an N turns concentrated coil (N turns installed in a pair of slots) is substituted by two coils of $N/2$ turns each associated in series. This approach is convenient for two reasons. Firstly, the manipulation of a coil of $N/2$ turns is easier than the manipulation of an N turns coil and secondly, the slot of a distributed coil has a half cross section of the elementary winding.

A reduction of *flux per pole* is observed comparing figures 8.10 and 8.11. That is the price to pay for the SM performance improvement.

Figure 8.12 shows a magnetic flux density distribution of a distributed AC winding with three coils of $N/3$ turns, each using the same span distribution. As the total turns of the winding are the same, the magnitude of the distribution does not change.

Although the *flux per pole* in the last distribution is less than the one of figure 8.11, the shape of the magnetic flux density distribution is more adjustable to the sinusoidal spatial distribution.

The Fourier series mapping is commonly used to identify how the shape of the magnetic field distribution fits the spatial sinusoidal distribution. As the distribution of magnetic field is a periodic function, the components of the Fourier series are

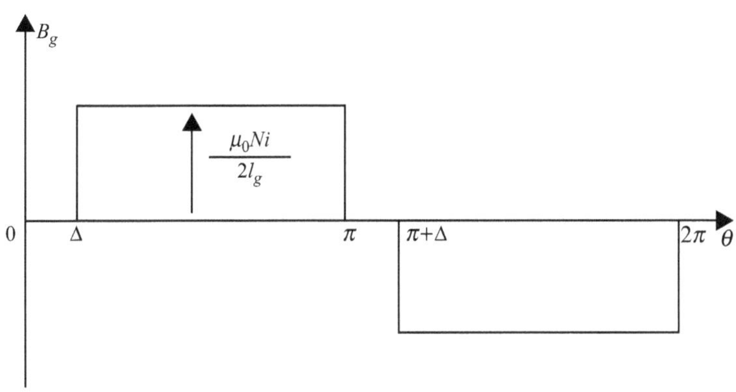

Figure 8.11. The magnetic flux density distribution along the air-gap. (+): north pole of the rotor. (−): south pole of the rotor.

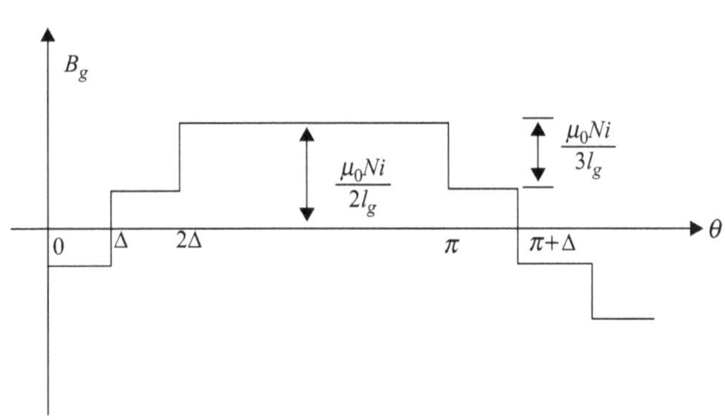

Figure 8.12. The magnetic flux density distribution-half period. (+): north pole of rotor. (−): south pole of rotor.

composed of a *fundamental*, that has the same period of the function under analysis and an infinite sum of terms of odd harmonics.

For the square wave of figure 8.9, the series Fourier expansion is given by:

$$B_g(\theta) = \frac{4}{\pi}\frac{\mu_0 Ni}{2l_g}\sin\theta + \frac{4}{3\pi}\frac{\mu_0 Ni}{2l_g}\sin 3\theta + \cdots + \frac{4}{h\pi}\frac{\mu_0 Ni}{2l_g}\sin h\theta + \cdots \tag{8.7}$$

$$B_g(\theta) = \sum_{\substack{h=1 \\ h \text{ odds}}}^{\infty} \frac{4}{h\pi}\frac{\mu_0 Ni}{2l_g}\sin(h\theta)$$

The magnitude of the *fundamental* is:

$$B_{1\max} = \frac{4}{\pi}\frac{\mu_0 Ni}{2l_g} \tag{8.8}$$

and the magnitude of the *h-harmonic* is:

$$B_{h\max} = \frac{4}{h\pi}\frac{\mu_0 Ni}{2l_g} \tag{8.9}$$

Comparing the expressions (8.8) and (8.9) yields:

$$B_{h\max} = \frac{1}{h}B_{1\max} \tag{8.10}$$

Figure 8.13 shows the fundamental and the third harmonic of the square wave of the magnetic flux density distribution, and figure 8.14 shows the half wave shape of the summation of fundamental, third and fifth harmonics.

Considering now that the winding is distributed in two coils with $N/2$ turns each, with the slots separated by angle of $\Delta = 30°$ (this is the case for a 2 poles triphasic winding), it is possible to demonstrate that:

$$B_{1\max} = 0.966 \times \frac{4}{\pi}\frac{\mu_0 Ni}{2l_g}$$

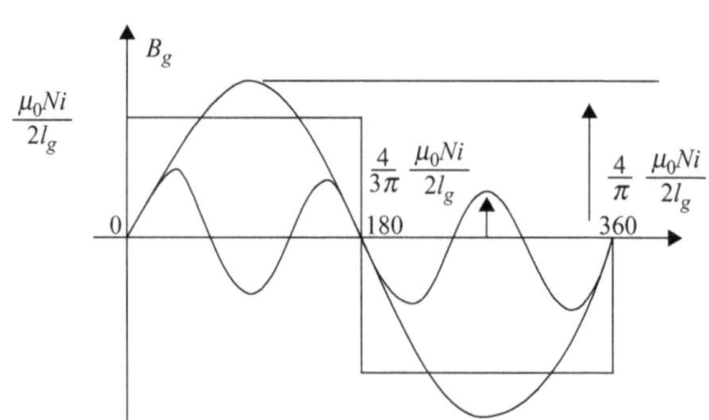

Figure 8.13. The square wave of magnetic flux density distribution accompanied by the fundamental + third harmonic.

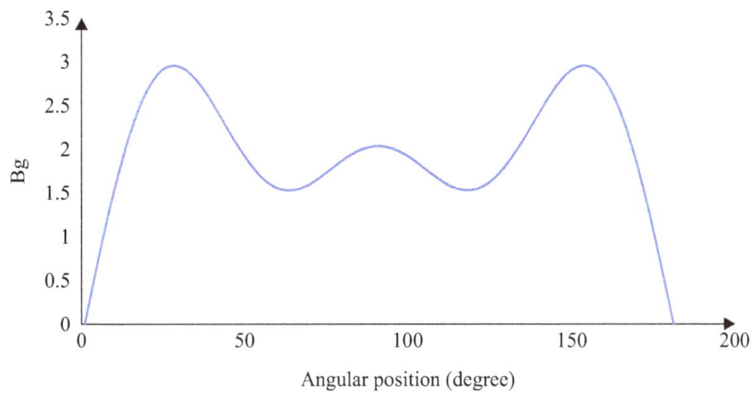

Figure 8.14. Half cycle wave of the fundamental + third harmonic + fifth harmonic resultant.

and, for $h = 3$

$$B_{3\max} = 0.707 \times \frac{4}{3\pi}\frac{\mu_0 Ni}{2l_g}$$

Considering now the general case where the winding is distributed in q coils with N/q turns each with the slots separated by angle of $\Delta = \left(\frac{60^\circ}{q}\right)$ for both 2 poles and triphasic winding, it is possible to demonstrate that:

$$B_{1\max} = \frac{4}{\pi}\frac{\mu_0 Ni}{2l_g}k_d \tag{8.11}$$

where:

$$k_d = \frac{\sin\left(\frac{q\Delta}{2}\right)}{q\,\sin\left(\frac{\Delta}{2}\right)} \tag{8.12}$$

This coefficient is called the *distribution factor* of the winding. It represents the reduction of the maximum value of the magnetic flux density caused by the distribution of the winding compared with the concentrated winding (where $k_d = 1$).

A similar coefficient is defined for the reduction of the maximum value of the h-harmonic of the magnetic flux density such that:

$$B_{h\max} = \frac{4}{h\pi}\frac{\mu_0 Ni}{2l_g}k_{dh} \tag{8.13}$$

where:

$$k_{dh} = \frac{\sin\left(\frac{hq\Delta}{2}\right)}{q\,\sin\left(\frac{h\Delta}{2}\right)} \tag{8.14}$$

Figure 8.15 shows the *fundamental* distribution factor for a triphasic winding as a function of q. As q increases the limit of k_d is about 0.955.

The $\lim_{q\to\infty} k_{dh}$ for the third and fifth harmonic is 0.637 and 0.191, respectively.

Those numbers show that the distribution winding presents a spatial wave of magnetic flux density distribution that is nearly sinusoidal. This can be considered because the *distribution factor* of the harmonics is low enough to neglect the harmonics of the magnetic flux density distribution.

8.3.1 The fractional pitch winding

Another way of improving the quality of the magnetic flux density distribution to obtain a better sinusoidal shape distribution consists of reducing the span of the winding.

The full pole pitch winding is the one where each side of a single coil is separated by 180°, as is shown in figure 8.7. It is also possible to make a winding where each side of a single coil is separated by an angle less than 180°, as is shown in figure 8.16. It is sometimes convenient to speak of an armature coil span as having a fractional pitch expressed as a fraction e.g., a $\frac{5}{6}$ pitch, or an $\frac{11}{12}$ pitch, etc.

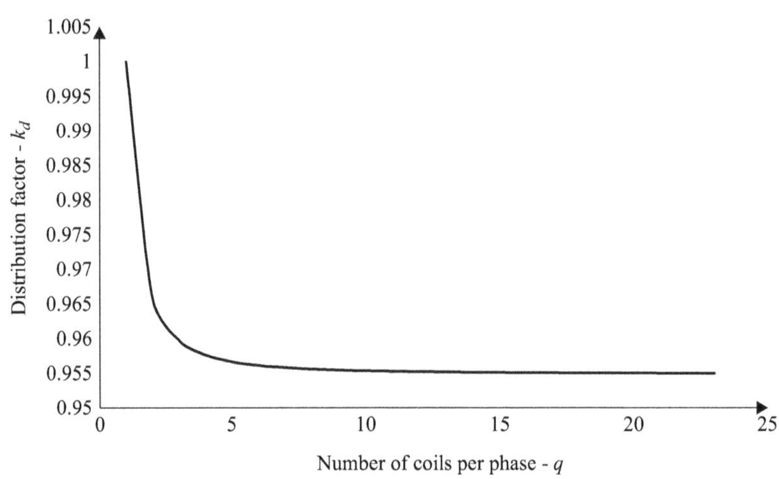

Figure 8.15. Distribution factor for the *fundamental* of the magnetic flux density distribution $\lim_{q\to\infty} k_d = 0.955$ for triphasic winding.

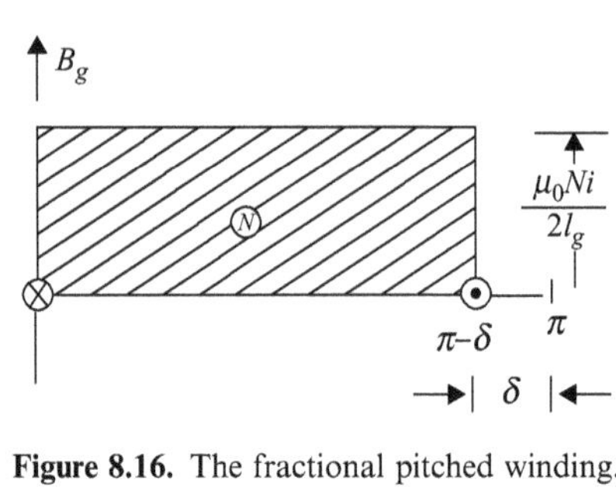

Figure 8.16. The fractional pitched winding.

Let δ be the short pitched angle of the coil. It is possible to demonstrate that the maximum value of the *fundamental* of the magnetic flux density distribution is reduced by a *pitch factor* k_p, given by:

$$k_p = \cos\left(\frac{\delta}{2}\right) \tag{8.15}$$

Therefore, the maximum value of the magnetic flux density distribution for a short pitched winding is:

$$B_{1\max} = \frac{4}{\pi}\frac{\mu_0 Ni}{2l_g}k_p \tag{8.16}$$

Applying the same approach for the *h-harmonic*, the *pitch factor* results in:

$$k_{ph} = \cos\left(\frac{h\delta}{2}\right) \tag{8.17}$$

Therefore, the maximum value of the *h-harmonic* of the magnetic flux density distribution yields:

$$B_{\text{hmax}} = \frac{4}{h\pi}\frac{\mu_0 Ni}{2l_g}k_{ph} \tag{8.18}$$

Supposing that not only is the winding distributed, but also fractioned, the penalty associated with the improved magnetic flux density distribution in the amplitude of its *fundamental* is given by the winding factor k_w given by:

$$k_w = k_d k_p \tag{8.19}$$

Therefore, the amplitude of the fundamental of the magnetic flux density distribution is:

$$B_{1\max} = \frac{4}{\pi}\frac{\mu_0 Ni}{2l_g}k_w \tag{8.20}$$

The same approach can be applied to the h-harmonic.

$$B_{\text{hmax}} = \frac{4}{h\pi}\frac{\mu_0 Ni}{2l_g}k_{wh} \tag{8.21}$$

where:

$$k_{wh} = k_{dh}k_{ph} \tag{8.22}$$

Example 8.1

The stator of a non-salient pole SM has a 40 cm inside diameter, an air-gap of 2 mm and a length of 50 cm (figure 8.17). Two slots separated by 180° house a 4 turns

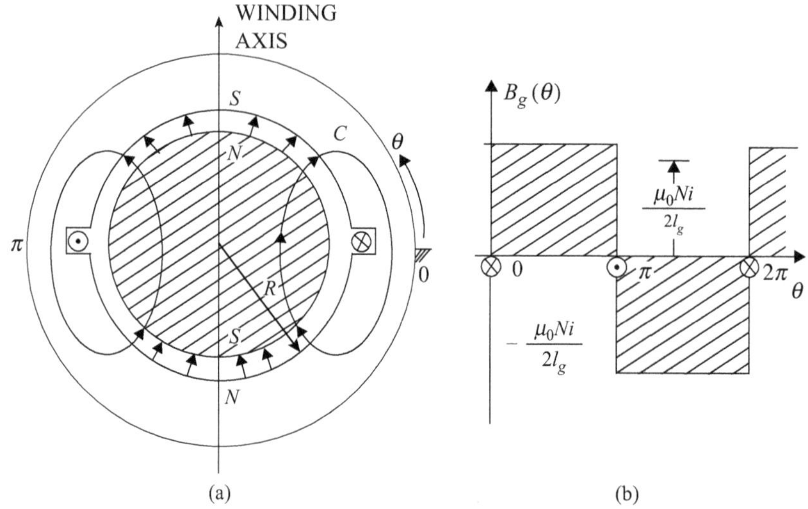

Figure 8.17. Stator of example 8.1.

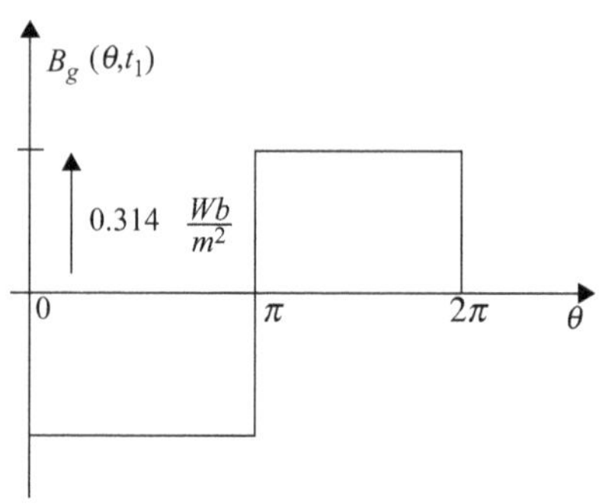

Figure 8.18. Magnetic flux density distribution for $\omega t_1 = \frac{5\pi}{4}$ rad s.

concentrated winding. The winding carries an electric current given by $i_s = \sqrt{2} \times 250 \cos(\omega t)$ A:

(a) sketch the magnetic flux density distribution in the air-gap when $\omega t_1 = \frac{5\pi}{4}$ rad s and $\omega t_2 = 2\pi$ rad s;

(b) considering the frequency 60 Hz, determine the time dependent function of the *flux per pole;*

(c) determine the time dependent function of both the *fundamental* and the 3-*harmonic* of the magnetic flux density distribution.

Solutions

(a) The magnitude of the square wave of magnetic flux density distribution (figure 8.18) is given by equation (8.3). Therefore, for $i_s = \sqrt{2} \times 250 \cos\left(\frac{5\pi}{4}\right) = -250$ A results in:

$$B_g = \frac{\mu_0 Ni}{2l_g} = \frac{4\pi \times 10^{-7} \times 4 \times (-250)}{2 \times 2 \times 10^{-3}} = -0.314\ (\text{Wb m}^{-2})$$

and for $i_s = \sqrt{2} \times 250 \cos(2\pi) = 353$ A results in (figure 8.19):

$$B_g = \frac{\mu_0 Ni}{2l_g} = \frac{4\pi \times 10^{-7} \times 4 \times 353}{2 \times 2 \times 10^{-3}} = 0.444\ (\text{Wb m}^{-2})$$

(b) The *flux per pole,* is given by equation (8.6). As

$$B_g = \frac{\mu_0 Ni}{2l_g} = \frac{4\pi \times 10^{-7} \times 4 \times \sqrt{2} \times 250 \cos(\omega t)}{2 \times 2 \times 10^{-3}} = 0.44 \cos(\omega t)$$

results in, for $\omega = 2\pi f = 377$ rad s^{-1}:

$$\phi_p = B_g \pi Rh = 0.44 \cos(377t) \times \pi \times 0.2 \times 0.5$$
$$\phi_p = 0.14 \cos(377t)\ \text{Wb}$$

(c) The maximum value of the *fundamental* of the magnetic flux density for a concentrated winding is given by:

$$B_{1\max} = \frac{4}{\pi}\frac{\mu_0 Ni}{2l_g} = \frac{4}{\pi} \times 0.44 = 0.56\ (\text{Wb m}^{-2})$$

Based on equation (8.10), the maximum value of the 3-*harmonic* is:

$$B_{3\max} = \frac{1}{3} B_{1\max} = \frac{1}{3} \times 0.56 = 0.19\ (\text{Wb m}^{-2})$$

8.3.2 Winding for more than two magnetic poles

The concentrated winding of figure 8.20 has its sides separated by 180°. When electric current is injected in the coil, two magnetic poles are produced. With reference to the rotor the north pole covers the range from $\theta = 0$ to $\theta = \pi$ and the south pole from $\theta = \pi$ to $\theta = 2\pi$.

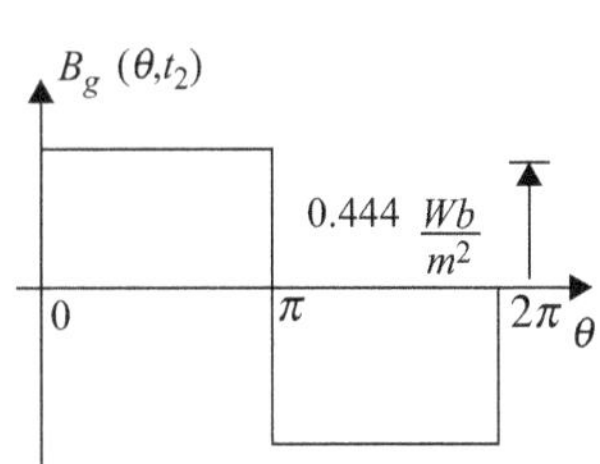

Figure 8.19. Magnetic flux density distribution for $\omega t_2 = 2\pi$ rad s.

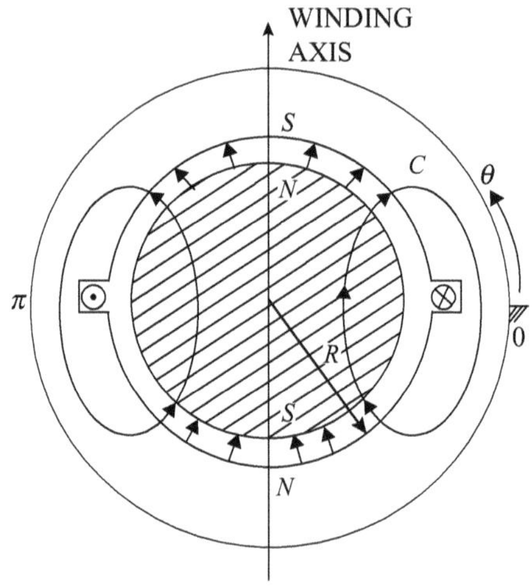

Figure 8.20. 2 poles winding. Only one coil with N turns spanning π rad s.

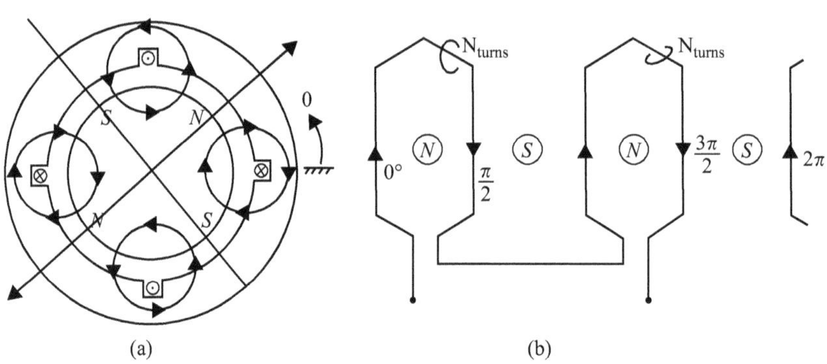

Figure 8.21. 4 poles winding. Two coils with N turns separated by $\frac{\pi}{2}$ rads each.

There are several reasons that makes the stator with more than 2 poles more convenient to have. For example, if we intend to have a 4 pole stator—two north poles and two south poles—it would be sufficient to implement two identical coils whose sides are 90° apart (as figure 8.21 shows). Those coils should be associated in series (or parallel) obeying the coil's polarity to assure the concordance of magnetic flux.

Figure 8.22 shows the magnetic flux density distribution issued from the 4 poles winding. The shape of this distribution is the same as the 2 poles except that each pole covers only 90° instead of the 180° observed in the 2 poles winding.

In conclusion, the magnetic flux density distribution under 2 poles of the 4 poles winding is the same as the 2 poles winding. Only the coil's span is reduced.

Using logical reasoning, if we intend to have a stator wound for $2p$ poles—p north poles and p south poles—it is enough to have p identical coils with N turns where the

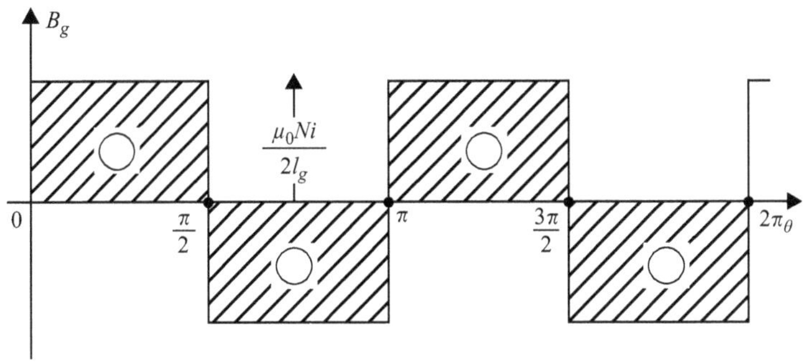

Figure 8.22. Magnetic flux density distribution of 4 poles winding. Each pole covers only $90°\left(\frac{\pi}{2}\text{ rads}\right)$.

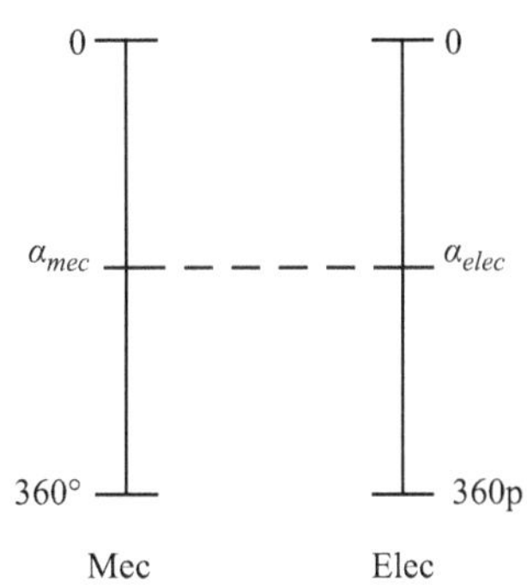

Figure 8.23. The relation of α_{mec} and α_{elec}.

sides span π/p rads. If all the these coils are associated in series, the total number of turns of the winding, is given by:

$$N_s = pN \tag{8.23}$$

(i) N_s: number of turns per coil associated in series;
(ii) p: number of pairs of poles;
(iii) N: number of turns per coil per pair of poles.

8.3.3 The electrical angle

As we observed previously, the magnetic flux density distribution repeats under every two stator poles and hence it is not necessary to represent the field distribution in all of the stator circumference. The representation of only 2 poles is enough for a complete understanding of the stator magnetic behavior.

It is convenient to suppose that the span angle of two single poles is 360 electrical degrees.

In conclusion, for a winding with p pairs of poles, the complete circumference of this stator will have '360p' electrical degrees. Through the figure 8.23, we can establish the relation of both the mechanical (α_{mec}) and the electrical angle (α_{elec}).

Using the proportionality:

$$\frac{\alpha_{\text{mec}} - 0}{360 - 0} = \frac{\alpha_{\text{elec}} - 0}{360p - 0}$$

resulting in:

$$\alpha_{\text{elec}} = p\alpha_{\text{mec}} \tag{8.24}$$

Therefore, the graphical representation of the magnetic flux density distribution in 2 poles, considering the distribution in the stator winding of a p pair of poles is presented in figure 8.24.

From (8.23), $N = N_s/p$,we can calculate the magnitude of the magnetic flux density distribution by:

$$B_g = \frac{\mu_0 N_s i}{2pl_g} \tag{8.25}$$

If the winding is both distributed and fractioned in q coils with (N_s/qp) turns each, the magnitude of the *fundamental* component of the magnetic flux density distribution is given by:

$$B_{1\text{max}} = \frac{4}{\pi}\frac{\mu_0 N_s i}{2pl_g}k_w \tag{8.26}$$

and the magnitude of the h-harmonic yields:

$$B_{\text{hmax}} = \frac{4}{h\pi}\frac{\mu_0 N_s i}{2pl_g}k_{wh} \tag{8.27}$$

Remarks: All angles involved in the evaluation of k_w and k_{wh}, Δ and δ, should be expressed in *electrical degrees*.

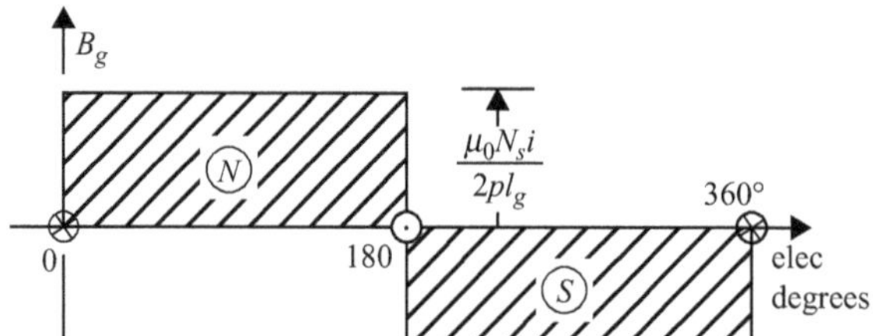

Figure 8.24. The magnetic flux density distribution of 2 poles in electrical degrees. The coil number of turns from (8.23) is $N = \frac{N_S}{p}$.

Example 8.2

A 4 poles SM stator with 24 slots uniformly distributed has an inner diameter of 30 cm and a length of 25 cm. Only eight slots will be used for installing a 4 poles distributed AC winding. The maximum value of the *fundamental* of the magnetic flux density distribution should be 0.7 Wb m^{-2}, when the winding carries 40 A.

(a) Describe how these coils are distributed in the stator slots for creating 4 poles (two north poles + two south poles) showing its connections (use a plan representation). Assume that each coil spans a pole pitch (180°).

(b) In part (a), each slot houses one side of the coil and only two groups of coils are enough to establish the 4 poles required. Redistribute these coils to have four groups showing how they are connected.

Tip: This kind of arrangement is called double-layer distributed winding.

(c) Evaluate the number of turns in series necessary to establish the maximum value of the *fundamental* component of the magnetic flux density distribution, specified by the current required. Under following constraints:

  (i) The winding span is six slot-pitches, also called *full-pitch.*

  (ii) The winding span is five slot-pitches.

(d) Evaluate the maximum value of both the third and the fifth harmonics of the magnetic flux density distribution.

(e) Compare the maximum value of the fundamental, third and fifth harmonics of magnetic flux density distribution in both cases of windings with the concentrated winding.

Solutions

(a) Describe the winding.

As we have 24 slots in which to place 4 poles, the result is that the full-pitch winding spans six slots. With eight sides of the coil (two per each pole), the winding must have 4 coils distributed in two groups.

Figure 8.25 shows the developed representation of the north poles group. Only slots 1, 2, 7, 8, 13, 14, 19 and 20 are occupied.

—First group:

The first coil side of the winding starts at slot number 1. As the pole pitch is 6 slots, the other side is housed at slots 7. The second coil starts at slot 2 and finishes at slot 8, completing the first group. Both coils are associated in series.

—Second group:

The first coil of the second group starts at slot 13—12 slots after slot 1—and finishes at slot 19. They are associated in series with the side of the second coil that starts at slot 14. The last side of the winding is located at slot 20.

Those two groups are associated in series and have the same polarity, as is shown in figure 8.25. If the winding has N_s turns in series in each coil and two pairs of poles, it should have $\frac{N_S}{4}$ turns per coil.

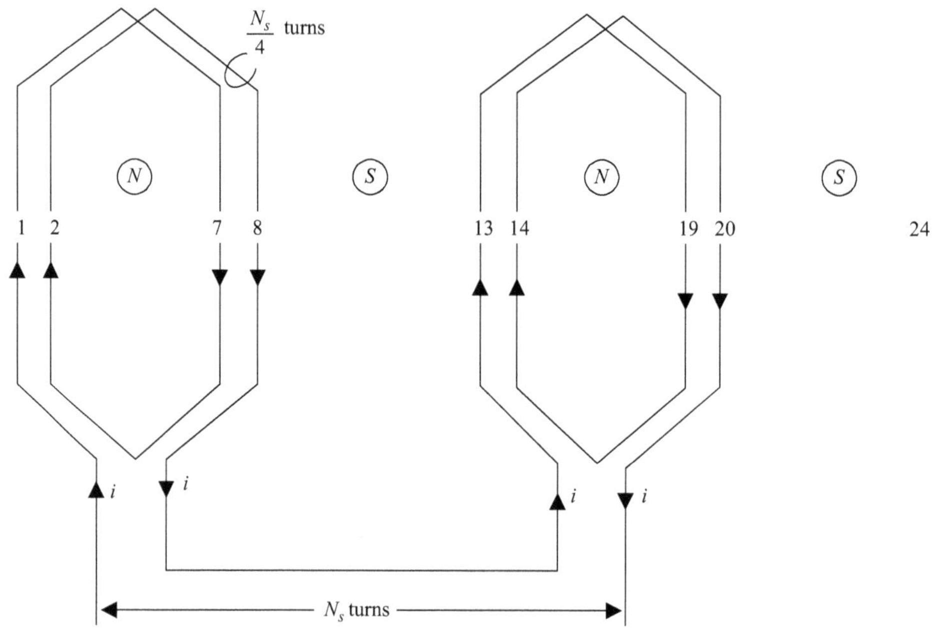

Figure 8.25. 4 poles distributed single layer winding (plan view of one group). One group forms 2 poles $\left(\frac{N_S}{2p}\frac{\text{turns}}{\text{coil}}\right)$.

(b) Double-layer distributed winding.

The double-layer winding is an evolution of the single-layer one. With this procedure two sides of coils are inserted in a slot, as shown figure 8.26. In this case, each coil of the single-layer configuration (with $\frac{N_S}{2p}$ turns) is divided into two coils with $\frac{N_S}{4p}$ turns. Half of them are mirrored.

Figure 8.27 shows how this kind of winding is built. The coil connections should obey the polarity for imposing the same sense of current inside the slot.

(c) Evaluation of the turns and number of the winding.

As we are working with a distributed winding, we need to evaluate its winding factor first.

In the first case, only the distribution factor is necessary, because the winding is a full-pitched span (six slot-pitches).

For 4 poles winding the total amount of electrical degrees in the circumference is:

$$360p = 360 \times 2 = 720° \text{ electrical degrees}$$

As the number of slots is 24 this results in:

$$\Delta = \frac{720}{24} = 30° \text{ electrical degrees}$$

that is, the angle between two consecutive slots.

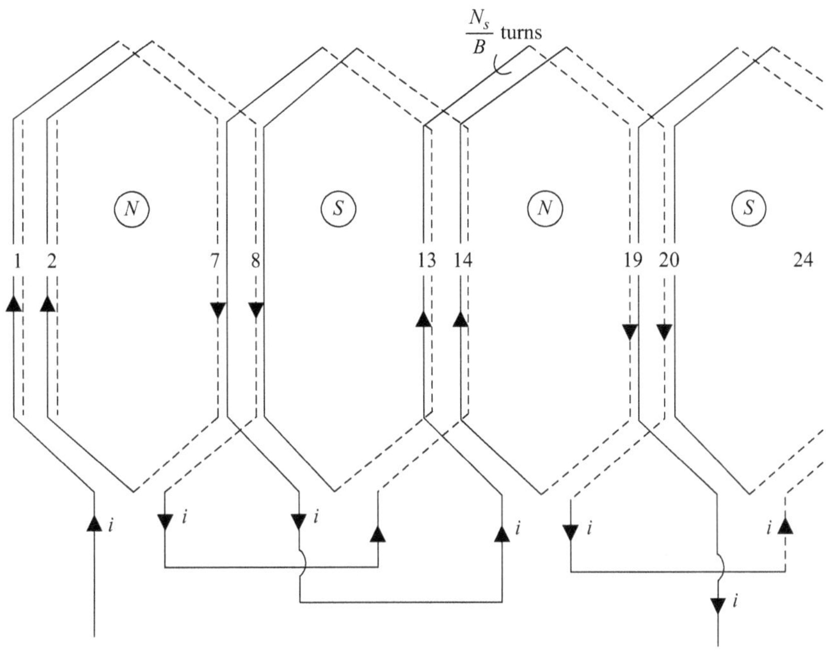

Figure 8.26. 4 poles distributed double-layer winding (plan view). The dotted line represents the side of the coil that is allocated at the bottom of the slot. One group forms 1 pole $\left(\frac{N_S}{4p}\frac{\text{turns}}{\text{coil}}\right)$.

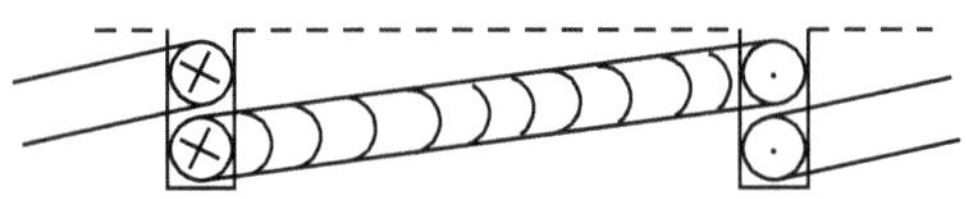

Figure 8.27. Double-layer winding. The current inside the slot must be in the same sense.

Therefore, the distribution factor given by the equation (8.12), yields:

$$k_d = \frac{\sin\left(\frac{q\Delta}{2}\right)}{q\sin\left(\frac{\Delta}{2}\right)} = \frac{\sin\left(\frac{2\times 30}{2}\right)}{2\times\sin\left(\frac{30}{2}\right)} = 0.966$$

Hence:

$$k_w = k_d$$

Applying equation (8.26),

$$B_{1\max} = \frac{4}{\pi}\frac{\mu_0 N_s i}{2pl_g}k_w$$

results in:

$$0.7 = \frac{4}{\pi}\frac{4\pi\times 10^{-7}\times N_s\times 40}{2\times 2\times 2\times 10^{-3}}\times 0.966$$

$$N_s = 91 \text{ turns}$$

For a single layer winding, we have two groups of two coils each, that results in:

$$N_{\text{coil}} = \frac{N_s}{4} = 22.75 \approx 23 \text{ turns}$$

For a double-layer winding, we have four groups of two coils each, that results in:

$$N_{\text{coil}} = \frac{N_s}{8} = 11.375 \approx 11 \text{ turns}$$

In the second case, the pitch coil is five slots that corresponds to:

$$\text{Span coil} = 5 \times 30 = 150° \text{ electrical degrees}$$

as a result:

$$\delta = 180 - 150 = 30° \text{ electrical degrees}$$

Note that in this case $\delta = \Delta = 30°$ because the winding is short pitched only by one slot.

Based on (8.15) the *pitch factor* is given by:

$$k_p = \cos\left(\frac{\delta}{2}\right) = \cos\left(\frac{30}{2}\right) = 0.966$$

Therefore, the winding factor becomes:

$$k_w = k_d \times k_p = 0.966 \times 0.966 = 0.933$$

Applying the same procedure, the number of turns in series results in:

$$N_s = 93 \text{ turns}$$

For a double-layer winding, where we have four groups of two coils each, that results in:

$$N_{\text{coil}} = \frac{N_s}{8} = 11.625 \approx 12 \text{ turns}$$

Remarks: This procedure avoids having slots of different sizes for a short pitched winding (non-null short pitched angle) and it is only applied in a double-layer winding.

(d) The maximum value of both the third and the fifth harmonic of the magnetic flux density distribution.

From (8.21), the maximum value of the h-harmonic of magnetic field distribution is given by:

$$B_{\text{hmax}} = \frac{4}{h\pi}\frac{\mu_0 Ni}{2l_g}k_{wh}$$

The winding factor is obtained applying (8.14) and (8.17). The distribution factor is:

$$k_{d3} = \frac{\sin\left(\frac{3q\Delta}{2}\right)}{q\sin\left(\frac{3\Delta}{2}\right)} = \frac{\sin\left(\frac{3\times 2\times 30}{2}\right)}{2\times\sin\left(\frac{3\times 30}{2}\right)} = 0.707$$

The pitch factor:

$$k_{p3} = \cos\left(\frac{3\delta}{2}\right) = \cos\left(\frac{3\times 30}{2}\right) = 0.707$$

and the winding factor for the third harmonic:

$$k_{w3} = k_{d3}\times k_{p3} = 0.5$$

Consequently,

$$B_{3\max} = \frac{4}{3\pi}\frac{4\pi\times 10^{-7}\times 12\times 8\times 40}{2\times 2\times 10^{-3}}\times 0.5 = 0.28\ \text{Wb m}^{-2}$$

Applying the same procedure for evaluating the maximum value of the fifth harmonic, results in:

$$B_{5\max} = 0.02\ \text{Wb m}^{-2}$$

(e) Comparing the magnitude of the both concentrate and distributed winding we obtain the following results.

For a concentrated winding, the maximum value of the fundamental is given by:

$$B_{1\text{maxconc}} = \frac{4}{\pi}\frac{\mu_0 N_s i}{2pl_g} \tag{8.28}$$

and the maximum value of the fundamental of a distributed winding is:

$$B_{1\max} = \frac{4}{\pi}\frac{\mu_0 N_s i}{2pl_g}k_w \tag{8.29}$$

Dividing (8.29) by (8.28), we have:

$$\frac{B_{1\max}}{B_{1\text{maxconc}}} = k_w = 0.933$$

Applying the same procedure for comparing the same effect in both the third and fifth harmonic, results in:

$$\frac{B_{3\max}}{B_{3\text{maxconc}}} = k_{w3} = 0.5$$

Remarks: The results of both the winding distribution and the short pitch span are effective in the reduction of the harmonic components of the magnetic flux density distribution. Although we must pay a penalty for doing it, the performance of the SM is greatly improved.

8.4 The triphasic winding

We can install more than one winding in the stator of the SM. The most common assemblage is the use of three identical windings uniformly separated in the circumference of the stator. Although, the number of windings that can be installed in the stator does not have a limit.

In a 3-phase winding, the total available slot space is divided equally between the phases, and all slots are filled. The separation of two consecutive phases is such that:

$$\text{phase separation} = \frac{360}{m} \text{ electrical degrees}$$

where m is the number of phases, $m = 3$ for a 3-phase winding hence the separation of two consecutive phases is 120°.

Figure 8.28, extracted from example 8.2, shows only one-phase of a 24 slots-4 poles double-layer winding.

As only 8 slots were used, it is possible to install two more identical windings applying the following procedure:

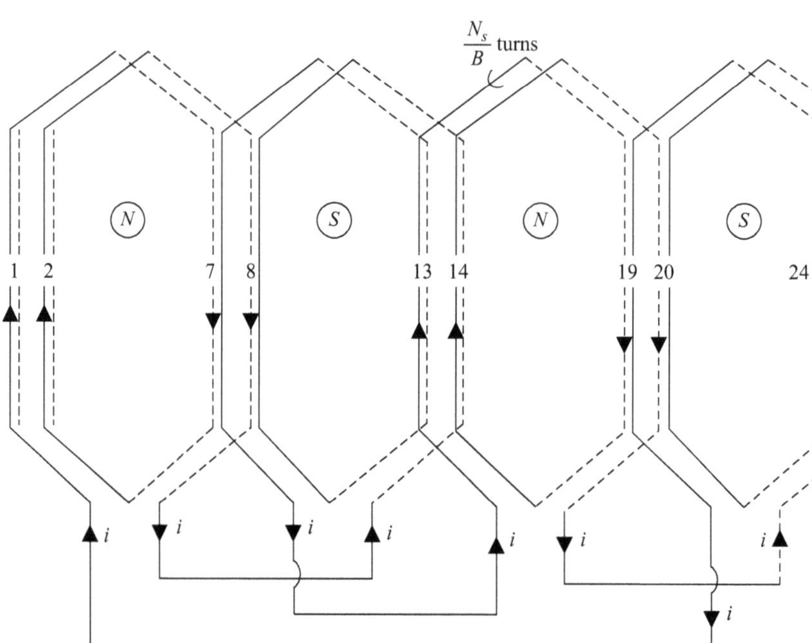

Figure 8.28. 4 poles distributed double-layer winding (plan view).

- As the angle between two consecutive slots is $\Delta = 30°$ electric, the second winding will start at 120° electric—4 slots forward—after the first phase.
- The first coil of the second winding starts at the slot number 5 and finish at slot 11.

We can observe that the position of the second winding is obtained by shifting the entire first winding 4 slots forward. The third winding is also obtained by shifting the entire second winding 4 slots forward. As the second winding starts at slot number 5, the first coil of the *third* winding will start at slot number 9.

Figure 8.29 shows a sketch of three windings for which the starts are A, B and C.

Remarks: For an m-phases winding it is easy to identify that:

$$q = \frac{N_r}{2pm} \tag{8.30}$$

where:

N_r: number of slots;
m: number of phases;
p: number of pair of poles.

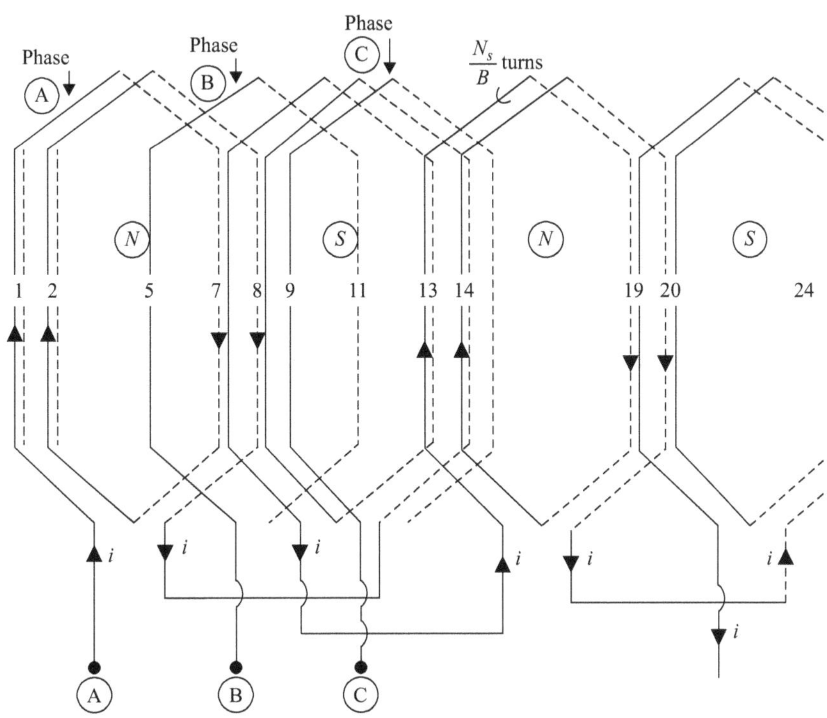

Figure 8.29. 4 poles, 3-phase distributed double-layer winding. Only the beginning of the windings B and C are shown, the winding A is fully represented.

8.4.1 The magnetic flux density distribution of a 3-phase winding

As we discussed previously, the magnetic flux density distribution for both a distributed and short pitched winding is virtually sinusoidal because the magnitude of harmonics is strongly damped with this technique.

Therefore, we can consider that the magnetic flux density distribution is composed only by its *fundamental*.

Figure 8.30 shows a representation of a 2 poles, 3-phase winding. i_a is the electric current of the winding of phase A and i_b and i_c are the currents of phases B and C respectively.

The magnetic flux density distribution produced by the phase A winding is given by:

$$B_a(\theta) = \frac{4}{\pi}\frac{\mu_0 N_s i_a}{2pl_g}k_w \sin\theta \tag{8.31}$$

A similar expression is obtained for the windings of phases B and C

$$B_b(\theta) = \frac{4}{\pi}\frac{\mu_0 N_s i_b}{2pl_g}k_w \sin(\theta - 120°) \tag{8.32}$$

$$B_c(\theta) = \frac{4}{\pi}\frac{\mu_0 N_s i_c}{2pl_g}k_w \sin(\theta - 240°) \tag{8.33}$$

As the magnetic flux density distribution is radially oriented, the final distribution in the air-gap is given by:

$$B_g(\theta) = B_a(\theta) + B_b(\theta) + B_c(\theta) \tag{8.34}$$

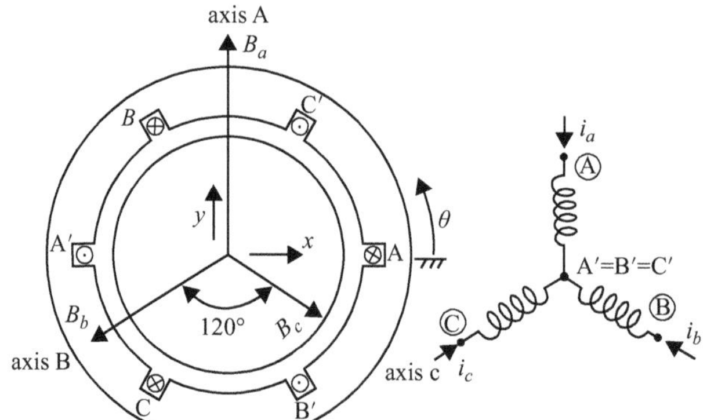

Figure 8.30. SM with 2 poles, 3-phase AC windings wye connected. The stator windings can be associated in either wye or delta.

If these windings are connected to a 3-phase system voltage in a wye or delta association, the resulting currents in each winding are such that:

$$i_a = \sqrt{2}I \cos \omega t,\ i_b = \sqrt{2}I \cos(\omega t - 120°) \text{ and } i_c = \sqrt{2}I \cos(\omega t - 240°)$$

Therefore, the magnetic flux density distribution in the air-gap is:

$$B_g(\theta, t) = \frac{4\sqrt{2}}{\pi}\frac{\mu_0 N_s I}{2pl_g}k_w[\cos \omega t \sin \theta + \cos(\omega t - 120°) \sin(\theta - 120°) + \cos(\omega t - 240°)\sin(\theta - 240°)]$$

After some mathematical manipulation we have:

$$B_g(\theta, t) = \frac{3\sqrt{2}}{\pi}\frac{\mu_0 N_{eff} I}{pl_g}\sin[\theta - \omega t] \quad (8.35)$$

with:

$$N_{\text{eff}} = N_s k_w \quad (8.36)$$

N_{eff}: the winding effective number of turns in series.

8.4.2 Physical analysis of $B_g(\theta, t)$

Expression (8.35) can be written as:

$$B_g(\theta, t) = B_{\max}\sin[\theta - \omega t] \quad (8.37)$$

where:

$$B_{\max} = \frac{3\sqrt{2}}{\pi}\frac{\mu_0 N_{eff} I}{pl_g} \quad (8.38)$$

Remarks: The maximum value of the final magnetic flux density distribution is such that:

$$\frac{B_{\max}}{B_{1\max}} = \frac{3}{2}$$

the magnetic flux density distribution issued from the action of the three AC currents is 1.5 times the magnetic distribution of only one winding excited by AC current ($B_{1\max}$).

The expression of the magnetic flux density distribution is a function of two variables, θ and t. To sketch $B_g(\theta, t)$ as a function of θ only we must specify t.

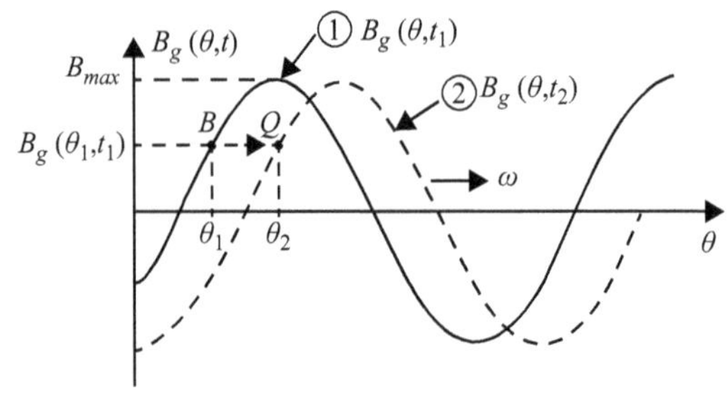

Figure 8.31. The wave of the magnetic field distribution. Curve (1): $B_g(\theta, t_1)$, curve (2): $B_g(\theta, t_2)$. Synchronous speed: $n_s = \frac{f}{p}$ rps.

Curve (1) of figure 8.31 represents the function $B_g(\theta, t_1)$, hence the coordinate of (P) is $B_g(\theta_1, t_1)$. Let us search the position θ_2 in the instant $t_2 > t_1$ such that $B_g(\theta_2, t_2) = B_g(\theta_1, t_1)$. A simple solution is to impose:

$$\theta_2 - \omega t_2 = \theta_1 - \omega t_1$$

Therefore:

$$\theta_2 - \theta_1 = \omega[t_2 - t_1] \tag{8.39}$$

We conclude there is a position $\theta_2 > \theta_1$ at $t_2 > t_1$ such that $B_g(\theta_2, t_2) = B_g(\theta_1, t_1)$. Point (Q) in figure 8.31 represents the value of $B_g(\theta_2, t_2)$.

Using the same approach for all points of curve (1), we can plot curve (2), that represents the function $B_g(\theta, t_2)$.

Comparing both curves, we identify that they have the same shape, indicating that the function $B_g(\theta, t)$ is a representation of a wave of a magnetic flux density distribution that travels in the air-gap with an angular speed given by:

$$\omega_s = \omega(\text{electric radians})/\text{s} \tag{8.40}$$

Representing (8.40) using mechanical angles, results in:

$$\omega_s = \frac{\omega}{p} \quad \text{radians/s} \tag{8.41}$$

ω_s is called the *synchronous angular speed* of the SM.

As $\omega_s = 2\pi n_s$ and $\omega = 2\pi f$, from (8.41), we can write:

$$n_s = \frac{f}{p} \quad \text{rps} \tag{8.42}$$

n_s is called the *synchronous speed* of the SM.

8.4.3 The induced electromotive force—emf

Let us evaluate the induced electromotive force (emf) developed in both a distributed and a short pitched 3-phase winding due to the injection of a 3-phase electrical current.

The emf is easily evaluated substituting the real winding with its equivalent concentrated winding that has an effective number of turns given by:

$$N_{\text{eff}} = N_s k_w$$

Figure 8.32 represents a sinusoidal wave that travels in the air-gap with an angular speed ω(electric radians)/s.

In this figure, we can identify:

(1) The sinusoidal wave of magnetic flux density distribution described by:

$$B_g(\theta, t) = B_{\max}\sin[\theta - \omega t] \tag{8.43}$$

This is the magnetic flux density distribution issued from a polyphase winding fed by a polyphase voltage system.

(2) The equivalent full-pitch winding with $N_{\text{eff}} = N_s k_w$.

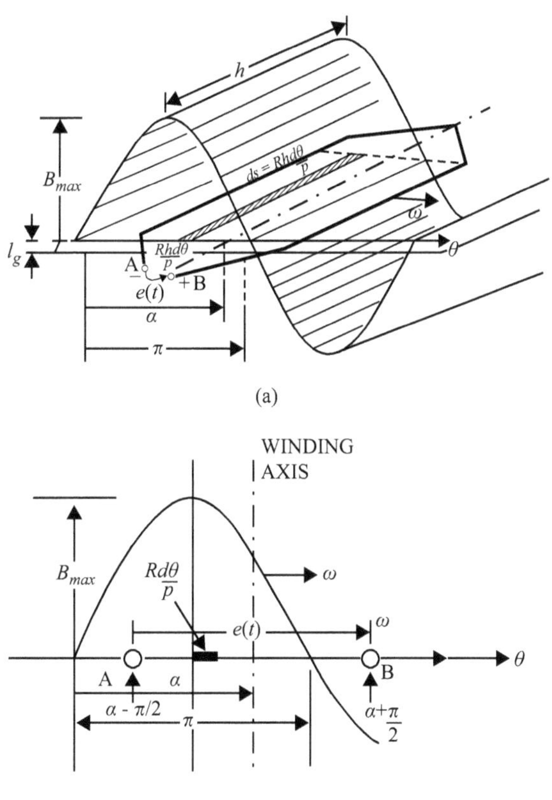

Figure 8.32. A concentrated winding under a rotating magnetic field. (a) The 3D geometry of the problem. (b) The 2D cross section.

The magnetic flux through the winding is given by:

$$\phi(\theta) = \int_{\alpha_1}^{\alpha_2} B_g(\theta, t) Rh d\theta_{\text{mec}}$$

as:

$$d\theta_{\text{mec}} = \frac{d\theta}{p}$$

resulting in:

$$\phi(\theta) = \int_{\alpha-\frac{\pi}{2}}^{\alpha+\frac{\pi}{2}} B_g(\theta, t) Rh \frac{d\theta}{p} \tag{8.44}$$

Substituting $B_g(\theta, t)$ by its value, we get:

$$\phi(\theta) = \int_{\alpha-\frac{\pi}{2}}^{\alpha+\frac{\pi}{2}} B_{\max} \sin[\theta - \omega t] Rh \frac{d\theta}{p} \tag{8.45}$$

After the integration the result is:

$$\phi(t) = -2B_{\max} \frac{Rh}{p} \sin(\omega t - \alpha)$$

The magnitude:

$$\phi_p = 2B_{\max} \frac{Rh}{p} \tag{8.46}$$

is called the *flux per pole*. So,

$$\phi(t) = -\phi_p \sin(\omega t - \alpha) \tag{8.47}$$

Based on Faraday's law the emf is given by:

$$e(t) = -N_{\text{eff}} \frac{d\phi(t)}{dt}$$

Therefore:

$$e(t) = \omega N_{\text{eff}} \phi_p \cos(\omega t - \alpha) \tag{8.48}$$

or:

$$e(t) = \sqrt{2} E_{\text{ph}} \cos(\omega t - \alpha) \tag{8.49}$$

where:

$$E_{\text{ph}} = \frac{1}{\sqrt{2}} \omega N_{\text{eff}} \phi_p$$

is the rms value of emf. As $\omega = 2\pi f$ we can write:

$$E_{\mathrm{ph}} = 4.44 f N_{\mathrm{eff}} \phi_p \tag{8.50}$$

We can conclude, a sinusoidal emf is generated by a rotating sinusoidal magnetic flux density distribution whose frequency depends on the angular speed of the field.

8.4.4 The synchronous reactance

We can rewrite (8.50) as:

$$E_{\mathrm{ph}} = 4.44 f N_{\mathrm{eff}} 2 B_{\max} \frac{Rh}{p}$$

Substituting $B_{\max}$ expressed in (8.38) results in:

$$E_{\mathrm{ph}} = 4.44 f N_{\mathrm{eff}} 2 \frac{3\sqrt{2}}{\pi} \frac{\mu_0 N_{\mathrm{eff}} I}{p l_g} \frac{Rh}{p}$$

The relation:

$$X_s = \frac{E_{\mathrm{ph}}}{I} = \frac{12 \times N_{eff}^2 \times f \times \mu_0 \times R \times h}{p^2 \times l_g} (\Omega) \tag{8.51}$$

is called the *synchronous reactance* of the winding.

The electric circuit of figure 8.33 is the equivalent electric circuit per phase of a polyphase distributed and short pitched AC winding.

Example 8.3

Figure 8.34 shows a cross section of a SM stator that has 36 slots uniformly distributed, a 30 cm inner diameter, a 25 cm length and a 2 mm air-gap.

A 4 poles, 3-phase, double-layer winding should be designed to develop at 60 Hz an induced emf such that the line-to-line voltage is 760 V wye connected. A short pitched technique should be applied to mitigate the seventh harmonic. Due to magnetic limitations, the maximum magnitude of the magnetic flux density could not be higher than 0.7 (Wb m^{-2}). Let us determine:

(a) the number of conductors inside each slot and the number of turns of each coil of the winding;
(b) the electric current for establishing the magnetic flux density required;
(c) the rotational speed of the magnetic field in rpm.

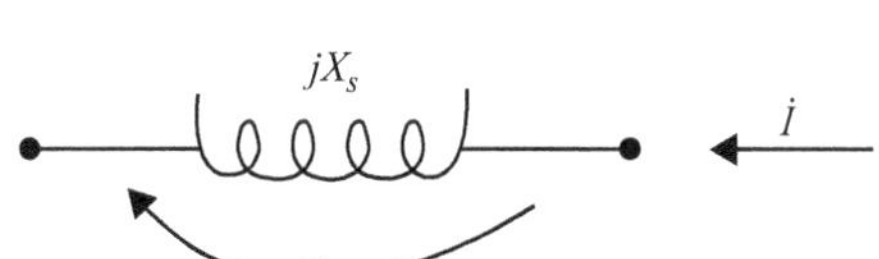

Figure 8.33. The synchronous reactance.

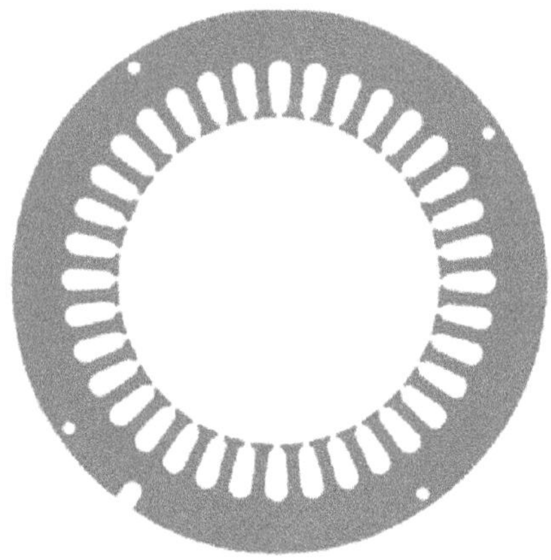

Figure 8.34. Stator of example 8.3.

Solutions

(a) The turns number of the winding

Based on (8.50), the induced emf is given by:

$$E_{ph} = 4.44 f N_{eff} \phi_p$$

As the line-to-line voltage is 760 V–60 Hz wye connected the induced voltage per phase will be:

$$E_{ph} = \frac{760}{\sqrt{3}} = 439 \text{ V}$$

The maximum admissible *flux per pole* should be:

$$\phi_p = 2B_{max}\frac{Rh}{p} = 2 \times 0.7 \times \frac{0.25 \times 0.15}{2} = 0.0265 \text{ Wb}$$

As the stator has 36 slots the slot pitch Δ is such that:

$$\Delta = \frac{360 \times 2}{36} = 20° \text{ elec.deg}$$

The winding short pitching for eliminating the seventh harmonic should be:

$$k_{p7} = \cos\left(\frac{7 \times \delta}{2}\right) = 0$$

that results in:

$$\delta = 25.7°$$

As $\delta = n\Delta$ the more suitable value is to use $= \Delta = 20°$ (20° = 1 slot). Therefore:

$$k_p = \cos\left(\frac{\delta}{2}\right) = \cos\left(\frac{20°}{2}\right) = 0.984$$

From (8.30), we get:

$$q = \frac{N_r}{2pm} = \frac{36}{2 \times 2 \times 3} = 3$$

therefore:

$$k_d = \frac{\sin\left(\frac{q\Delta}{2}\right)}{q\sin\left(\frac{\Delta}{2}\right)} = \frac{\sin\left(\frac{2 \times 20}{2}\right)}{3 \times \sin\left(\frac{20}{2}\right)} = 0.958$$

and the winding factor yields:

$$k_w = k_d \times k_p = 0.944$$

from:

$$E_{\text{ph}} = 4.44 f N_{\text{eff}} \phi_p$$

resulting in:

$$N_{\text{eff}} = \frac{E_{\text{ph}}}{4.44 f \phi_p} = \frac{439}{4.44 \times 60 \times 0.0265} = 621 \text{ turns}$$

Consequently, the series turns coil is such that:

$$N_{\text{eff}} = N_s k_w$$

that results in:

$$N_s = \frac{N_{\text{eff}}}{k_w} = \frac{621}{0.944} = 657 \text{ turns}$$

This is the first approach for the number of series turns of the winding. As we have 12 slots per phase, the number of conductors inside each slot should be:

$$N_{\text{cond}} = \frac{2 \times 657}{12} = 109.5 \text{ conductors/slot}$$

but N_{cond} must be a whole even number. So, we must choose:

$$N_{\text{cond}} = \frac{2 \times 657}{12} = 110 \text{ conductors/slot}$$

As we have two sides of coil inside each slot, the number of turns of each partial coil of the winding will be 55 conductors and the number of series turns of the winding, considering these constraints, should be:

$$N_s = \frac{N_{\text{eff}}}{k_w} = \frac{621}{0.944} = 660 \text{ turns}$$

We should correct the peak of magnetic flux density making:

$$B_{1\max} = \frac{657}{660} \times 0.7 = 0.697 \text{ Wb m}^{-2}$$

(b) The electric current in the winding

Based on (8.38), the electric current for establishing the magnetic flux density required should be:

$$I = \frac{\pi p l_g}{3\sqrt{2}\mu_0 N_{\text{eff}}} B_{\max} = \frac{\pi \times 2 \times 2 \times 10^{-3}}{3\sqrt{2} \times 4\pi \times 10^{-7} \times 660 \times 0.944} \times 0.697 = 2.6 \text{ A}$$

(c) The rotational speed of the field

From (8.42), we obtain:

$$n_s = \frac{f}{p} = \frac{60}{2} = 30 \text{ rps (1800 rpm)}$$

8.5 The rotor winding

As we discussed previously, the winding located in the rotor of SM is fed by a DC electrical source. As occurred in the stator, any mmf produced is associated with a magnetic flux density distribution in the air-gap.

8.5.1 The field winding of a non-salient pole machine

Due to its cylindrical geometry, a non-salient pole produces a magnetic flux density distribution that has the same shape as the mmf distribution imposed by the carried electric current. In this case, we can write that:

$$B_g = \mu_0 H_g = \frac{\mu_0}{2l_g}\mathcal{F} \tag{8.52}$$

Figure 8.35 shows the rotor winding of a 2-non-salient pole SM. The stator windings are omitted for convenience. All conductors of the right side carry electric currents going perpendicularly into (⊗) the sheet and all conductors of the left side carry electric currents going out of (⊙) the sheet.

Supposing that each coil under the pole pitch has a different number of turns (N_1, N_2, N_3), as figure 8.35(a) is showing, the shape of the mmf distribution is the one shown in figure 8.35(c).

To have a distribution as near as possible to sinusoidal shape, the number of each partial coil of the rotor should be carefully chosen. It is also possible but not common to have an almost mmf sinusoidal distribution with the same number of conductors inside the slots but with different angle separation of the slots.

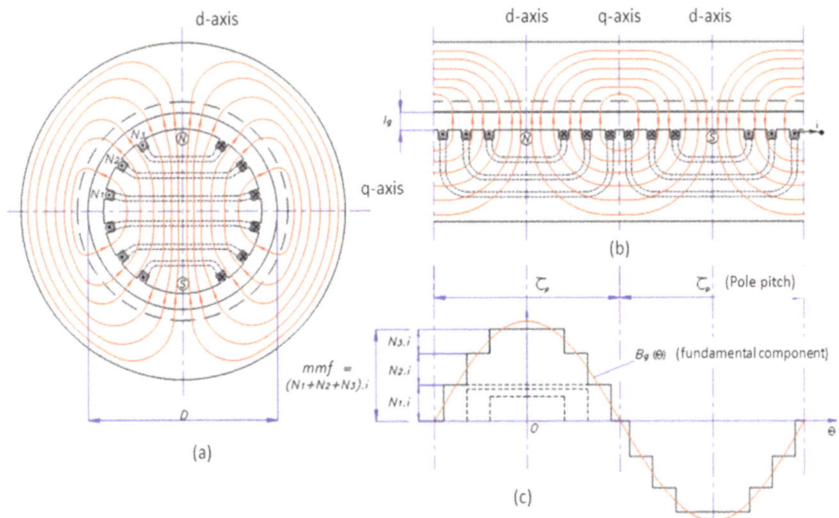

Figure 8.35. The rotor winding in a 2-non-salient pole SM. (a) Cross section of SM omitting the stator windings. (b) Developed representation of both mmf and B_g distribution. (c) Plot representation of both mmf and the fundamental of B_g in the air-gap. (Courtesy of Professor Ivan Eduardo Chabu).

8.5.2 The field winding of a salient pole machine

In the salient pole SM, the field coil is concentrated in the pole core producing an mmf distribution like the one shown figure 8.36(c). To have a magnetic flux density distribution as near as possible to a sinusoidal shape, the geometry of the pole piece should introduce a variable air-gap in the circumference. The minimum air-gap is located at the center of the pole. At the border of the pole piece, the length of the air-gap is big enough to establish as closely as possible a sinusoidal magnetic flux density distribution.

8.5.3 The effect of the rotor winding in the SM operation

As the field winding is fed by DC electric current, the magnetic flux density distribution is static regarding the rotor. As static field does not induce any emf in the stator windings, the rotor should rotate imposing a time dependent magnetic flux density distribution like the one established by 3-phase winding fed by triphasic voltage system.

The rms value of the induced emf in the AC winding due to the magnetic flux density distribution produced in the rotor is given by:

$$E_{\mathrm{ph}} = 4.44 f N_{\mathrm{eff}} \phi_p \tag{8.53}$$

with

$$f = p n_s \tag{8.54}$$

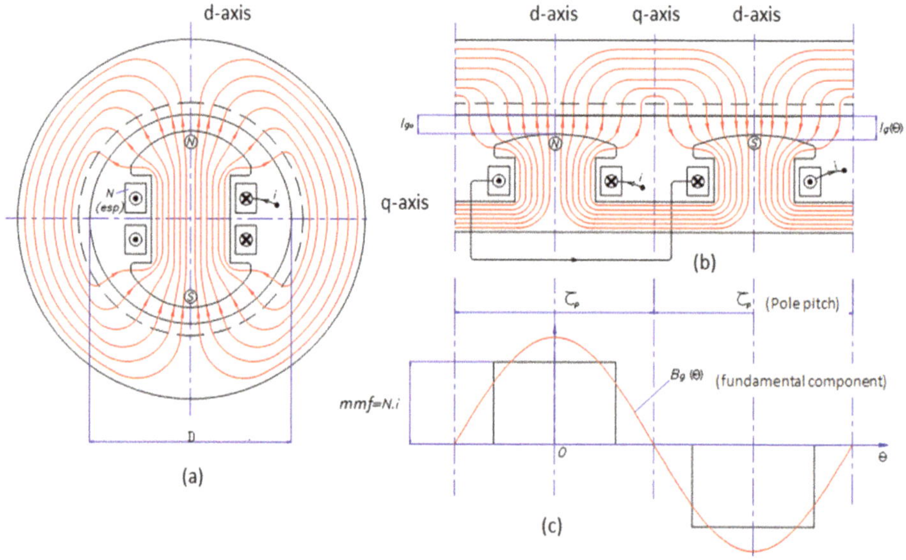

Figure 8.36. The rotor winding in a 2-salient pole SM. (a) Cross section of a SM omitting the stator windings. (b) Developed representation of B_g distribution. (c) Plot representation of both mmf and the fundamental of B_g in the air-gap. (Courtesy of Professor Ivan Eduardo Chabu).

Example 8.4

A cylindrical rotor of a 2-non-salient pole SM has 12 uniformly distributed slots (figure 8.37). Determine the number of turns of each coil laid in these slots in such a way that the distribution of conductor around the rotor periphery approximates as closely as possible the ideal sinusoidal magnetic flux density distribution (figure 8.38). N_s is the total number of turns of the field winding.

Solution

The sinusoidal mmf distribution is described by:

$$\mathcal{F}(\theta) = N_p i_f \sin\theta \tag{8.55}$$

where

N_p: the field winding number of turns per pole;
i_f: the DC field electric current.

As the slots are uniformly distributed around the cylindrical rotor, the first slot is located at the position $\theta = 15°$ and the next one at $\theta = 45°$. Figure 8.39 shows both the sinusoidal mmf distribution and the idealized mmf distribution from $\theta = 15°$ to $\theta = 45°$.

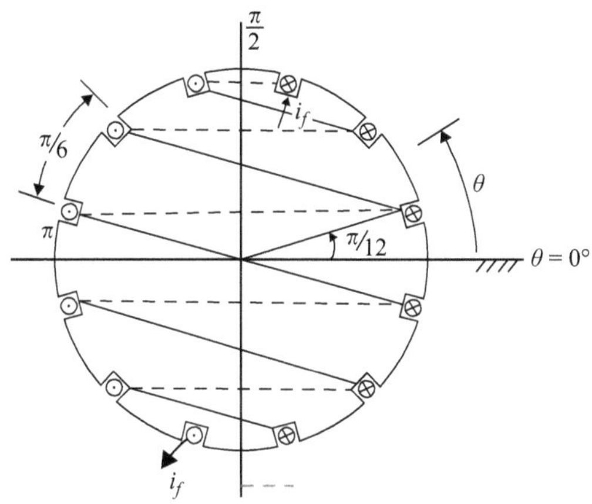

Figure 8.37. The rotor winding of the 12 slot 2-non-salient pole SM. The stator is omitted.

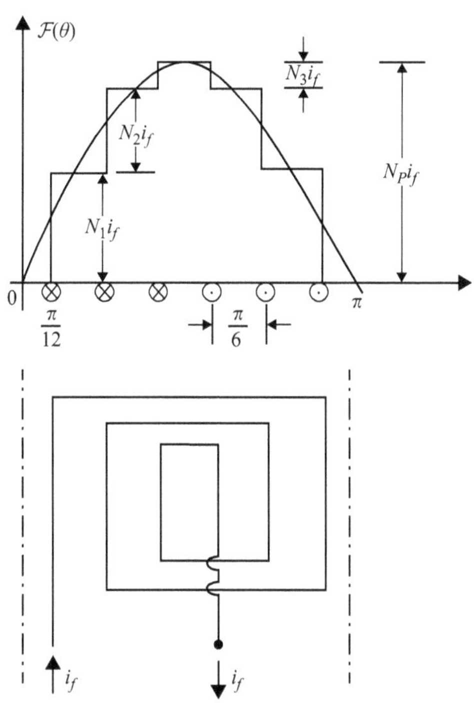

Figure 8.38. Developed 12 slot 2-non-salient pole SM.

The number of conductors inside the slot located at $\theta = 15°$ should be the one that establishes the same mmf distribution equivalent to the average value of the sinusoidal one in the range from $\theta = 15°$ to $\theta = 45°$. Therefore, we can write:

$$N_1 i_f = \frac{6}{\pi} \int_{15°}^{45°} N_p i_f \sin \theta d\theta$$

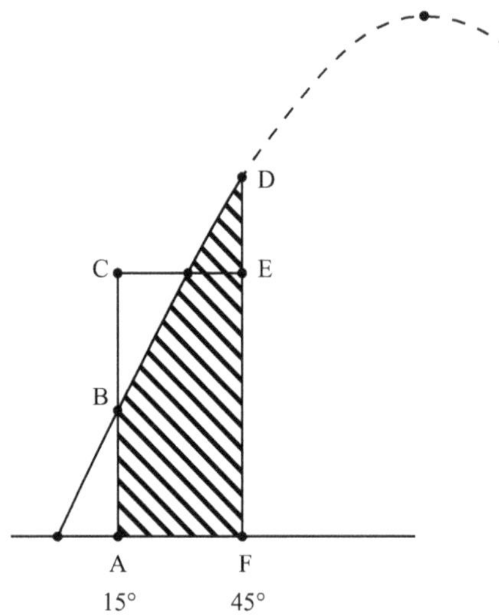

Figure 8.39. The average mmf from $\theta = 15°$ to $\theta = 45°$. Area ABDFA = area ACEFA.

that results in:

$$N_1 = 0.494N_p$$

For the slot located at $\theta = 45°$, we can write:

$$(N_1 + N_2)i_f = \frac{6}{\pi}\int_{45°}^{75°} N_p i_f \sin\theta d\theta$$

that results in:

$$N_2 = 0.361\ N_p$$

Finally, for the slot located at $\theta = 75°$ that results in:

$$(N_1 + N_2 + N_3)i_f = N_p i_f$$

therefore,

$$N_3 = 0.144\ N_p$$

8.6 Summary

This is the first chapter introducing the synchronous machine (SM). We started with the design aspects of a SM where we have made an analysis of its components: the stator that hosts an AC polyphase winding and the rotor that presents two kinds of construction, namely non-salient or salient pole. The non-salient pole is suitable for the high-speed SM, such as the one used in nuclear plants and the salient pole that is largely used in hydroelectric plants. The fundamentals of SM operation were shown when we discussed how the SM works, that showed the interaction of magnetic flux density distribution and the movement of the rotor. The performance of the SM is highly dependent on the shape of the magnetic flux density distribution in the air-gap. Therefore, all care should be taken for making an AC winding that produces a (virtually) sinusoidal distribution of magnetic field.

The most important point of the chapter is the production of a sinusoidal magnetic flux density distribution that moves in the air-gap at constant speed, called *synchronous speed*, that depends on the frequency of the AC currents.

A static sinusoidal magnetic flux density distribution produced by the rotor winding, excited by DC electric current, is also discussed at the end of the chapter. The influence of the pole shape in the magnetic flux density distribution created by the rotor winding in a non-salient pole machine and the salient pole SM were detailed. The static magnetic field distribution needs a movement of the rotor to induce an emf in the stator AC windings.

8.6.1 Project

The stator of an industrial non-salient synchronous motor exhibits the following data:

Internal diameter: 760 mm
Air-gap: 5 mm
Net length: 650–660–670–680–690 mm (choose one option)
Number of slots: 96
Number of poles: 8
Frequency of the 3-phase AC voltage source: 60 Hz
Maximum of magnetic flux density in the air-gap: $0.8T$ ($\pm 5\%$)
Design a wye 3-phase winding for this stator considering that the line-to-line applied voltage is 4800 V and all coils are series connected.

Figure 8.40 shows the details of the stator stamp and the complete information of the slot dimensions.

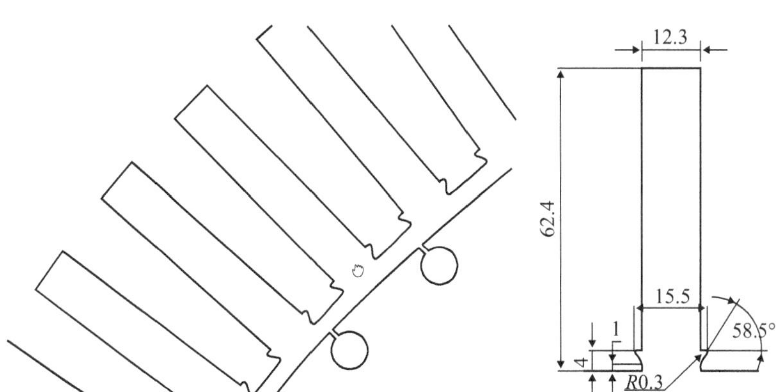

Figure 8.40. Draft for project. Left: cut view of the stator slots. Right: slot dimensions.

The project requires:

(1) Number of turns per phase of the winding short pitched designed to mitigate the effects of the fifth harmonic. All parameters of the winding should be presented: chord factor, distribution factor, winding factor, number of effective turns, magnetic flux per pole.

(2) The evaluation of the no-load current of the motor, that is the current for establishing the required magnetic flux density in the air-gap.

(3) Supposing that the slot occupation factor is defined by:

$$k_{oc} = \frac{\text{Copper area in the slot}}{\text{Area of the slot}}$$

is around 0.45 (for low voltage this factor can reach 0.6), you are asked to estimate the rated power of the motor. Consider that the current density is around 6 A mm^{-2}.

Problems

8.1 Find a synchronous machine and learn all you can about it from observation. Any kind of synchronous machine may be used: electric generator, electric motor, small synchronous machine, etc. Write about two pages of description, including a diagram. Write as if for an engineering report. Give its purpose, its general size, shape, construction and the provision for cooling. Give information about its electrical operation that is available from the name plate, with any deductions that you are able to make. Possible data to include are voltage, current, power ratings; connections of windings, number of phases, electric material and core laminations, insulation; electrical parameters, etc.

8.2 The stator of a synchronous machine is such that:

3-phase; 4 poles; 36 slots; winding double-layered; 2 slots short pitched and 2-turn for each single coil.

Determine both the winding factor and the effective turns per phase.

Sketch the winding over a pole pitch, indicating the location of the conductor of each phase.

Answer: 0.9; 21.6 turns.

8.3 Supposing that the three-phase winding of problem 8.2 carries a three-phase current whose magnitude is 50 A (rms). Compute:

(a) the maximum value of fundamental of mmf wave;

(b) the maximum value of the third harmonic of the mmf wave;

(c) compare both values with the one issued from a three-phase full-pitch double-layered winding.

Answers: (a) 1460 A; (b) 18 A; (c) 6.5% and 99%.

8.4 The three-phase winding of problem 8.3 is built in a stator of a non-salient synchronous machine with the following dimensions:

diameter 40 cm, length 25 cm and air-gap of 2 mm.

Compute both the magnitude of the fundamental component of the magnetic flux density and the magnetic flux per pole.

Answer: 0.46 Wb m^{-2}; 230 × 10^{-4} Wb.

8.5 Supposing that all coils of each phase of the winding of problem 8.4 are series connected, compute the induced emf (rms) generated by the current of 50 A–60 Hz.

Answer: 132 V.

8.6 A cylindrical rotor of a 4-non-salient pole SM that has 24 uniformly distributed slots. Determine the number of turns of each coil laid in theses slots in such a way that the distribution of conductor around the rotor periphery approximates as closely as possible the ideal sinusoidal magnetic flux density distribution. Let 193 be the total number of turns of the field winding.

Answer: 52; 89; 52 turns.

8.7 The cylindrical rotor of problem 8.5 is installed in a non-salient synchronous machine whose internal diameter is 40 cm and the air-gap is 2 mm. Compute the DC current that should be injected in the rotor winding to establish a sinusoidal magnetic flux density that has a maximum value of 0.46 Wb m^{-2}.

Answer: 7.6 A.

8.8 The cylindrical rotor of problem 8.6 is 25 cm long. Compute the induced emf (rms) generated by the field current evaluated in problem 8.7 when the rotor is spinning at 1500 rpm

Answer: 110 V.

8.9 A 2-pole 3-phase winding is installed in a stator of a linear synchronous machine of a magnetic levitation train. This stator is a developed round stator, as figure 8.41 shows. The length of the device is 1500 mm.

The winding is fed by a 3-phase–50 Hz-AC voltage source.

Evaluate the linear speed in km h^{-1} of the magnetic flux density distribution.

Answer: 270 km h^{-1}.

8.10 Determine the length of a 4-pole linear stator of a magnetic levitation train that is fed by an AC voltage source of 140 Hz and develops a linear speed of 200 km h^{-1}.

Answer: 80 cm.

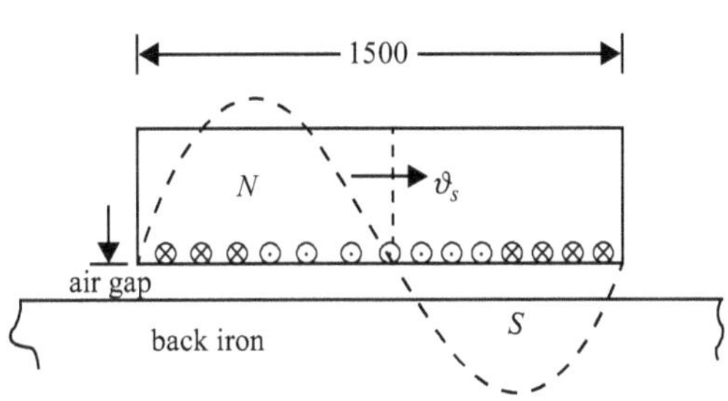

Figure 8.41. Stator of a linear synchronous machine. Dimensions in millimeters.

Further reading

[1] Langsdorf A S 1955 *Theory of Alternating Current Machinery* 2nd edn (New York: McGraw-Hill)
[2] Fitzgerald A E, Kingsley C Jr and Umans S D 1992 *Electric Machinery* 5th edn (New York: McGraw-Hill)
[3] Meisel J 1984 *Principles of Electromechanical-Energy Conversion* (Malabar, FL: R.E. Krieger)
[4] Kuhlmann J H 1940 *Design of Electrical Apparatus* 2nd edn (New York: Wiley)
[5] Nasar S A and Unnewehr L E 1983 *Electromechanics and Electrical Machines* 2nd edn (New York: Wiley)
[6] Slemon G R and Straughen A 1980 *Electric Machines* (Reading, MA: Addison-Wesley)
[7] Skilling H H 1962 *Electromechanics: A First Course in Electromechanical Energy Conversion* (New York: Wiley)
[8] Krause P, Wasynczuk O and Pekarek S 2012 *Electromechanical Motion Devices* 2nd edn (Hoboken, NJ: Wiley)
[9] Hughes A and Drury B 2013 *Electric Motors and Drives: Fundamentals Types, and Applications* 4th edn (Amsterdam: Elsevier)
[10] Jordao R G 1980 *Maquinas Sincronas* (Sao Paulo: LTC/EDUSP) (in Portuguese)
[11] Lipo T A 2017 *Introduction to AC Machine Design* 1st edn (New York: Wiley)
[12] Boldea I and Syed A N 2010 *The Induction Machines Design Handbook* 2nd edn (Boca Raton, FL: CRC Press)
[13] Gieras J F 2016 *Electrical Machines: Fundamentals of Electromechanical Energy Conversion* 1st edn (Boca Raton, FL: CRC Press)
[14] Gonen T 2012 *Electrical Machines with MATLAB®* 2nd edn (Boca Raton, FL: CRC Press)
[15] Mohan N 2012 *Electric Machines and Drives: A First Course* 1st edn (New York: Wiley)
[16] Pyrhonen J, Jokinen T and Hrabovcova V 2013 *Design of Rotating Electrical Machines* 2nd edn (New York: Wiley)

IOP Publishing

Electromechanical Energy Conversion Through Active Learning

J R Cardoso, M B C Salles and M C Costa

Chapter 9

Synchronous machine: operation

9.1 Introduction

The previous chapter introduced the basic principles of SM operation emphasizing not only its components, but also the magnetic effect issued by the electrical current carried by its winding combining with the movement of the rotor. Two independent phenomena can be identified when the SM is running.

The first is the induction of a sinusoidal time dependent electromotive force (emf) in the stator winding. This emf is induced due to the combination of the magnetomotive force (mmf) produced by the DC electric current injected in the field winding that produces a magnetic field distribution and the movement of the rotor.

The second is the self-induction of an emf in the stator winding due to the circulation of currents in its own coils in an *m-phase* system (commonly 3-phase).

When the emf is induced by the moving field winding, the frequency is given by:

$$f = pn_s \tag{9.1}$$

where:

n_s is the rotor speed in r s^{-1},
p: number of pairs of poles.

When the emf is self-induced by the AC electrical current carried by its own coils, the frequency of induced emf is the same as the current, and the rotational speed of the magnetic field wave will be:

$$n_s = \frac{f}{p} \tag{9.2}$$

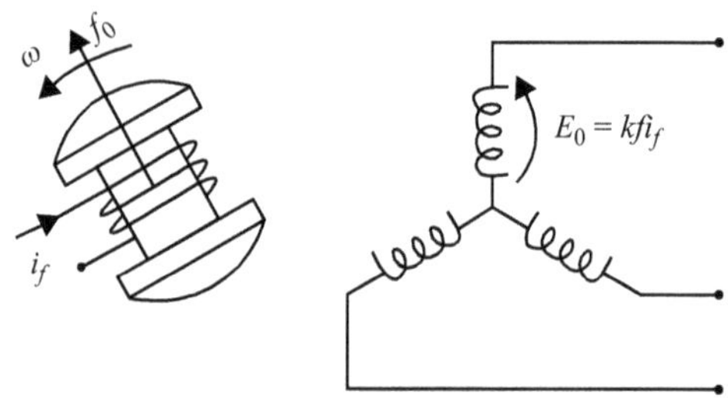

Figure 9.1. Synchronous machine at no-load. On the left, the field winding is fed by DC voltage source. On the right, the phase induced voltage is shown at no-load $E = kfi$.

9.2 No-load operation

The rotor winding, also called the field winding (figure 9.1(left)), is fed by a DC electrical voltage source that establishes a sinusoidal static magnetic flux density distribution related to the rotor itself. Therefore, as the rotor is running at a rotational speed n_s, the stator winding is exposed to a rotating magnetic field distribution. At the ends of the stator windings (or its terminals, shown in figure 9.1(right)), the induced emf is giving by:

$$E_0 = 4.44 f N_{\text{eff}} \phi_0 \tag{9.3}$$

where:

$\phi_0 = 2B_0\left(\frac{hR}{p}\right)$: flux per pole produced by the field winding in Wb,
f: frequency of induced voltage in Hz,
N_{eff}: number of effective turns in series per phase.

Supposing that the SM magnetic circuit is linear results in ϕ_0 being directly proportional to the excitation current i_f. Therefore, we can write:

$$E_0 = kfi_f \tag{9.4}$$

Figure 9.2 shows the spatial distribution of both, the rotor mmf and the induced emf in each stator conductor, considering the rotor moving in an 'anti-clockwise' sense. To plot the direction of conductor's emf, it was considered that the rotor is 'stalled', and the stator was moving in a 'clockwise' direction, so, the relative motion is the same when the stator is 'stalled' and the rotor is moving in an 'anti-clockwise' sense. After that, we can observe that applying the left-hand rule (e, B and v), the emf direction is such that the conductors located from $\theta = 0$ to $\theta = \pi$ are pointing up out of the plane of the figure. From $\theta = \pi$ to $\theta = 2\pi$, the emf direction is down into the plane of the figure.

Applying the right-hand rule to the emf distribution results in the emf E_0 lagging behind the mmf $\mathcal{F}_0$ by $\pi/2$.

9.3 On-load operation

Connecting a 3-phase load at the terminals of the SM, an AC electric current is injected into the load and a power transfer from the SM to the load starts.

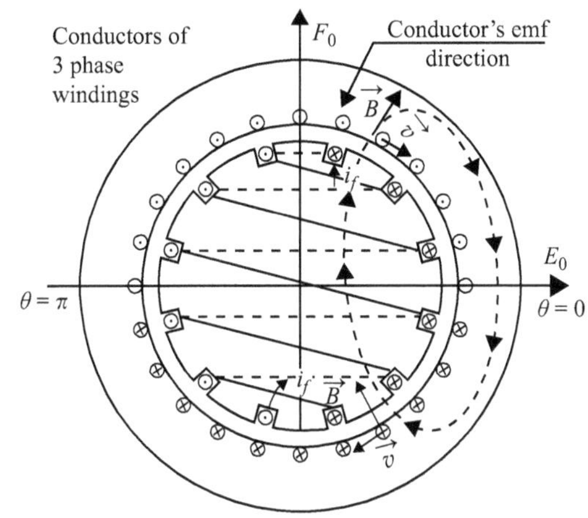

Figure 9.2. Both the rotor mmf and the stator emf distribution at no-load operation. ⊗: the positive direction down into the plane of the figure. ⊙: the negative direction up out of the plane of the figure.

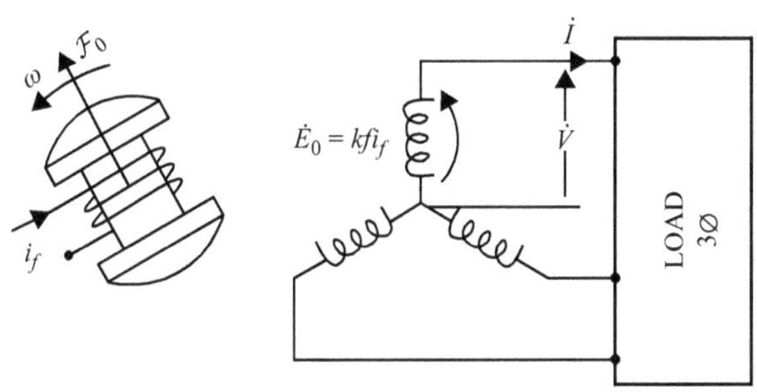

Figure 9.3. SM operating as generator.

This is what occurs when the SM is operating as a generator feeding an isolated load (figure 9.3).

The main electromagnetic quantities involved in this operation are:

$\mathcal{F}_0$: the mmf distribution produced by the DC electric current carried by the field winding. Regarding the rotor frame, $\mathcal{F}_0$ is given by:

$$\mathcal{F}_0(\theta) = N_p i_f \sin\theta \tag{9.5}$$

$\mathcal{F}_0$ is also the source of the magnetic flux density distribution B_0. Therefore:

$$B_0(\theta) = \frac{\mu_0}{2l_g}\mathcal{F}_0(\theta) = \frac{\mu_0}{2l_g}N_p i_f \sin\theta \tag{9.6}$$

If the rotor is running at a rotational speed n_s (r s^{-1}), this magnetic field distribution is observed by the stator frame as a rotative magnetic field, such that:

$$B_0(\theta, t) = \frac{\mu_0}{2l_g}\mathcal{F}_0(\theta, t) = \frac{\mu_0}{2l_g}N_p i_f \sin(\theta - \omega t) \tag{9.7}$$

where $\omega = 2\pi f = 2\pi p n_s$ rad s^{-1}

All sinusoidal time dependent magnetic flux density distributions cause an emf in the stator winding. In such cases, the rms no-load emf developed in the stator winding *per phase* will be:

$$E_0 = 4.44 f N_{\text{eff}} \phi_o = k f i_f \ (V/\text{phase}) \tag{9.8}$$

where:

$$\phi_o = 2B_{0\max}\frac{hR}{p} = \frac{\mu_0}{l_g}N_p i_f \frac{hR}{p} \tag{9.9}$$

with:

$$B_{0\max} = \frac{\mu_0}{2l_g}N_p i_f \tag{9.10}$$

From figure 9.2, we can establish that E_0 lags behind $\mathcal{F}_0$ by $\pi/2$.

The load current (I) is also the source of a rotational mmf distribution $\mathcal{F}_{ar}$, also called the *armature reaction mmf* given by:

$$\mathcal{F}_{ar}(\theta, t) = \frac{2l_g}{\mu_0}B_g(\theta, t) \tag{9.11}$$

From (8.35) we have:

$$B_g(\theta, t) = \frac{3\sqrt{2}}{\pi}\frac{\mu_0 N_{\text{eff}} I}{p l_g}\sin[\theta - \omega t]$$

that results in:

$$\mathcal{F}_{ar}(\theta, t) = \frac{6\sqrt{2}}{\pi}\frac{N_{\text{eff}} I}{p}\sin[\theta - \omega t] \tag{9.12}$$

Due to the self-induction effect, the *armature reaction mmf* is the source of an induced emf in the stator winding, also called the *armature reaction emf,* such that:

$$E_{\text{ar}} = 4.44 f N_{\text{eff}} \phi_{ar} \ (V/\text{phase}) \tag{9.13}$$

where:

$$\phi_{\text{ar}} = 2B_{\text{gmax}}\frac{hR}{p} = \frac{6\sqrt{2}}{\pi}\frac{\mu_0 N_{\text{eff}} I h R}{p^2 l_g} \tag{9.14}$$

with:

$$B_{\text{gmax}} = \frac{3\sqrt{2}}{\pi} \frac{\mu_0 N_{\text{eff}} I}{p l_g} \tag{9.15}$$

As ϕ_{ar} depends on the load current, as we saw in the previous chapter, we can write:

$$E_{\text{ar}} = X_s I \tag{9.16}$$

X_s: is the synchronous reactance as we saw in the previous chapter.

As E_0 lags behind $\mathcal{F}_0$ by $\pi/2$, we also can establish that E_{ar} lags behind $\mathcal{F}_{\text{ar}}$ by $\pi/2$.

Remarks: As the air gap is uniform, it results in:

$$\frac{\mathcal{F}_{0\,\text{max}}}{E_0} = \frac{\mathcal{F}_{\text{armax}}}{E_{\text{ar}}} \tag{9.17}$$

Supposing an inductive load, that is, I lags behind E_0 by an angle α, the vectors of both mmfs and emfs are plotted in figure 9.4.

As those vectors are turning at an angular speed ω, they can also represent the phasors of those quantities. Therefore, using complex notation, we can write:

$$\dot{E} = \dot{E}_0 + \dot{E}_{\text{ar}} \tag{9.18}$$

If we do not consider any additional voltage losses, the net emf $\dot{E}$ will be the phase voltage applied to the load. Considering the resistance of the stator winding the equation (9.18) becomes:

$$\dot{V} = \dot{E} - r_s \dot{I} \tag{9.19}$$

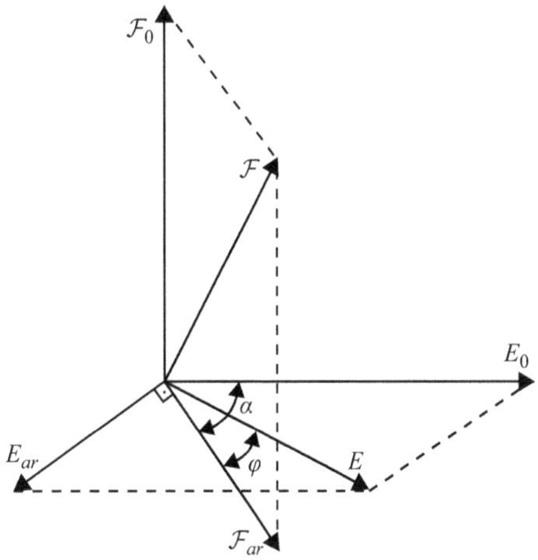

Figure 9.4. Mmf and emf for SM at load. $\mathcal{F}_0$: field winding mmf—E_0: induced emf due to $\mathcal{F}_0$. $\mathcal{F}_{\text{ar}}$: armature reaction mmf—E_{ar}: induced emf due to $\mathcal{F}_{\text{ar}}$. $\mathcal{F} = \mathcal{F}_0 + \mathcal{F}_{\text{ar}}$: net mmf—$E = E_0 + E_{\text{ar}}$: net emf.

Writing $\dot{E}_{\text{ar}}$ as a drop voltage in the synchronous reactance, such that:

$$\dot{E}_{\text{ar}} = -jX_s\dot{I} \tag{9.20}$$

The final voltage equation for the synchronous generator yields:

$$\dot{V} = \dot{E}_0 - jX_s\dot{I} - r_s\dot{I} \tag{9.21}$$

Based on equation (9.21), we can model the synchronous generator (SG) steady-state operation by the electric circuit of figure 9.5.

9.3.1 The rated quantities of synchronous generator

The main rated values (or the full-load values) of a SG are:

S_{FL}: the rated power (kVA, MVA)—the maximum apparent power that the generator can deliver continuously to the load without reducing its life cycle;

V_{FL}: the rated voltage (V, kV)—the maximum line voltage that a SG can operate continuously at without reducing its life cycle;

I_{FL}: the rated current (A, kA)—the maximum line current that a SG can operate continuously at without reducing its life cycle;

n_s: the rated speed (rpm) or the synchronous speed—the maximum rotational speed that a SG can operate continuously at without reducing its life cycle; the rated speed is also called the synchronous speed;

PF = cos φ: the power factor that limits the developed active power delivered to the load.

For a 3-phase SG, using the line values, we can write:

$$S_{FL} = \sqrt{3}\, V_{FL} I_{FL}\ (VA) \tag{9.22}$$

The maximum active power is such that:

$$P_{FL} = \sqrt{3}\, V_{FL} I_{FL} \cos\varphi\ (W) \tag{9.23}$$

The rated active power is related to the maximum torque that should be introduced in the SG shaft by the equation:

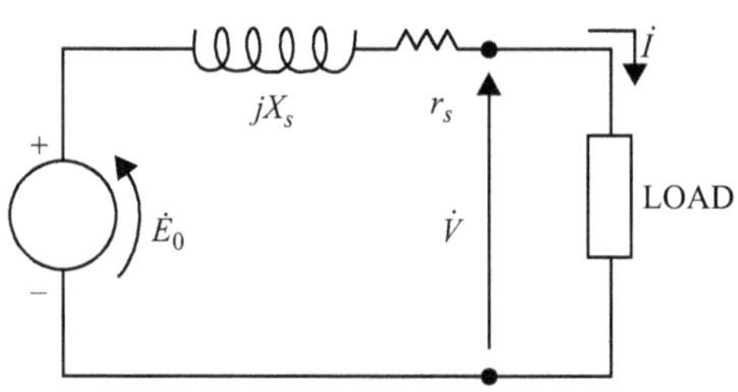

Figure 9.5. Equivalent electric circuit of SG. r_s: phase resistance of stator winding (Ω); X_s: synchronous reactance (Ω); $\dot{E}_0$: induced emf due to $\mathcal{F}_0$ (V); $\dot{V}$: phase voltage at load (V); $\dot{I}$: load current (A).

$$T_{FL} = \frac{P_{FL}}{2\pi n_s}(\mathrm{Nm}) \tag{9.24}$$

9.3.2 Drawing the phasor diagram

The phasor diagram is very useful for extracting the most important quantities involved in SG operation. If you wonder what will happen if some steady-state quantity varies, the drawing of the phasor diagram is an ideal tool.

The phasor diagram requires inputs. In the SG operation, the user normally knows its end voltages, apparent power, the nature of the load (inductive/capacitive) and power factor and wants to find both the emf E_0—also called the intern emf—and the excitation current that should be injected at the field winding. Here:

V_L: the line-to-line voltage (V),
S: the apparent power (VA),
$\cos\varphi$: the power factor.

The inputs for drawing the phasor diagram are:

$V = \frac{V_L}{\sqrt{3}}$: the phase voltage at the end of SG,
$I = \frac{S}{\sqrt{3}\,V_L}$: the load current
$\varphi = \arccos(\varphi)$.

Remarks: $\varphi < 0$ inductive load; $\varphi > 0$ capacitive load and $\varphi = 0$ resistive load.

The phasor diagram is based on the voltage equation:

$$\dot{E}_0 = \dot{V} + jX_s\dot{I} + r_s\dot{I} \tag{9.25}$$

Choosing $\dot{V} = V\angle 0°$ results in $\dot{I} = I\angle\varphi$. Therefore, the first step for drawing these two phasors is shown in figure 9.6, where an inductive load was chosen as an example.

The second step is to add the phasor of $r_s\dot{I}$ to $\dot{V}$, that is in phase with $\dot{I}$ phasor (see figure 9.7). Therefore, the phasor $r_s\dot{I}$ is parallel the phasor $\dot{I}$.

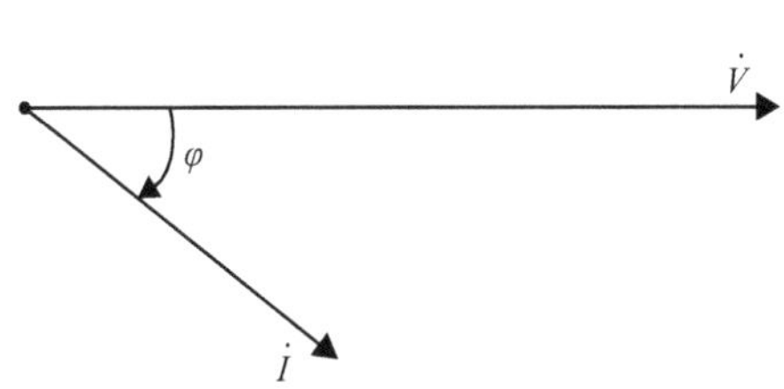

Figure 9.6. First step of the phasor diagram. Inductive load.

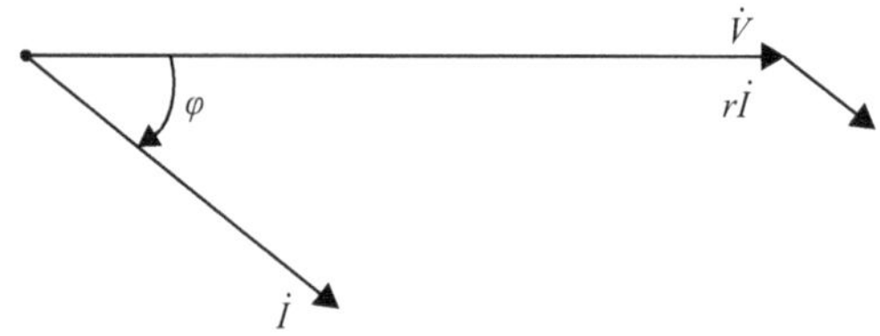

Figure 9.7. Second step of the phasor diagram: $\dot{V} + r_s\dot{I}$.

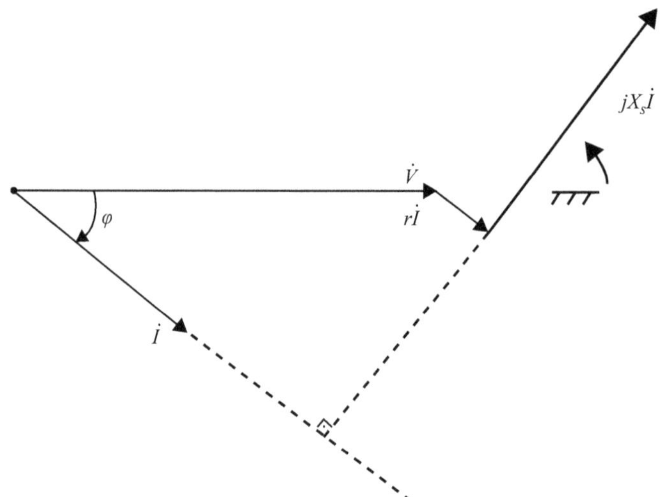

Figure 9.8. Third step of the phasor diagram: $\dot{V} + jX_s\dot{I} + r_s\dot{I}$.

The third step is to add the phasor of $jX_s\dot{I}$ to $\dot{V} + r_s\dot{I}$. The phasor $jX_s\dot{I}$ is normal to the phasor $\dot{I}$, because the current is lagging behind $jX_s\dot{I}$ by 90°. Figure 9.8 shows the third step.

Figure 9.9 shows the final step of the phasor diagram of SG, placing its necessary internal voltage E_o to feed an inductive load at the required voltage.

The angle δ is called the *power angle* for reasons that we will see later.

In standard SM, the resistances of stator windings are negligible. So, equation (9.21) can be simplified as:

$$\dot{V} = \dot{E}_0 - jX_s\dot{I} \tag{9.26}$$

and the final phasor diagram is shown in figure 9.10.

9.3.3 Power and torques

As we have seen in previous chapters, the interactions between both the magnetic field and electric current develop power and electromagnetic torque. If a SM operates as a generator, the mechanical power introduced in its shaft is converted into electrical power that is delivered to the load.

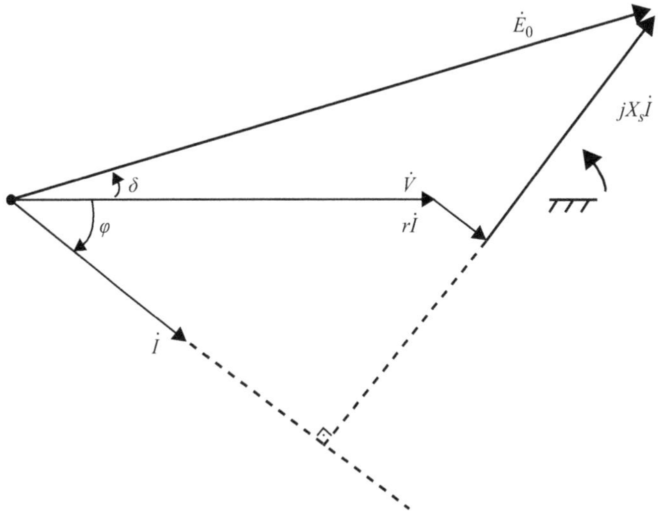

Figure 9.9. The final step of the phasor diagram $\dot{E}_0 = \dot{V} + jX_s\dot{I} + r_s\dot{I}$.

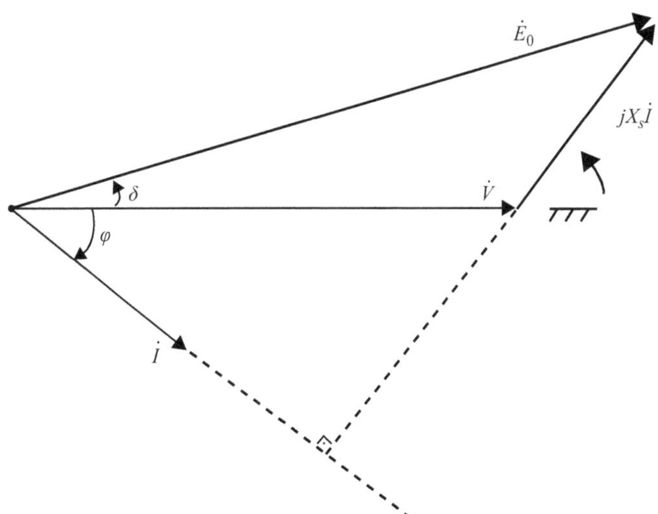

Figure 9.10. The phasor diagram neglecting the windings resistances. In a generator operation $\delta > 0$ and V is lagging behind E_o by δ.

Evaluating the electric power from both the electric values and SM parameters is an important task in analyzing the performance of the machine.

The apparent power *per phase* delivered to the load is expressed by:

$$\dot{S} = \dot{V}\dot{I}^* \text{ (VA)} \tag{9.27}$$

From equation (9.26) we can write:

$$\dot{I} = \frac{\dot{E}_0 - \dot{V}}{jX_s} \tag{9.28}$$

From the phasor diagram of figure 9.10, choosing $\dot{V} = V\angle 0°$ results in $\dot{E}_0 = E_o\angle\delta$. Therefore:

$$\dot{I} = \frac{E_o\angle\delta - V}{jX_s}$$

or,

$$\dot{I} = \frac{E_o \sin\delta}{X_s} - j\frac{E_o \cos\delta - V}{X_s} \tag{9.29}$$

So, the conjugate complex of $\dot{I}$ will be:

$$\dot{I}^* = \frac{E_o \sin\delta}{X_s} + j\frac{E_o \cos\delta - V}{X_s} \tag{9.30}$$

Therefore, the apparent power can be written as:

$$\dot{S} = \dot{V}\dot{I}^* = \frac{VE_o \sin\delta}{X_s} + j\frac{V(E_o\cos\delta - V)}{X_s} \tag{9.31}$$

Identifying $\dot{S}$ with $\dot{S} = P + jQ$, where P and Q are, respectively, the active and reactive power, results in:

$$P = \frac{VE_o \sin\delta}{X_s}(\mathrm{W}) \tag{9.32}$$

$$Q = \frac{V(E_o\cos\delta - V)}{X_s}(\mathrm{VAr}) \tag{9.33}$$

The mechanical torque applied by the prime mover to the SG shaft will be:

$$T_{\mathrm{mec}} = \frac{3P}{2\pi n_s}(\mathrm{Nm}) \tag{9.34}$$

9.3.4 The voltage regulation

The measure of how well a SG maintains the terminal voltage constant over a range of load currents is called the SG's *voltage regulation*. It can be calculated from the following formula:

$$\mathcal{R} = \frac{V_0 - V}{V} \times 100\% \tag{9.35}$$

where:

V: on-load SG terminal voltage (V),
V_0: no-load SG terminal voltage (V).

Remarks: The no-load terminal voltage V_0 is the internal emf E_0 (V). That is:

$$\mathcal{R} = \frac{E_0 - V}{V} \times 100\% \tag{9.36}$$

The voltage regulation is strongly affected by the load nature. It can be positive under both resistive and inductive load and negative under a capacitive one.

Example 9.1

A 3-phase (500 kVA; 380 V; 4 poles; 1800 rpm) SG has a synchronous reactance $X_s = 0.45\ \Omega$. The winding resistances are negligible. Under rated speed, this machine presents rated voltage when the field current is 20 A.

(a) determine the field current required when the generator is supplying rated output at 0.8 lagging power factor, that is, the load on the generator is inductive;
(b) the regulation in this condition;
(c) determine the required field current when the generator is supplying the same active power as part (a) for a resistive load under rated voltage;
(d) the regulation for the condition in part (c);
(e) determine the required field current when the generator is supplying the same active power as parts (a) and (c) for a 0.8 capacitive power factor load under rated voltage;
(f) the regulation for the condition in part (e);
(g) determine the mechanical torque applied to the SG shaft in all conditions.

Solution

(a) Field winding current for a full-load operation

Based on equation (9.22), the armature current is given by:

$$I = \frac{S_{FL}}{\sqrt{3}\,V_{FL}} = \frac{500\ 000}{\sqrt{3} \times 380} = 760\ \text{A}$$

Adopting the voltage per phase as:

$$\dot{V} = \frac{380}{\sqrt{3}}\angle 0^\circ = 220\angle 0^\circ\ (\text{V})$$

and considering that the power factor is $\cos\varphi = 0.8$ inductive, results in $\varphi = -36.87^\circ$. Therefore:

$$\dot{I} = 760\angle - 36.87^\circ\ (\text{A})$$

The internal emf for such a condition is evaluated from:

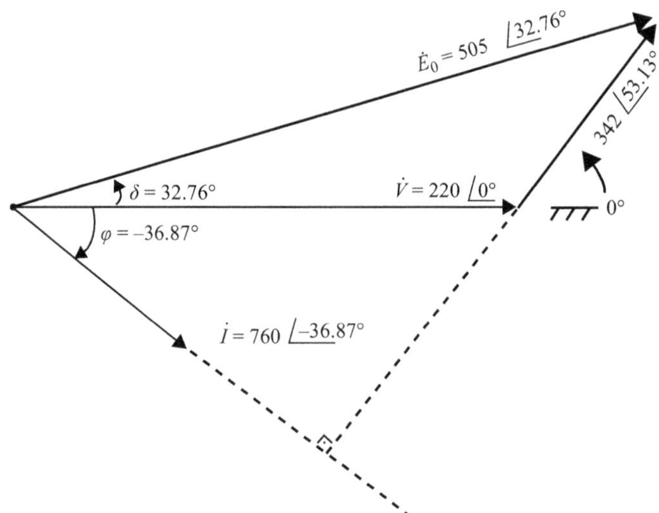

Figure 9.11. The phasor diagram. On inductive load $E_0 > V$.

$$\dot{E}_0 = \dot{V} + jX_s\dot{I} = 220\angle 0^\circ + j0.45 \times 760\angle - 36.87^\circ$$
$$\dot{E}_0 = 505\angle 32.76^\circ\ (V)$$

That is: $E_o = 505$(V) and $\delta = 32.76^\circ$

As we saw previously, the internal emf is directly dependent on the field current. Consequently, as the field current for establishing the rated voltage is 20 A the new field current for establishing 505 V will be:

$$i_f = \frac{505}{220} \times 20 = 46\ \text{A}$$

Figure 9.11 shows the phasor diagram for this condition.

(b) The regulation in this condition

Based on equation (9.36), the regulation of the SG in the operating condition of part (a) is given by:

$$\mathcal{R} = \frac{E_0 - V}{V} \times 100\% = \frac{505 - 220}{220} \times 100\%$$

Therefore:

$$\mathcal{R} = 129\%$$

(c) Field current for a resistive load under rated voltage

From part (a), the phase-active power is such that:

$$P = \frac{1}{3} S_{FL} \cos\varphi = 133.33\ \text{kW}$$

For a resistive load with the same active power, the phase-current is given by:

$$I = \frac{P}{V\cos\varphi} = \frac{133\,330}{220 \times 1} = 606 \text{ A}$$

consequently,

$$\dot{I} = 606\angle 0^\circ \text{ A}$$

The new value of the internal emf results in:

$$\dot{E}_0 = \dot{V} + jX_s\dot{I} = 220\angle 0^\circ + j0.45 \times 606\angle 0^\circ$$
$$\dot{E}_0 = 350\angle 51.13^\circ \text{ V}$$

that is:

$$E_0 = 350 \text{ V and } \delta = 51.13^\circ$$

The phasor diagram for this condition is the one shown in figure 9.12.

The field current results in:

$$i_f = \frac{350}{220} \times 20 = 32 \text{ A}$$

(d) The regulation for the resistive load results in:

$$\mathcal{R} = \frac{E_0 - V}{V} \times 100\% = \frac{350 - 220}{220} \times 100\%$$

therefore:

$$\mathcal{R} = 59\%$$

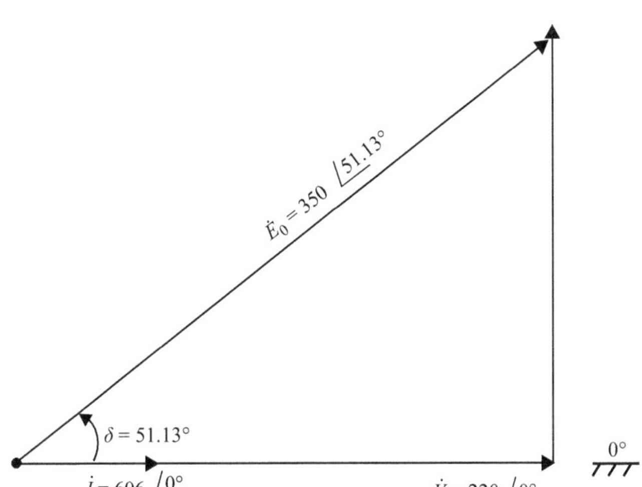

Figure 9.12. Phasor diagram for resistive load. On resistive load $E_0 > V$.

(e) Field current for a capacitive load under rated voltage
From part (a), the phase-active power is such that:

$$P = \frac{1}{3} S_{FL} \cos \varphi = 133.33 \text{ kW}$$

For a 0.8 capacitive power factor load with the same active power, the phase-current is given by:

$$I = \frac{P}{V \cos \varphi} = \frac{133\,330}{220 \times 0.8} = 758 \text{ A}$$

consequently,

$$\dot{I} = 758\angle 36.87^\circ \text{ A}$$

Remarks: $\varphi > 0$ for a capacitive load.

The new value of the internal emf results in:

$$\dot{E}_0 = \dot{V} + jX_s\dot{I} = 220\angle 0^\circ + j0.45 \times 760\angle 36.87^\circ$$
$$\dot{E}_0 = 273.4\angle 86.78^\circ \text{ V}$$

that is:

$$E_0 = 273.3 \text{ V and } \delta = 86.78^\circ$$

The phasor diagram for this condition is the one shown in figure 9.13.
The field current results in:

$$i_f = \frac{273.3}{220} \times 20 = 25 \text{ A}$$

(f) The regulation for the resistive load results in:

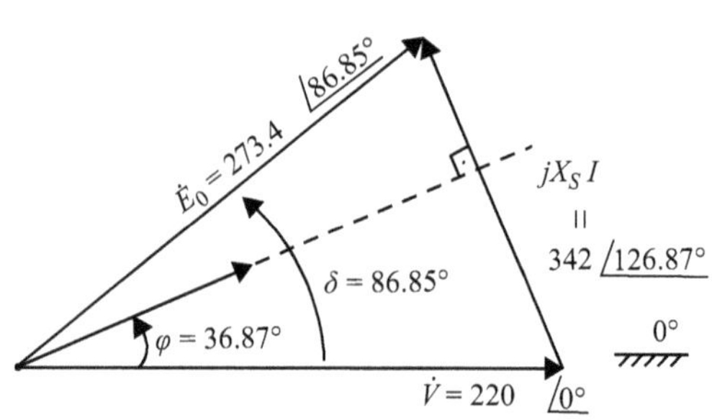

Figure 9.13. Phasor diagram for resistive load. On capacitive load E_0 can be $\cong V$.

$$\mathcal{R} = \frac{E_0 - V}{V} \times 100\% = \frac{273.3 - 220}{220} \times 100\%$$

therefore:

$$\mathcal{R} = 24.2\%$$

(g) The mechanical torque applied to the SG shaft in all conditions

As the active power is the same in all conditions, the full-load torque is such that:

$$T_{FL} = \frac{3P}{2\pi n_s} = \frac{400000}{2\pi \frac{1800}{60}}$$

that results in:

$$T_{FL} = 2122 \text{ Nm}$$

Final remarks: It is important to observe that for the *same* active power, the regulation is such that: $\mathcal{R}_L > \mathcal{R}_R > \mathcal{R}_C$. In all these three cases, the SG was able to regulate its field current to adapt to the load characteristics, that is exactly what happens in the grid or in an isolated system.

9.4 Motor operation

Although a SM is commonly used as a generator, we can also find in some industrial applications that the SM operates as a motor. Currently, the synchronous motor (SMot) is being used in the powertrain of hybrid vehicles with excellent performance for this kind of application.

Figure 9.14 shows the electric scheme of SMot connection at a 3-phase AC power supply.

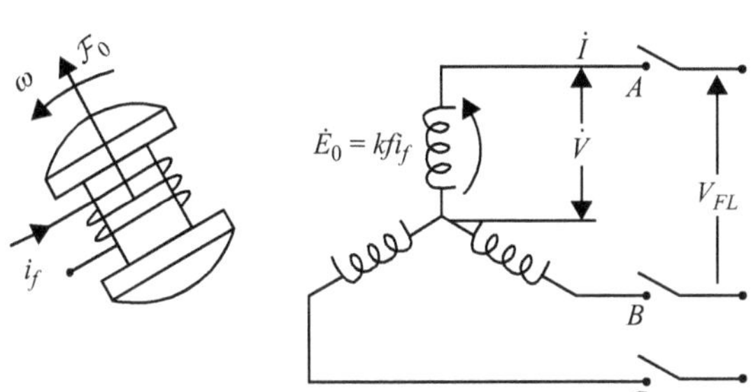

Figure 9.14. Synchronous motor. V_{FL}: line-to-line rated voltage.

As we discussed previously, the synchronous motor only develops electromagnetic torque when it is running at the synchronous speed n_s. Consequently, no starting torque is available for taking the load from inertia. This is the main reason for not using a SM as a motor.

Nowadays, power electronics converters allied with some construction features have overcome this issue; however, study of this technology is not the objective of this book.

Our purpose is to present the steady-state operation of a SMot supposing that it is running at its rated speed n_s.

The equivalent electric circuit of the SMot is the same as the SG when changing the direction of the electric current (figure 9.15).

As before, neglecting the winding resistance, the voltage equation becomes:

$$\dot{V} = \dot{E}_0 + jX_s\dot{I} \tag{9.37}$$

Analyzing the phasor diagram of a SMot operating under several conditions of load, we can identify from figure 9.16 the following:

(1) The SMot over-excited ($E_0 > V$) is a capacitive load for AC power supply. That is very interesting for industrial application, because we can use this feature to avoid low inductive power factor in the plant.

(2) The SMot under-excited ($E_0 < V$) is an inductive load for AC power supply. It is not an interesting operation point because it is necessary to have enough reactive power capacity to supply the required reactive power by the machine.

(3) Based on statements (1) and (2), the SMot power factor can be controlled by the excitation current. This feature is powerful and can improve the performance of the machine and the plant.

(4) In the motor operation of a SM, $\delta < 0$, that is E_o lags behind V by δ.

9.4.1 Properties of synchronous motor phasor diagram

The synchronous motor is fed by an AC voltage source that keeps both the frequency and the voltage input constant. As we have total control over the

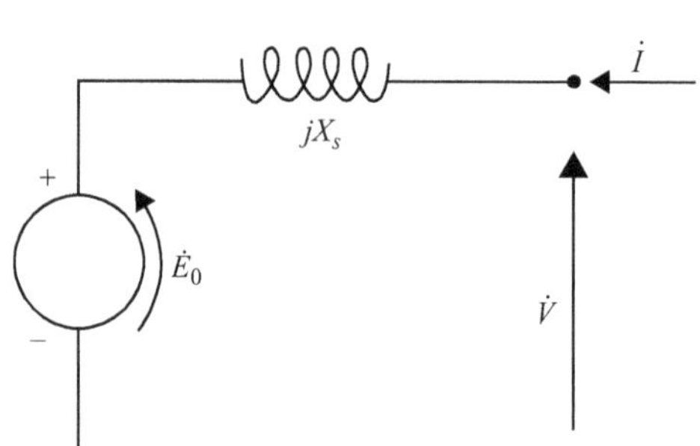

Figure 9.15. Equivalent electric circuit for SMot. Comparing with a SG, only the direction of current is changed.

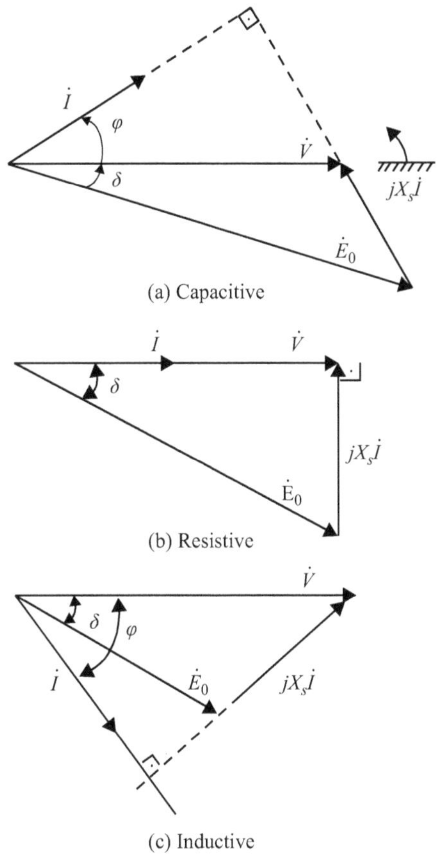

Figure 9.16. Phasor diagrams for a SMot ($\delta < 0$). (a) Over-excited ($E_0 > V$) operation—capacitive load. (b) Unit power factor operation—resistive load. (c) Under-excited ($E_0 < V$) operation—inductive load.

excitation current (i_f), we can improve its performance by choosing a suitable value of i_f to achieve this target.

Based on equation (9.31), both the active power delivered to the load (since all losses are neglected) and the reactive power delivered by the AC power supply are given by:

$$P = \frac{VE_o \sin \delta}{X_s} \tag{9.38}$$

$$Q = \frac{V(E_o \cos \delta - V)}{X_s} \tag{9.39}$$

If we suppose that the active power delivered to the mechanical load is constant—that is a common motor operation—and observing that V imposed by the AC source is also constant, we can establish:

$$E_o \sin \delta = \text{constant}$$

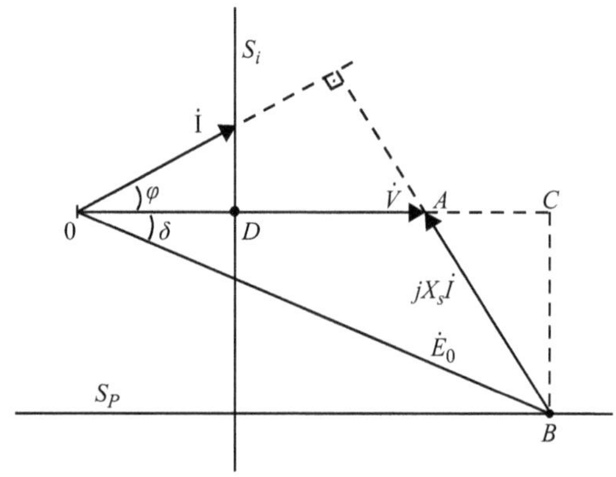

Figure 9.17. Loci of $\dot{E}_0$ and $\dot{I}$ for both P and V constants.

Figure 9.17 shows the phasor diagram drawn for an over-excited operation that clearly presents the previous equation.

The segment BC is the projection of E_o in a vertical straight line to V, that is $BC = E_o \sin\delta$. As a result, this segment is proportional to the power delivered to load. We can write this evidence as:

$$P \propto BC = E_o \sin \delta \tag{9.40}$$

Therefore, the straight line s_p is the *loci* of $\dot{E}_0$ for a constant active power independent of i_f.

Noticing that the segment OD is the projection of I over the straight line of V, resulting in $OD = I \cos \varphi$. As $P = VI\cos \varphi$ and V is constant, we can also write that:

$$P \propto OD = I\cos \varphi \tag{9.41}$$

As a result, the straight line s_i is the *loci* of $\dot{I}$ for a constant active power independent of i_f.

9.4.2 'V' curves

Based on the properties of the SMot discussed previously through the phasor diagram, we can improve our analysis of the synchronous motor operation elaborating curves that show the variation of the armature current (I) due to the variation of the field current (i_f).

Figure 9.18 shows the phasor diagram of a SMot for three different operation conditions. Figure 9.18(a) shows an over-excited operation where the motor behaves like a capacitive load. The field current is proportional to E_o, with i_c being the field current in this condition.

Figure 9.18(b) also shows an over-excited operation where the motor behaves like a resistive load, with i_R being the field current in this condition.

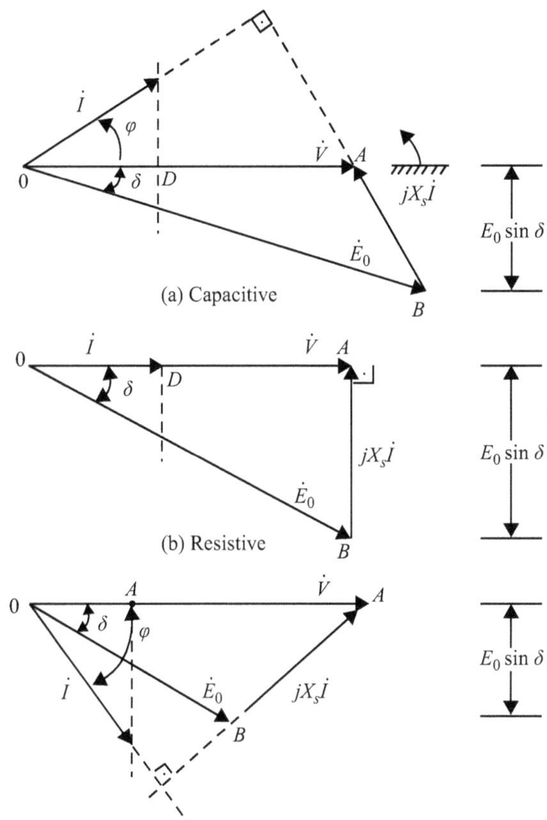

Figure 9.18. Phasor diagram for motor operation in three different conditions.

In both conditions the active power is supposed to be the same. As a result, in both cases not only the projection of E_o over a vertical straight line of V, but also the projection of I over V should be the same.

Therefore, comparing the magnitude of both phasors E_o and of I, issued from the capacitive to resistive operation, we conclude that as the field current decreases the armature current also decreases.

Figure 9.18(c) shows the phasor diagram for an under-excited operation, the SMot behaves like an inductive load for the AC supply, with i_L being the field current. Comparing both, the field and the armature currents from the resistive to inductive load we identify that as the field current decreases the armature current increases. Figure 9.19 shows these results graphically.

The 'V' curve of figure 9.19 was drawn supposing that the active power is constant. If we plot it for several different values of active power, a family of 'V' curves will be generated, like the one shown by figure 9.20.

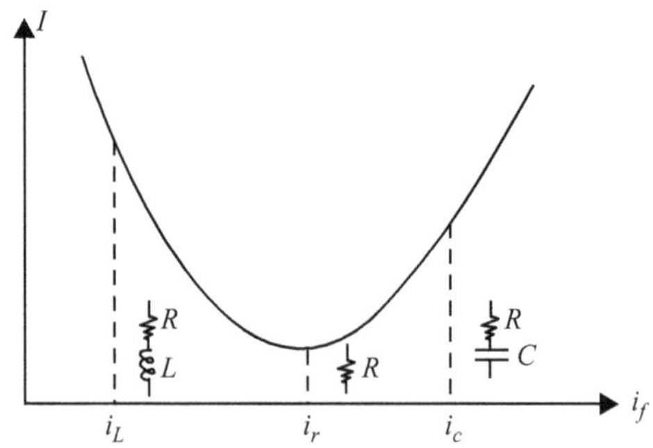

Figure 9.19. 'V' curve. $i_c > i_r$: capacitive current—power factor angle positive. $i_r > i_L$: inductive current—power factor angle negative. i_c: resistive current—unitary power factor.

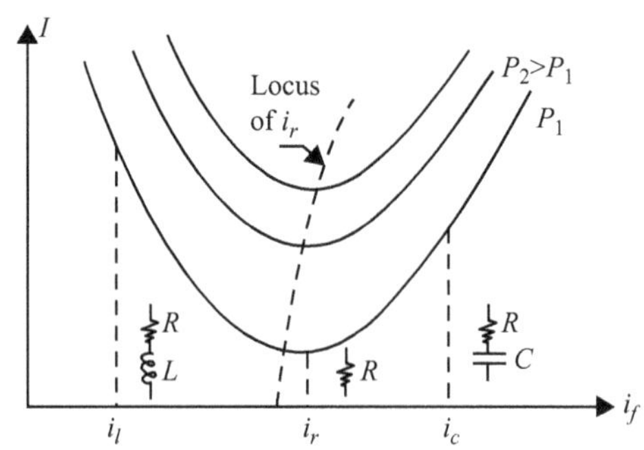

Figure 9.20. Family of 'V' curves.

Example 9.2

A 3-phase SMot, 500 kW–660 V–4 poles–60 Hz, has a synchronous reactance $X_s = 1.5\ \Omega$. The winding resistances are negligible. Running at rated speed with 25 A carried by the field winding results in the intern *emf* equal to the rated voltage.

(a) Determine the field current required when the motor is delivering its rated power to the load under a power factor 0.8 capacitive. Plot the phasor diagram for this condition.

(b) Determine both the active power delivered to the load and the new power factor when the field current is reduced 10% of the previous value of part (a) without changing any other operational condition.

(c) Due to operational condition, the power delivered to the mechanical load was decreased 10%, maintaining a constant field current. Determine the new armature current, the new power factor and the torque delivered to the load.

Solution

Initial condition: for both $n_s = \frac{f}{p} = \frac{60}{2} = 30$ rps(1800 rot s^{-1}) and the field current is $i_f = 25$ A the internal emf is such that $E_0 = \frac{660}{\sqrt{3}} = 380$ V.

(a) Field current for rated power and power factor 0.8 capacitive

From the equation $P = VI \cos\varphi$ we obtain:

$$I = \frac{P}{V\cos\varphi} = \frac{167 \times 10^3}{380 \times 0.8} = 549 \text{ A}$$

Choosing $\dot{V} = 380\angle 0°$, it results in $\dot{I} = 549\angle 36.87°$ A, therefore:

$$\dot{E}_0 = \dot{V} - jX_s\dot{I} = 380\angle 0° - j1.5 \times 549\angle 36.87°$$
$$\dot{E}_0 = 1094\angle -37° \text{ V}$$

that is:

$$E_0 = 1094 \text{ V and } \delta = -37°$$

As for $i_f = 25$ A we have $E_0 = 380$ V resulting in:

$$i_f' = \frac{1094}{380} \times 25 = 72 \text{ A}$$

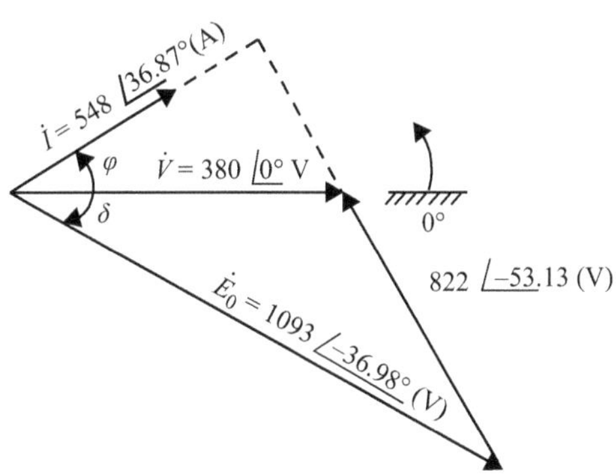

Figure 9.21. Answer of (a), the phasor diagram.

(b) Active power delivered to the load and the new power factor when the field current is reduced 10%

Reducing the field winding current without changing any other operational condition, the active power delivered to the load remains constant. As the active power is the same, we can write:

$$E_o \sin\delta = E_0' \sin\delta'$$

where E_o and δ are both the internal emf and the power angle of the previous operational condition, E_0' and δ' are the same quantities for the new operational condition. As a result:

$$1094 \times \sin(-37°) = 0.9 \times 1094 \times \sin\delta'$$

therefore:

$$\sin\delta' = -0.65 \text{ and } \delta' = -40.57°$$

so, the new internal emf is:

$$\dot{E}_0' = 984.6\angle - 40.57° \text{ (V)}$$

Based on the equation $\dot{E}_0 = \dot{V} - jX_s\dot{I}$ we obtain:

$$\dot{I} = \frac{\dot{V} - \dot{E}_0'}{jX_s} = \frac{380\angle 0° - 984.6\angle - 40.57°}{j1.5}$$

that is:

$$\dot{I} = \frac{\dot{V} - \dot{E}_0'}{jX_s} = \frac{380\angle 0° - 984.6\angle - 40.57°}{j1.5}$$

$$\dot{I} = 492\angle 29.9° \text{ (A)}$$

the new power factor is such that:

$$\cos\varphi = \cos 29.9° = 0.87 \text{ cap.}$$

(c) The new armature current and the new power factor due to the decreasing power.

The new power is: $P' = 0.9 \times 167\,k = 150$ kW. As the field current remains constant, it result in $E_0' = 984.6$ V. Based on the equation:

$$P = \frac{VE_o \sin\delta}{X_s}$$

the new value of the power angle is:

$$\sin\delta = \frac{PX_s}{VE_o} = \frac{150 \times 10^3 \times 1.5}{380 \times 984.6} = 0.601$$

Therefore, $\delta = -37°$ (negative for motor operation) and the new current is obtained from:

$$\dot{I} = \frac{\dot{V} - \dot{E}_0'}{jX_s} = \frac{380\angle 0° - 984.6\angle -37°}{j1.5}$$

that is:

$$\dot{I} = 479\angle 34.4° \text{ A}$$

So, the new current and the new power factor are:

$$I = 479 \text{ A and } \cos\varphi = \cos 34.4° = 0.82 \text{ cap.}$$

The torque delivered to the load will be:

$$T_L = \frac{3 \times P}{2\pi n_s} = \frac{3 \times 150 \times 10^3}{2\pi \times 30} = 2837 \text{ Nm}$$

9.5 Summary

This is the second chapter on the subject of the synchronous machine (SM). This chapter covered the aspects of SM related to its operation as a generator and as a motor. The operation as a generator focused on the isolated systems and it feeds three types of loads: inductive, resistive and capacitive. We have developed the phasor diagram for the SG to analyze the impact of the different types of load and the field current. We have seen that by changing the field current it is possible to regulate the terminal voltage and feed all the loads at the rated voltage.

The motor mode focused on the operation connected to a power source with fixed voltage and frequency. We have discovered that it is possible to provide the mechanical load demanded and to inject or consume reactive power, or even operate at unitary power factor. The 'V' curves provide a kind of operation map that helps to not only understand the idea, but also to set the best field current value to regulate the terminal voltage.

9.5.1 Project

The rotor of the industrial non-salient synchronous motor of chapter 8 has 64 slots uniformly distributed. The maximum field current should be less than 50 A due to DC voltage source limitation.

Design a concentric field winding following the procedure presented in chapter 8 to establish the specified magnetic flux density distribution required.

Based on the results obtained from the previous project, determine the synchronous reactance of the machine.

Considering the machine parameters, you should evaluate the field current enough for a rated power operation under 0.8 lagging power factor. Neglect the resistance of the 3-phase winding.

Finally, compute the mass of both iron and copper contained in the active part of the stator of the machine. Compare it with the one obtained for the other students or groups in your classroom.

Problems

9.1 Describe how the synchronous machine works at no-load operation. What is required to have an induced emf of a required frequency? What is the influence of both the number of poles and the rotational speed for acquiring a specific frequency?

9.2 How does a synchronous generator work at load? How many mmf distributions are presented during this kind of operation? What is the role of each mmf distribution?

9.3 A 3-phase source of 400 Hz is required in an aviation application. The only utility available is a 3-phase at 60 Hz. The frequency conversion is to be accomplished using a synchronous motor driving a synchronous generator. A variation of about $\pm 3\%$ in the frequency of 400 Hz is permissible. Determine a suitable number of poles for each of the synchronous machines and the speed at which the set will run.

Answer: 6 and 40 poles.

9.4 Draw a phasor diagram for a generator supplying:
   (a) a purely capacitive load;
   (b) a purely inductive load;
   (c) a purely resistive load.

9.5 A 3-phase synchronous motor operates at its rated voltage and frequency and delivers rated mechanical power output. The field current is adjusted so that the power factor is 1 (unitary). Draw the phasor diagram for this condition.

9.6 Referring to the motor of example 9.4.3, assume that the load requires a torque independent of speed. Describe quantitatively the effects such as power input, power factor and current in response to the reduction of 10% in the AC voltage source, maintaining constant both the frequency and the field current.

Answer: $\delta > \delta'$; $\varphi > \varphi'$; $I' > I$.

9.7 Returning to the motor in example 9.4.3, describe quantitatively the effects such as power input, power factor, and current in response to the reduction at the same time of both 10% in the AC voltage source and 10% in the frequency, maintaining a constant field current.

9.8 When a synchronous generator is supplying a capacitive load (as shown in figure 9.13), E_0 may be less than V. Suppose the terminal voltage V is to be held constant. The load is almost pure capacitance. Can the capacitance of the load be so great that E_0 becomes zero? What then is the correct field current? Explain how the generator operates in such a case. If the capacitance of the load is even greater than this value, can the field current reversed? Is the generator then stable? (Note: This is a problem of practical importance when a long transmission line without load is connected to a generator.)

9.9 A non-salient pole 115 V–60 Hz, 4-pole, 3-phase synchronous machine is operating in a steady-state as a motor. The power input to the AC windings on the rotor is 800 W and the current is 6.5 A. The resistance of 3-phase winding is negligible, and its reactance is $X_s = 4.5\ \Omega$. Suppose that the power input is kept constant and the DC field current can be varied:
   (a) determine the power angle δ for a field current corresponding to generated voltage of 50.75 and100 V, respectively;
   (b) compute the AC line current on the stator corresponding to a generated voltage of 100 V.

Answers: (a) 11.8°; 6°; (b) 4.16 A; p.f. 0.56 ind.

9.10 A 15 kVA, 220 V, 60 Hz, 4 poles synchronous generator is designed to operate at a full-load lagging power factor of 0.8. The synchronous reactance of the machine is $X_s = 2.23\ \Omega$/phase. Injecting a DC current of 2.27 (A) in the field winding, the difference of the potential at the terminal for no-load operation at rated rotational speed is 220 (V). Compute:
   (a) the required field current for a rated power operation;

(b) maintaining the DC field current constant as the one evaluated in part (a), compute the difference of the potential at the terminal when the generator is supplying rated current at unitary power factor.

Answers: (a) 3.45 A; (b) 297 V.

Further reading

[1] Fitzgerald A E, Kingsley C Jr and Umans S D 1992 *Electric Machinery* 5th edn (New York: McGraw-Hill)
[2] Slemon G R and Straughen A 1980 *Electric Machines* (Reading, MA: Addison-Wesley)
[3] Skilling H H 1962 *Electromechanics: A First Course in Electromechanical Energy Conversion* (New York: Wiley)
[4] Hughes A and Drury B 2013 *Electric Motors and Drives: Fundamentals, Types, and Applications* 4th edn (Amsterdam: Elsevier)
[5] Jordao R G 1980 *Maquinas Sincronas* (Sao Paulo: LTC/EDUSP) (in Portuguese)
[6] Gieras J F 2016 *Electrical Machines: Fundamentals of Electromechanical Energy Conversion* 1st edn (Boca Raton, FL: CRC Press)
[7] Gonen T 2011 *Electrical Machines with MATLAB®* 2nd edn (Boca Raton, FL: CRC Press)
[8] Mohan N 2012 *Electric Machines and Drives: A First Course* 1st edn (New York: Wiley)
[9] Sen P C 2013 *Principles of Electric Machines and Power Electronics* 3rd edn (New York: Wiley)

IOP Publishing

Electromechanical Energy Conversion Through Active Learning

J R Cardoso, M B C Salles and M C Costa

Chapter 10

Asynchronous machine: operation

10.1 Introduction

The asynchronous machine is the most frequently used electrical machine worldwide. It can be found in all industries operating as an electrical motor, driving several manufacturer machines and process. An asynchronous machine operating like a motor is known as an *induction motor* due to the electromagnetic phenomena responsible for its operation.

The induction motor occupies an important space in electrical vehicle technology. Its versatility, combined with both its simpler construction and low price, gives the induction motor the unbeatable position of the most used electrical machine in the world.

The asynchronous machine is also used as a generator, not only in unassisted small hydroelectric plants, but also in some industrial processes, but the most modern application is in wind energy generator technology.

The developments of the power electronics devices has substantially improved both the performance and the diversity of application of induction motors in recent decades. That is the reason why a good understanding of this machine in worldwide use is necessary.

10.2 Design aspects of an asynchronous machine

Figure 10.1 is a picture of an induction motor that is the union of two parts: the rotor—which rotates when the power is delivered—and the stator—which stays in place.

Stator: Built with Fe–Si sheet-steel uniformly slotted, the stator houses the conductors of polyphase windings (normally triphasic). It is very close to the construction of the stator in synchronous machines (figure 10.2).

Rotor: It is also built with Fe–Si sheet-steel uniformly slotted that houses the rotor conductors. The rotor can be constructed in two forms that depend on the winding.

(1) *Slip ring rotor*: The slots of the rotor house the conductors of polyphase winding like those used in the stator. The number of poles of both windings

doi:10.1088/978-0-7503-2084-9ch10

Figure 10.1. The induction motor.

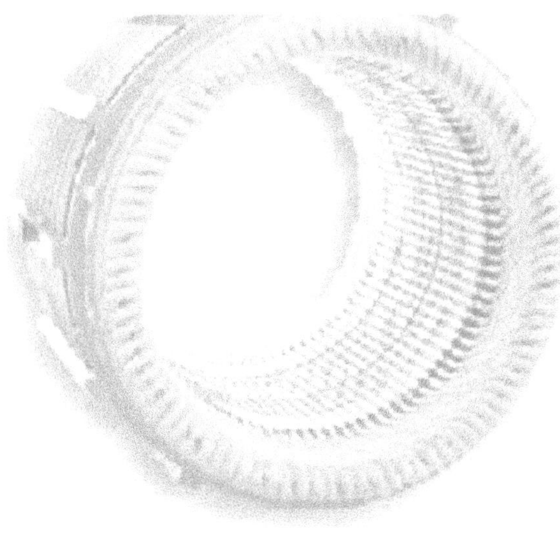

Figure 10.2. Induction motor stator wound with a triphasic winding. Source: Courtesy from EQUACIONAL LTDA Brazil.

are the same. The rotor winding terminals can be accessed via static brushes, which are in contact with the slip rings that rotate with the shaft. This rotor type is also called a wound rotor (figure 10.3).

(2) *Squirrel-cage rotor*: In this construction, the rotor winding is composed of bars made of conductive materials, normally cast aluminum, short-circuited by two conductor rings at its extremities (figure 10.4).

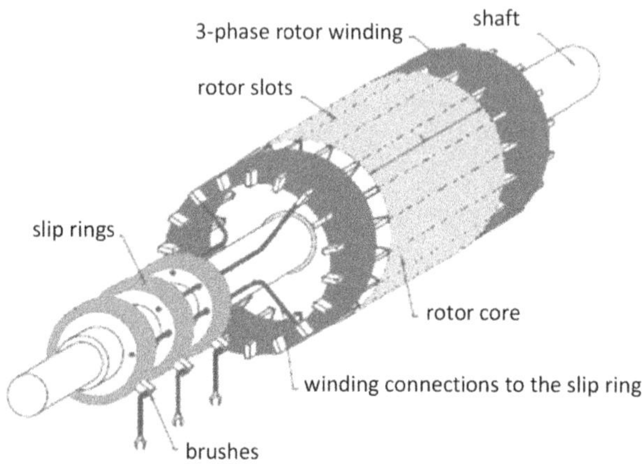

Figure 10.3. The slip ring rotor (wound rotor). Source: Courtesy from EQUACIONAL LTDA Brazil.

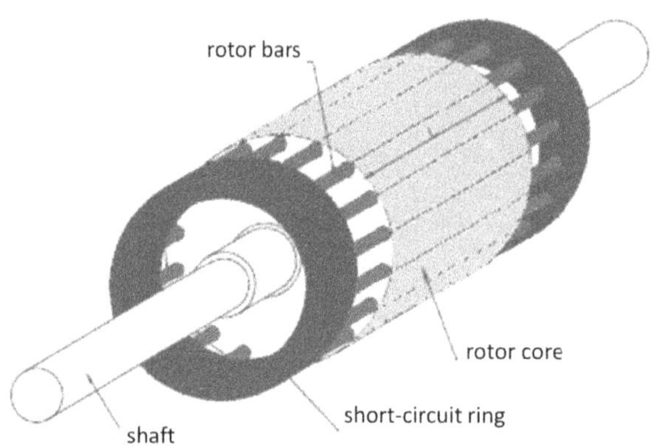

Figure 10.4. The squirrel-cage rotor. Source: Courtesy from EQUACIONAL LTDA Brazil

10.3 How the asynchronous machine works

Is the asynchronous machine a transformer? The answer to this question is: it is almost a transformer. Except for its capacity to convert electrical energy to mechanical energy (and vice-versa), the asynchronous machine operates exactly as a transformer.

Figure 10.5 shows the windings of a slip ring triphasic induction motor. The stator winding is fed by an AC 3-phase voltage source (Vs) that injects 3-phase electric currents. The rotor windings are connected to the brushes that can achieve different types of operation, in this case the rotor windings are opened.

In chapter 9, we demonstrated that a set of 3-phase currents carried by a 3-phase winding produces a magnetic field distribution that rotates at the synchronous speed, given by:

$$n_s = \frac{f}{p}\text{rps} \tag{10.1}$$

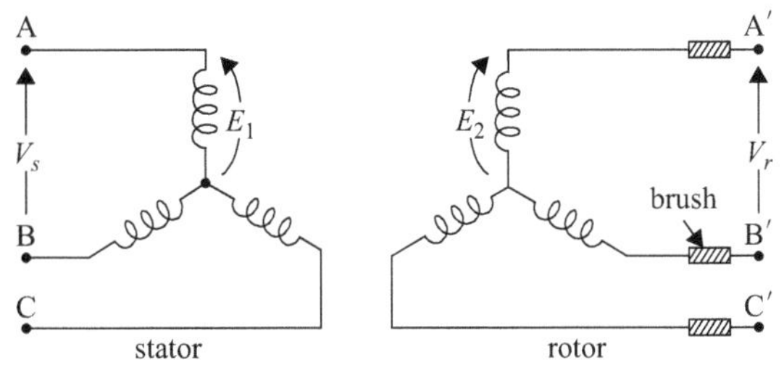

Figure 10.5. Induction motor stalled with the winding rotor opened.

This magnetic field distribution induces an internal emf in the stator winding such that:

$$E_1 = 4.44\, f N_{1\mathrm{eff}} \phi \tag{10.2}$$

In equation (10.2), ϕ is the magnetic flux that crosses the air gap and concatenates with the rotor winding, that can also be a 3-phase winding like the stator one.

If the rotor is stalled, the conductors of its windings *feel* a rotating magnetic field with the same speed *felt* by the stator winding. As a result, an induced emf is generated in the rotor winding and is given by:

$$E_2 = 4.44\, f N_{2\mathrm{eff}} \phi \tag{10.3}$$

therefore:

$$\frac{E_1}{E_2} = \frac{N_{1\mathrm{eff}}}{N_{2\mathrm{eff}}} = a \tag{10.4}$$

The relation (10.4) is the *transformer relation* of the induction motor.

As the rotor winding is opened, no current is carried by its conductors. Supposing that no additional losses are presented, we can write that the same transformer relation is applied to the terminal voltages:

$$\frac{V_s}{V_r} = \frac{E_1}{E_2} = \frac{N_{1\mathrm{eff}}}{N_{2\mathrm{eff}}} = a \tag{10.5}$$

As a transformer, the relation (10.5) is not obeyed because effects like leakage flux, winding resistances and iron losses are not neglected. In a real induction motor with the rotor winding opened, we can establish that:

$$\frac{V_s}{V_r} \approx \frac{E_1}{E_2} = \frac{N_{1\mathrm{eff}}}{N_{2\mathrm{eff}}} = a \tag{10.6}$$

Remarks: The frequencies of the induced emfs of both stator and rotor winding are the same when the induction motor is stalled, working as a cylindrical transformer.

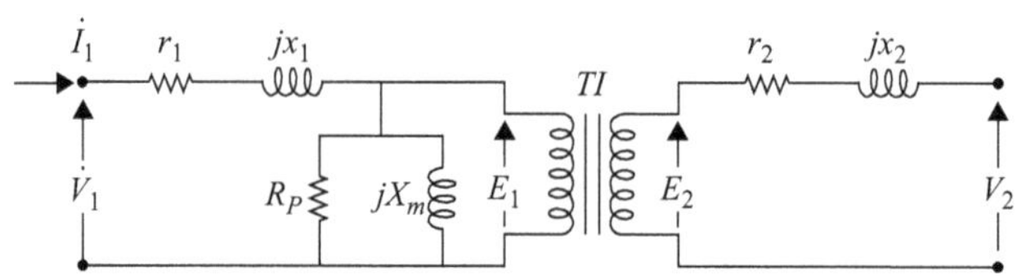

Figure 10.6. Phase-equivalent electric circuit. Induction motor stalled—rotor winding opened.

The phase-equivalent electric circuit for an induction motor stalled and with the rotor winding opened is very similar to the transformer, as shown in figure 10.6.

The circuit parameters are:

r_1 and r_2: the winding resistances of both stator and rotor;
x_1 and x_2: the leakage reactance at frequency of the AC voltage source;
X_m: magnetization reactance;
R_p: iron losses resistance.

Let us suppose now that the rotor is running at a generic speed $n \neq n_s$. Just for simplicity, we are going to consider that the rotor is running in the same direction as the synchronous speed.

As a result, the induced emf of the rotor winding will change proportionally to the change in relative velocity between the rotor conductors and the rotating magnetic field of the stator. To characterize this change, we define a parameter, called 'slip', given by:

$$s = \frac{n_s - n}{n_s} \tag{10.7}$$

That is, the *slip* is the relative velocity between the rotor conductors and the rotating magnetic flux distribution represented as a fraction of the synchronous speed. When the rotor is stalled ($s = 1$), the frequency of both stator and rotor winding are the same. But if ($s \neq 1$), the frequency of the rotor windings is not the same as the frequency of the stator winding.

We can easily identify that the rotor winding frequency is such that:

$$f_r = sf \tag{10.8}$$

where f is the frequency of the AC electric source.

The frequency changes not only alter the induced emf in the rotor winding but also affect the electrical parameter that depends on frequency, as the reactance.

Concerning the new induced emf in the rotor winding, we can write:

$$E_2' = 4.44 f_r N_{2\text{eff}} \phi \tag{10.9}$$

Based on both equations (10.3) and (10.8) we get:

$$E_2' = sE_2 \tag{10.10}$$

Concerning the electric parameters of the rotor electric circuit, the only parameter that should be correct with the frequency variation is the leakage reactance. That is:

$$x_2' = 2\pi f_r l_2 = 2\pi s f l_2 = s x_2 \tag{10.11}$$

No variation is observed in the electrical resistance of the rotor winding. Figure 10.7 shows the phase-equivalent circuit for the induction motor running at speed $n \neq n_s$, with the rotor winding opened.

If the rotor winding terminals are short-circuited, a 3-phase electric current will flow in the windings that will also promote an increment of the stator current.

The induced current in the rotor winding is a reaction due to the magnetic flux variation imposed by the stator winding current. As the electrical parameters of both windings are different, the rotor current and the stator current are not in phase. Because of that, a spatial displacement appears in the relative position of both mmf distributions, promoting the development of an electromagnetic torque. Figure 10.8 shows a *snapshot* of the directions of the electric current in each conductor of both stator and rotor highlighting this spatial displacement by a power angle δ. Both mmf distributions are rotating with a synchronous speed.

We can imagine that the two sets of 3-phase windings are equivalent to two coils magnetically coupled, with its magnetomotive force (mmf) misaligned of a power angle δ, as represented on the right of figure 10.8.

Based on the circuit of figure 10.9, the rms value of the rotor current results in:

$$I_2 = \frac{sE_2}{\sqrt{r_2^2 + (sx_2)^2}} \tag{10.12}$$

Multiplying numerator and denominator by $1/s$, equation (10.12) can be written as:

$$I_2 = \frac{E_2}{\sqrt{\left(\frac{r_2}{s}\right)^2 + x_2^2}} \tag{10.13}$$

We can interpret I_2 as the current obtained from the circuit of figure 10.10. The advantage of this mathematical approach is that we convert the rotor frequency to the same frequency used in the stator ($f_r = f$) for any point of operation of the machine—note that the leakage reactance is not affected by the slip—and the relation of emf is the transformer relation:

$$\frac{E_1}{E_2} = \frac{N_{1\text{eff}}}{N_{2\text{eff}}} = a$$

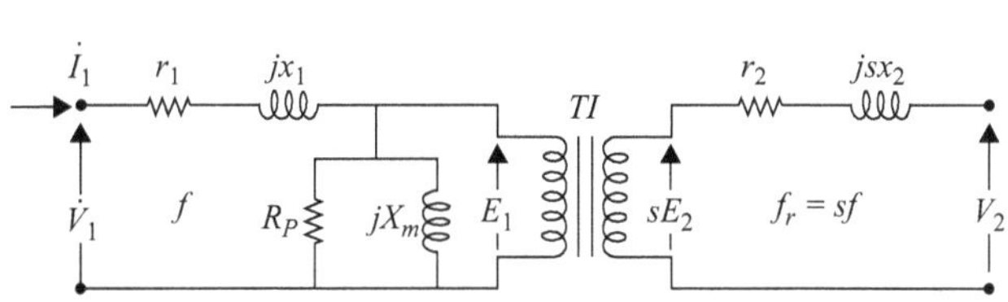

Figure 10.7. Phase-equivalent electric circuit. Induction motor at $n \neq n_s$ -> rotor winding opened. Stator frequency f -> rotor frequency $f_r = sf$.

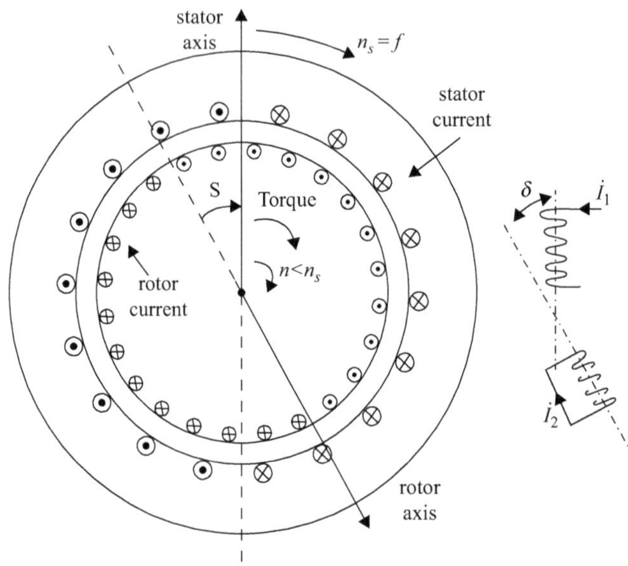

Figure 10.8. Directions of electric current in each conductor. Right: two misaligned equivalent coils.

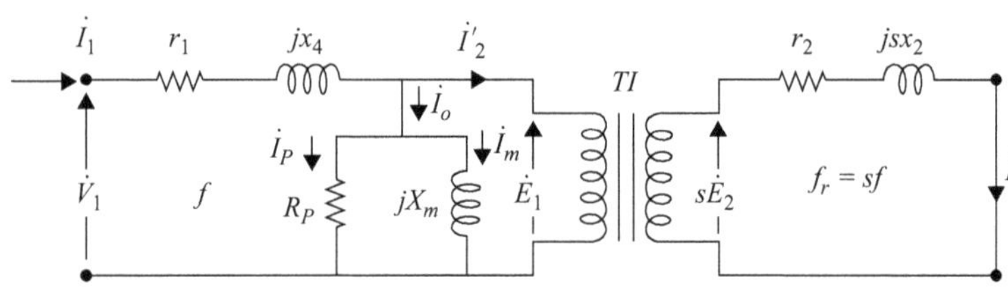

Figure 10.9. Phase-equivalent electric circuit for an induction motor. Rotor winding short-circuited and running at generic slip. Stator frequency f—rotor frequency $f_r = sf$.

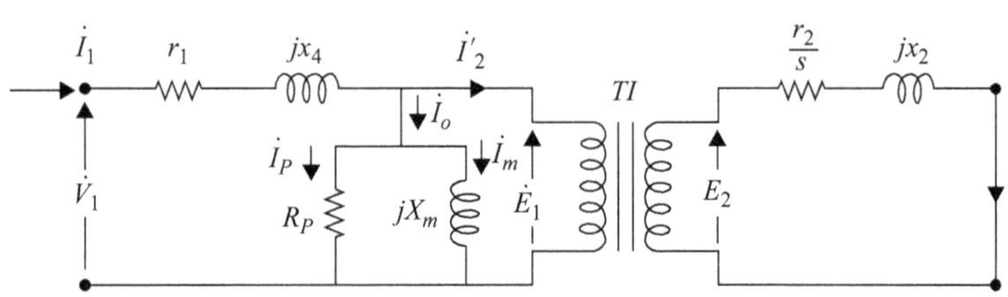

Figure 10.10. The modified equivalent circuit for an induction motor.

Splitting the equivalent resistance $\frac{r_2}{s}$ such that:

$$\frac{r_2}{s} = r_2 + \frac{r_2(1-s)}{s} \tag{10.14}$$

The equivalent circuit of figure 10.10 can be redrawn as shown in figure 10.11.

Referencing all rotor parameters to the stator, we achieve the final equivalent circuit of the induction motor, as shown in figure 10.12.

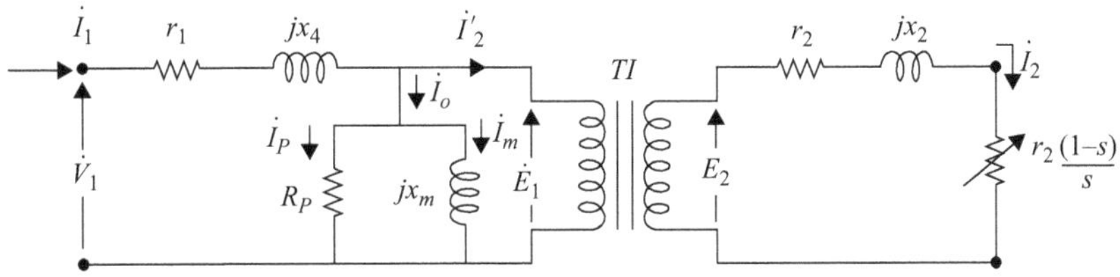

Figure 10.11. The modified equivalent circuit for an induction motor. $\frac{r_2}{s} = r_2 + \frac{r_2(1-s)}{s}$.

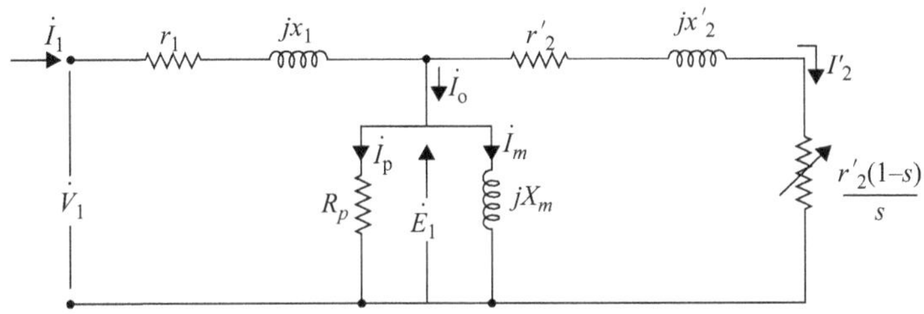

Figure 10.12. The modified equivalent circuit for an induction motor, with reference to the stator.

The energy balance for induction motors

The power flow is a very important tool for understanding the performance of the induction motor, not only for identifying the losses but also for evaluating the power that is converted from electrical to the mechanical and vice-versa.

(1) The input power—P_{in}

P_{in}

The power delivered to the motor by the polyphase AC voltage source. This quantity is expressed by:

$$P_{in} = mV_1 I_1 \cos\varphi \text{ (W)} \tag{10.15}$$

where:

m: is the number of phases;

V_1: is the phase voltage of the AC voltage source;

I_1: is the phase current of the AC voltage source;

$\cos\varphi$: is the power factor.

(2) The Joule losses at stator winding—P_{js}

$$P_{js} = mr_1 I_1^2 \text{ (W)} \tag{10.16}$$

(3) The iron losses—P_{il}

The iron losses are represented in the equivalent circuit by the losses in the iron resistance R_p.

$$P_{il} = m\frac{E_1^2}{R_p} \text{ (W)} \tag{10.17}$$

(4) The power transferred to the rotor—P_{12}

The remaining power $P_{12} = P_{in} - P_{js} - P_{il}$ is the one that is transferred from the stator to the rotor by the electromagnetic effect. It is also called the power that crosses the air gap.

Expressing this power using the rotor parameters, we can write:

$$P_{12} = m\frac{r_2'}{s}I_2'^2 \text{ (W)} \tag{10.18}$$

because the total equivalent resistance of the rotor is $\frac{r_2'}{s}$.

(5) The Joule losses at rotor winding P_{js}

$$P_{jr} = mr_2'I_2'^2 \text{ (W)} \tag{10.19}$$

(6) The developed electromagnetic power—P_{em}

The remaining power:

$$P_{em} = P_{12} - P_{jr} = m\frac{r_2'(1-s)}{s}I_2'^2 \text{ (W)} \tag{10.20}$$

is the one that is converted from electrical to mechanical by the electromagnetic effect. This power is used not only for driving the load attached the shaft, but also for overcoming mechanical losses like friction and ventilation.

(7) The useful power

The useful power is that delivered to the mechanical load. It can be evaluated by:

$$P_u = P_{em} - P_{fv} \tag{10.21}$$

where P_{fv} is the friction power dissipated in rolling bearings added to the power used for ventilation to avoid the motor over-heating.

The efficiency of the induction motor is calculated by:

$$\eta\% = \frac{P_u}{P_{in}} \times 100 \tag{10.22}$$

Figure 10.13 shows the flow chart of the energy balance.

The flow chart of the energy balance can be changed based on the statement that the iron losses are virtually independent of the load. In addition, the losses due to friction and ventilation, are also constant because the motor speed does not change enough from no-load to the full load operation. As a result, we can consider the sum of these losses to be constant during normal operation of the motor.

Therefore, the flow chart can be simplified, as is presented in figure 10.14. In this figure, the *rotational losses* are:

$$P_{rot} = P_{il} + P_{fv} \tag{10.23}$$

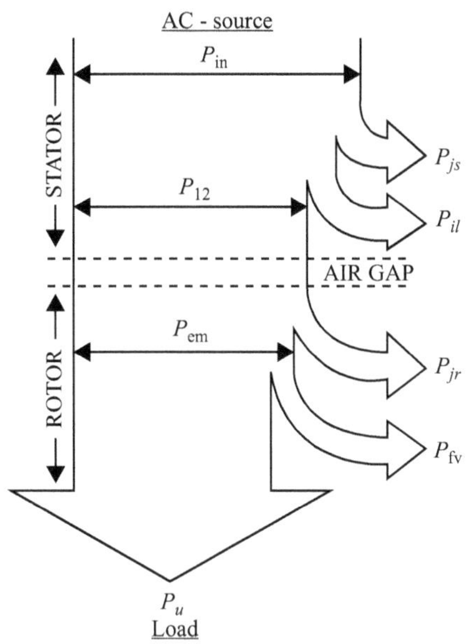

Figure 10.13. Flow chart of the energy balance.

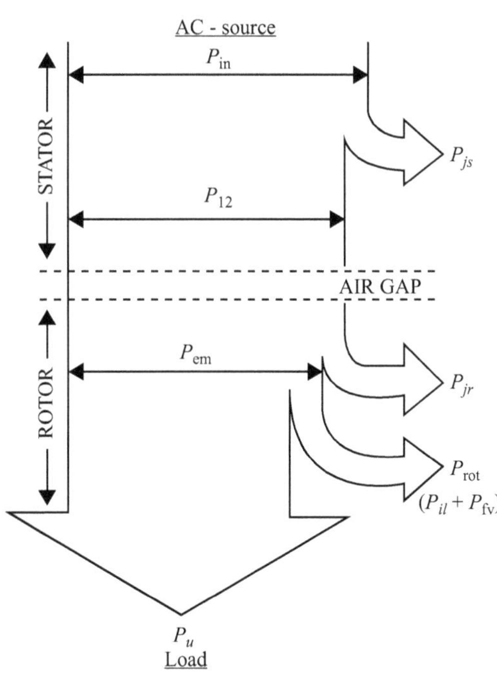

Figure 10.14. The modified flow chart of energy balance. $P_{rot} = P_{il} + P_{fv}$.

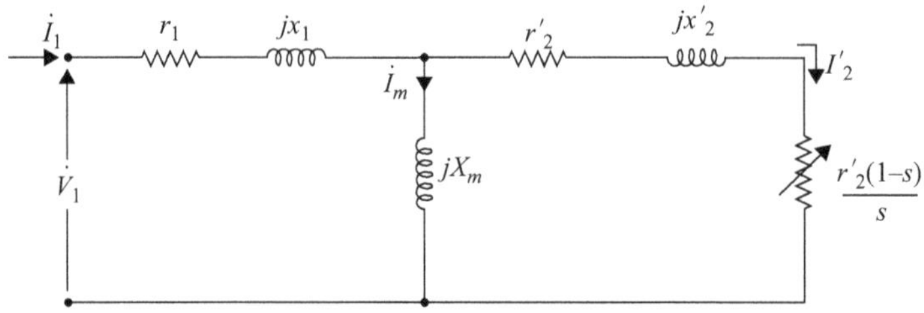

Figure 10.15. The final equivalent circuit based on the modified flow chart.

As the rotational losses are extracted from the energy balance after the developed electromagnetic power, we need to modify the equivalent electric circuit. Figure 10.15 can represent all power fractions, from the input power to the developed electromagnetic one.

Example 10.1

A 3-phase induction motor (60 Hz, wye connected, 6 poles) delivers to the mechanical load a useful power equal to 6.5 hp (1 hp = 746 W) at the rated speed of 1170 rot m^{-1}. The parameters obtained from tests are:

$r_1 = 0.294\ \Omega$; $x_1 = 0.503\ \Omega$; $X_m = 13.25\ \Omega$; $r'_2 = 0.144\ \Omega$; $x'_2 = 0.209\ \Omega$.

The rotational losses can be considered constant and equal to 410 W. Determine:

(a) the rated slip;
(b) draw the equivalent electric circuit for a normal induction motor operation;
(c) draw the flow chart of the energy balance evaluating the efficiency of the induction motor in this condition;
(d) the mechanical torque delivered to the load;
(e) the line-to-line AC voltage applied to the ends of the stator.

Solutions

(a) The slip is given by:

$$s = \frac{n_s - n}{n_s}$$

where: $n_s = \frac{f}{p} = \frac{60}{3} = 20$ rps and $n = \frac{1170}{60} = 19.5$ rps,
therefore:

$$s = \frac{n_s - n}{n_s} = \frac{20 - 19.5}{20} = 0.025$$

(b) The equivalent electric circuit (figure 10.16):
Based on the final circuit shown in figure 10.15.

(c) The energy balance
In this case, the data available is the useful power at the shaft of the motor. The useful power is:

$$P_u = 6.5 \times 746 = 4849\ (\text{W})$$

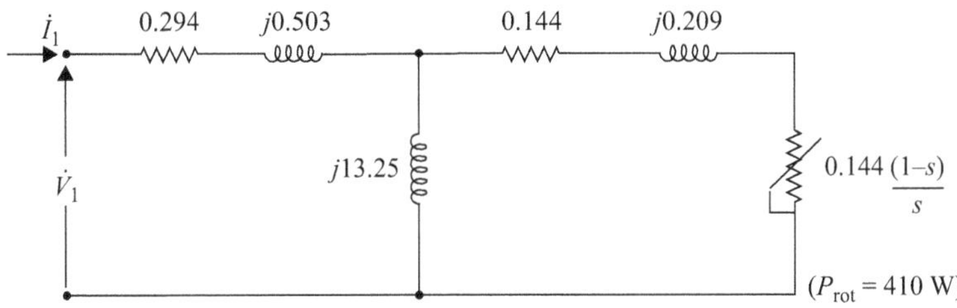

Figure 10.16. Equivalent electric circuit.

The developed electromagnetic power can be evaluated by the expression:

$$P_{em} = P_u + P_{rot} = 4849 + 410 = 5259 \text{ (W)}$$

The rotor current is determined from:

$$P_{em} = m\frac{r_2'(1-s)}{s}I_2'^2$$

So,

$$5259 = 3 \times \frac{0.144 \times (1 - 0.025)}{0.025}I_2'^2$$

Or:

$$I_2' = 17.7 \text{ (A)}$$

The rotor Joule losses:

$$P_{jr} = mr_2'I_2'^2 = 3 \times 0.144 \times (17.7)^2 = 135 \text{ (W)}$$

The power transferred to the rotor is such that:

$$P_{12} = \frac{P_{jr}}{s} = \frac{135}{0.025} = 5400 \text{ (W)}$$

Remarks*:* Observe that $P_{12} = P_u + P_{rot} + P_{jr}$

To evaluate the stator Joule losses, we must first evaluate the stator current.

Based on the equivalent electric circuit we have figure 10.17.

The equivalent impedance of the rotor is:

$$z_2 = \frac{0.144}{0.025} + j0.209 = 5.764\angle 2.08^\circ \ \Omega$$

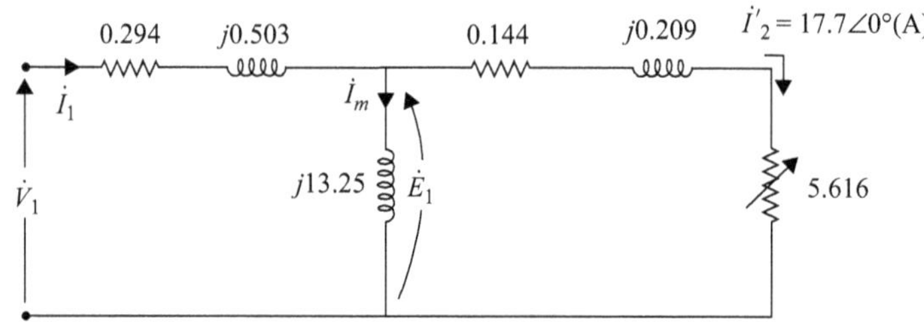

Figure 10.17. Equivalent electric circuit for $s = 0.025$. Operational condition: motor.

The internal emf is such that:

$$\dot{E}_1 = 5.764\angle 2.08° \times 17.7\angle 0° = 102.02\angle 2.08° \text{ (V)}$$

Remarks: The phase of the current was arbitrarily chosen.

The magnetization current is given by:

$$\dot{I}_m = \frac{102.02\angle 2.08°}{j13.25} = 7.7\angle -87.92° \text{ (A)}$$

Therefore, the stator current yields:

$$\dot{I}_1 = 17.7\angle 0° + 7.7\angle -87.92° = 19.56\angle -23.17° \text{ (A)}$$

As a result, the stator power losses are:

$$P_{js} = 3 \times 0.294 \times 19.56^2 = 337.4 \text{ (W)}$$

The input power is such that:

$$P_{\text{in}} = P_{12} + P_{js} = 5400 + 337.4 = 5737.4 \text{ (W)}$$

The efficiency in such a condition is:

$$\eta\% = \frac{P_u}{P_{\text{in}}} \times 100 = \frac{4849}{5737.4} \times 100 = 84.5\%$$

(d) Mechanical torque
The mechanical torque delivered to the load is such that:

$$T_L = \frac{P_u}{2\pi n} = \frac{4849}{2\pi \times 19.5} = 39.6 \text{ (Nm)}$$

(e) The line-to-line AC voltage
The phase voltage of the AC source is such that:

$$\dot{V}_1 = (0.294 + j0.503) \times 19.56\angle -23.17° + 102.02\angle 2.08°$$

or:

$$\dot{V}_1 = 111.6\angle 5.39° \text{ (V)}$$

The line-to-line voltage is:

$$V_l = \sqrt{3} \times V_1 = 193 \text{ (V)}$$

The generator operation

When the asynchronous machine is driving beyond the synchronous speed, it behaves like a generator. To do that, it is necessary to have a prime mover driving it to develop enough mechanical torque that will be converted in electrical energy. Note that for the asynchronous machine operating like a generator not only is its slip negative, but also it should be connected to an external AC voltage system that will receive the generated power by the machine.

As a result, the developed electromagnetic power $\left(P_{em} = m\frac{r_2'(1-s)}{s}I_2'^2\right)$ is also negative, that means that the direction of the electromagnetic power is from the prime mover to the machine and not from the machine to the mechanical load, as would occur during the motor operation.

The wind power plant is a suitable installation for an asynchronous generator. The wind passes through the blades, rotating the low speed shaft that is connected to a gearbox that multiplies the speed about one hundred times. The rotor of an asynchronous generator is connected to the high-speed shaft and is driven beyond the synchronous speed to generate electrical energy. The generated power is delivered directly to the AC system that is connected to the wind power plant.

Example 10.2

We now suppose that the asynchronous machine of Example 10.1 is driven by a prime mover at 1230 rot m^{-1} connected to an 220 (V) AC power system. For the purposes of this example, the effect of the magnetization reactance can be neglected. Determine:

(a) the slip in this condition;
(b) draw the equivalent electric circuit for this condition;
(c) draw the flow chart of the energy balance evaluating the efficiency of the asynchronous generator in this condition;
(d) the mechanical torque delivered to the generator;
(e) the line-to-line AC voltage applied to the ends of the stator.

Solution

(a) Slip is given by:

$$s = \frac{n_s - n}{n_s}$$

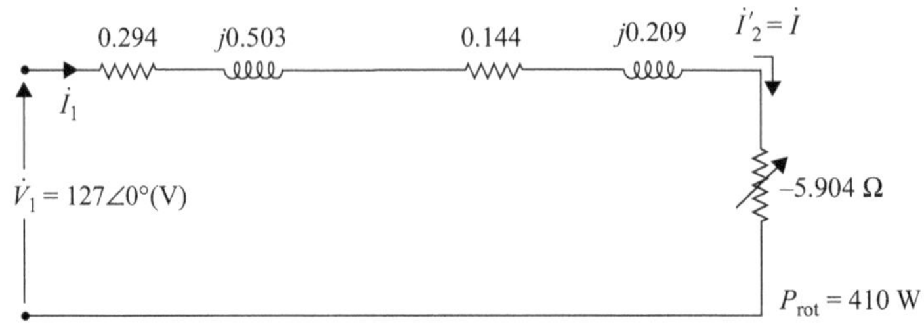

Figure 10.18. Equivalent electric circuit for $s = -0.025$. Effect of magnetization reactance is neglected. Operational condition: generator.

as $n = \dfrac{1230}{60} = 20.5$ rps. Therefore:

$$s = \frac{n_s - n}{n_s} = \frac{20 - 20.5}{20} = -0.025$$

(b) The equivalent electric circuit

Based on the final circuit shown in figure 10.15, it results in figure 10.18.

Remarks: Because $s = -0.025$, the resistance associated to the electromagnetic power $\frac{r_2'(1-s)}{s}$, that is the power that will be converted from mechanical to electrical, is negative. As a result, the power *dissipated* in this resistance is also negative, that means that the flow of the electromagnetic power is from the prime mover to the asynchronous machine.

(c) The energy balance

From the circuit of figure 10.18, the stator current yields:

$$I_1 = \frac{127}{\sqrt{(0.294 + 0.144 - 5.904)^2 + (0.503 + 0.209)^2}} = 23\ (\text{A})$$

The prime mover that injects mechanical power to the asynchronous machine is attached to the shaft of the rotor. Consequently, we start the flow energy balance diagram by the mechanical power input (P_{mech}).

As a result, $P_{em} = 3 \times 5.904 \times 23^2 = 9369$ W. The mechanical power input in the rotor shaft is:

$$P_{mech} = P_{em} + P_{rot} = 9369 + 410 = 9779\ (\text{W})$$

The Joule losses in the rotor winding are evaluated by;

$$P_{jr} = mr_2' I_2'^2 = 3 \times 0.144 \times (23)^2 = 229\ (\text{W})$$

Therefore, the power transferred from the rotor to the stator is given by:

$$P_{21} = P_{\text{em}} - P_{jr} = 9369 - 229 = 9140\ \text{(W)}$$

as $P_{js} = 3 \times 0.294 \times 23^2 = 467$ (W), the electrical power delivered to the AC system yields:

$$P_{\text{u}} = P_{21} - P_{js} = 9140 - 467 = 8673\ \text{(W)}$$

Figure 10.19 represents the flow energy balance diagram in this operational condition.

The efficiency in such a condition is:

$$\eta\% = \frac{P_u}{P_{\text{in}}} \times 100 = \frac{8673}{9779} \times 100 = 88.7\%$$

(d) Mechanical torque

The mechanical torque delivered to the rotor shaft is such that:

$$T_L = \frac{P_{\text{mech}}}{2\pi n} = \frac{9779}{2\pi \times 20.5} = 75.9\ \text{(Nm)}$$

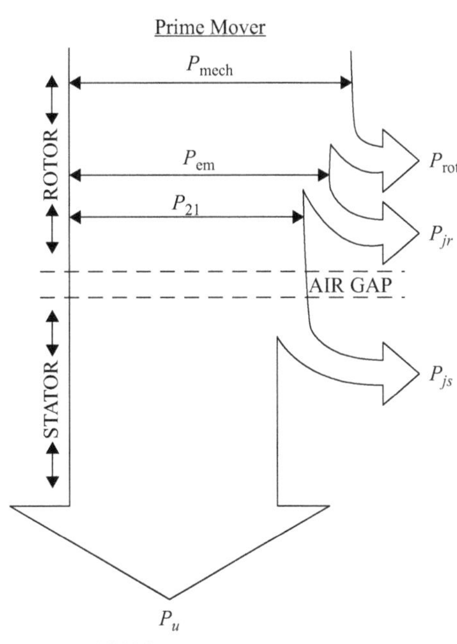

Figure 10.19. Flow energy balance chart; operational condition: generator. The input power is injected in the rotor.

The break operation

The break operation is a not common operation of an asynchronous machine. In this condition, not only is the electrical energy injected in the stator of the machine from the AC source, but also the mechanical power is an input in the rotor shaft from the natural deceleration of the rotor.

That occurs when the rotor is driving in the opposite direction to the rotating magnetic field. This is possible only if a prime mover is attached to the rotor shaft. Therefore, if the rotor is running in the opposite direction to the rotating magnetic field, its slip is such that $s > 1$. As a result, the resistance associated with the electromagnetic power $\frac{r_2'(1-s)}{s}$ is negative but the power transferred to the rotor is positive, because $\frac{r_2'}{s} > 0$. So, if both the mechanical energy and the electrical energy are injected in the rotor shaft at the same time, both are converted to Joule and rotational losses. No useful power is delivered and the only effect is the strong and fast heating process that must stop in a short time.

Example 10.3

We now suppose that the asynchronous machine in example 10.1 is driven by a prime mover at 30 rot m^{-1} in the opposite direction to the rotating magnetic field and connected to a 220 (V) AC power system. For the purposes of this example, the effect of the magnetization reactance can be neglected. Determine:

(a) the slip in this condition;
(b) draw the equivalent electric circuit for this condition;
(c) draw the flow chart of the energy balance evaluating the efficiency of the asynchronous generator in this condition.

Solutions

(a) Slip is given by:

$$s = \frac{n_s - n}{n_s}$$

as $n = -\frac{30}{60} = -0.5$ rps. The negative sign is due to the fact that the speed is in the opposite direction to the rotating magnetic field. Therefore:

$$s = \frac{n_s - n}{n_s} = \frac{20 + 0.5}{20} = 1.025$$

(b) The equivalent electric circuit

Based on the final circuit shown in figure 10.15, it results in figure 10.20.

Remarks: Because $s = 1.025$, the power that will be converted from mechanical to electrical is negative. On the other hand, the resistance associated with the

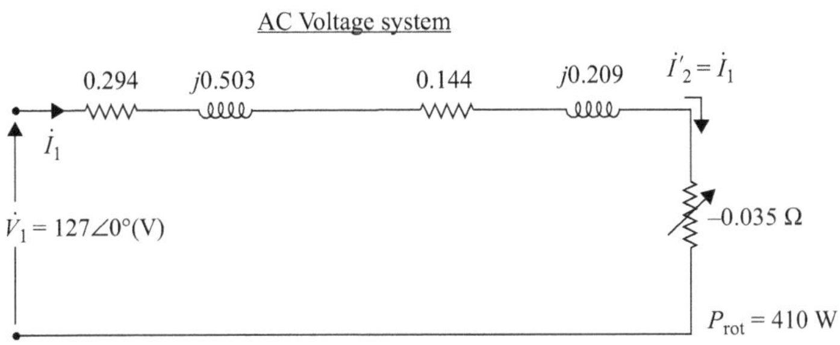

Figure 10.20. Equivalent electric circuit for $s = 1.025$. Effect of magnetization reactance was neglected. Operational condition: break.

electromagnetic power is $\frac{r_2'(1-s)}{s}$ and the power *dissipated* in this resistance is also negative. That means that the flow of the electromagnetic power is also from the prime mover to the asynchronous machine, as in the generator operation.

(c) The energy balance

From the circuit of figure 10.20, the stator current yields:

$$I_1 = \frac{127}{\sqrt{(0.294 + 0.144 - 0.035)^2 + (0.503 + 0.209)^2}} = 155\ (\text{A})!!$$

Remarks: This kind of operation normally involves a high current value.

The electromagnetic power that is converted from mechanical to electric energy is given by (the negative sign was omitted):

$$P_{em} = 3 \times 0.035 \times 155^2 = 2522\ \text{W}$$

The mechanical power input in the rotor shaft is:

$$P_{mech} = P_{em} + P_{rot} = 2522 + 410 = 2932\ (\text{W})$$

The Joule losses in the rotor and stator windings are evaluated by:

$$P_{jr} = 3 \times 0.144 \times (155)^2 = 10\,379\ (\text{W})$$
$$P_{js} = 3 \times 0.294 \times (155)^2 = 21\,190\ (\text{W})$$

As the power factor of the equivalent circuit of figure 10.20 is $\cos\varphi = 0.492$, the result is that the electric power injected in the break will be:

$$P_{in} = 3V_1I_1\cos\varphi = 3 \times 127 \times 155 \times 0.492$$

or:

$$P_{in} = 29\,055\ (\text{W})$$

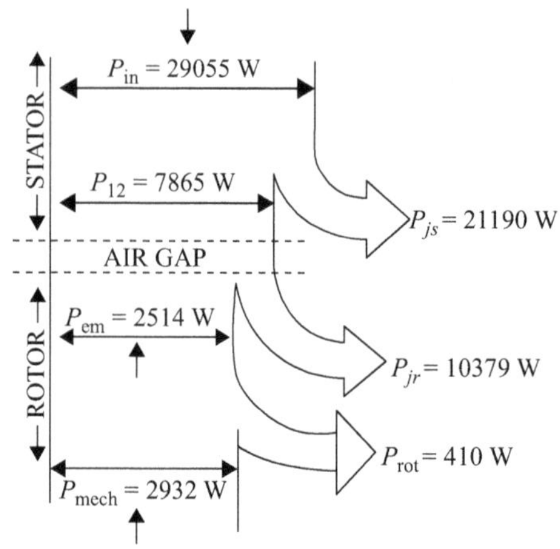

Figure 10.21. Flow energy balance chart. Operational condition: break s = 1.025. Input mechanical power is injected in the rotor. Input electrical energy power injected in the stator.

Figure 10.21 shows the energy balance with two energy inputs.

There is no sense talking aobut efficiency in this operational condition. All the energy inputs in the break are converted into heat. No work is developed in the break operation.

Frequency converter operation

The asynchronous machine also works as a frequency converter. So, the AC voltage system feeds the stator winding with the standard frequency and an AC voltage system with another frequency and voltage magnitude can be extracted from the rotor winding, depending on the rotor speed.

In this kind of operational condition, two kinds of energy are injected in the machine. The first one is the electrical energy supplied by the AC voltage system, connected in the stator winding and second one is the mechanical energy introduced at the rotor shaft. Both energies are transferred to the rotor winding that deliver it to the electric load.

An interesting exercise is to find how these energies are shared to supply the load.

To find it, let us consider an ideal asynchronous machine where all series parameters and the effect of the magnetization reactance are neglected.

Figure 10.22 shows the connections of the windings with both the AC voltage system and the load. The figure also shows a representation of the prime mover attached to the shaft of the machine.

Let $\dot{Z} = R + jX$ be the load impedance that will be fed by the rotor induced AC voltage system with a frequency $f_r = sf$. Neglecting all kinds of losses, the phase-equivalent circuit is shown in figure 10.23.

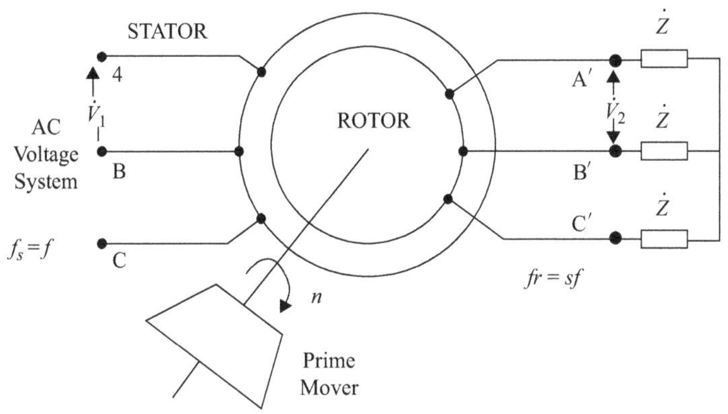

Figure 10.22. Asynchronous machine working as a frequency converter. The rotor speed is dependent on the rotor speed imposed by the prime mover. The sense of the rotor is opposite to the rotating magnetic field.

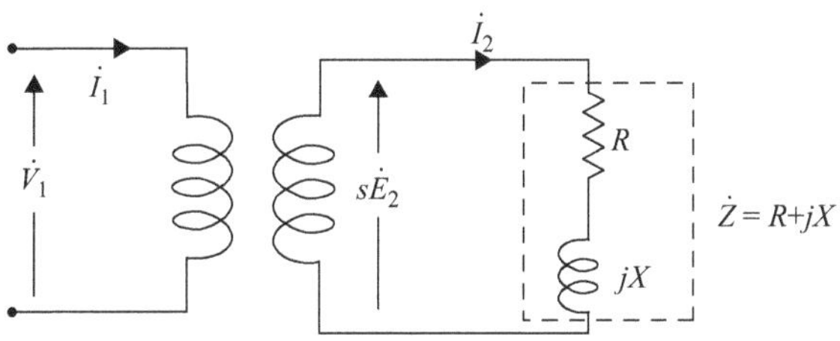

Figure 10.23. The equivalent electric circuit for a frequency converter. No losses are considered and the stator frequency is f. The load impedance is referred to the rotor frequency f_r. Rotor frequency is $f_r = sf$.

The magnitude of the electric current in the rotor winding is given by:

$$I_2 = \frac{sE_2}{\sqrt{R^2 + X^2}} \tag{10.24}$$

Dividing both by the slip s, results in:

$$I_2 = \frac{E_2}{\sqrt{\left(\frac{R}{s}\right)^2 + \left(\frac{X}{s}\right)^2}} \tag{10.25}$$

From equation (10.25), we can establish that the current I_2 is the one obtained from the circuit of figure 10.24, where the frequency is the stator frequency f.

Splitting the impedance $\frac{\dot{Z}}{s}$ in two such that we can separate the power involved, we get the circuit shown in figure 10.25.

The power delivered by the prime mover is the one dissipated in the representative resistance $\frac{R(1-s)}{s}$ that is negative, since the slip $s > 1$. The power dissipated in the

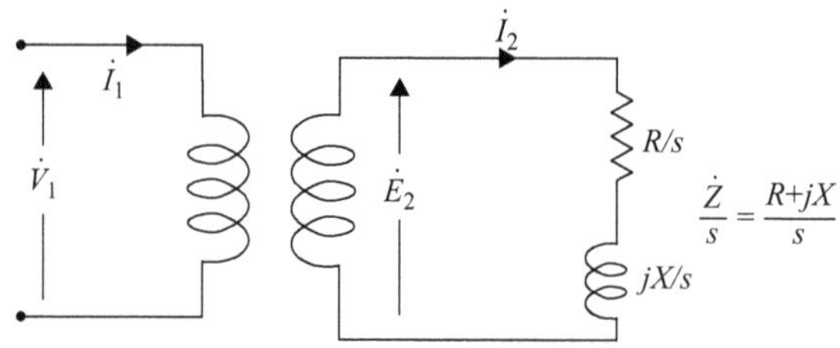

Figure 10.24. The modified equivalent electric circuit. The frequency of the circuit is the stator frequency.

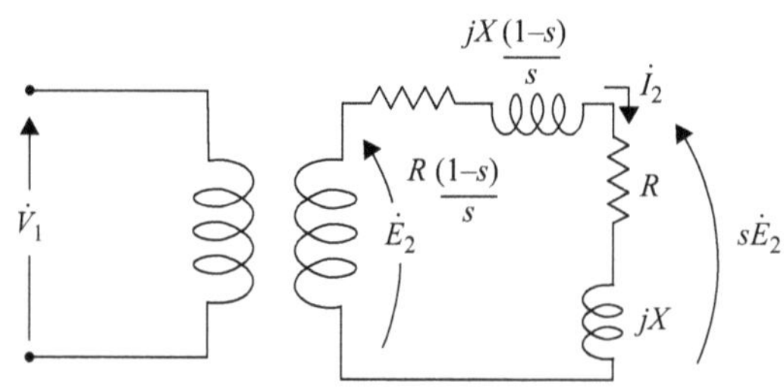

Figure 10.25. The modified equivalent electric circuit. The frequency of the circuit is the stator frequency.

resistance R is the active power delivered to the load and the power supplied by the AC voltage system is the one dissipated in the equivalent rotor resistance $\frac{R}{s}$.

Example 10.4

A portable polishing machine for professional applications is fed by a 220 V AC system—120 Hz. A frequency converter is normally applied as a power supply to the machine. Suppose that the stator is fed by a 60 Hz AC system, determine how this power is shared by both the AC system and the prime mover that drives the rotor.

Solution

As the rotor frequency is 120 Hz, the slip should be = 2. Therefore, the representative resistance of the power delivered by the prime mover is $\frac{R(1-s)}{s} = -\frac{R}{2}$.

As the electric current is the same as the load, we can establish that the power delivered by the prime mover is 50% of the power required by the polishing machine.

The resultant resistance offered to the 60 Hz AC system is such that $\frac{R(1-s)}{s} + R = \frac{R}{s}$ or $\frac{R}{s} = \frac{R}{2}$ in this case. So, we can also establish that 50% of the power required by the load is obtained from the 60 Hz AC system.

10.4 The torque speed characteristics

It is important to have information about the torque speed characteristics when choosing a suitable motor for driving a load. Dependent on the characteristics of the load, it is possible to find a specific motor to accomplish all the requirements that are necessary for good performance.

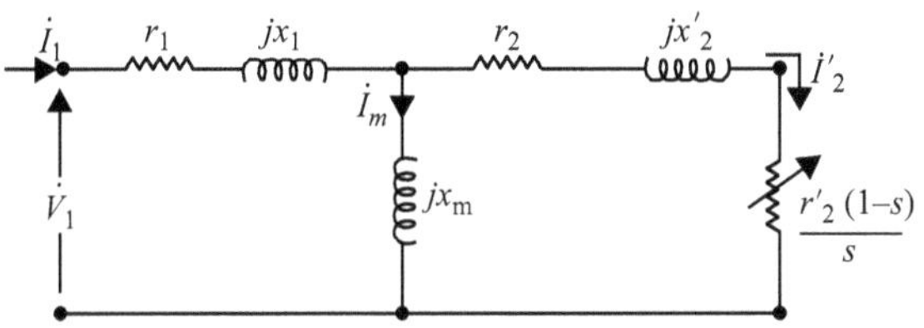

Figure 10.26. The modified equivalent circuit for an induction motor referred to the stator.

The equivalent circuit of the asynchronous motor provides a good starting point to discuss and calculate the developed electromagnetic torque. Figure 10.26 revisits the equivalent circuit of figure 10.15.

The developed electromagnetic power is the one that is converted from electrical to mechanical energy, and is given by:

$$P_{em} = P_{12} - P_{jr} = m\frac{r_2'(1-s)}{s}I_2'^2 \tag{10.26}$$

This power includes the rotational losses P_{rot}. The developed electromagnetic torque associated is written as:

$$P_{em} = C_{dev} \times \omega \tag{10.27}$$

where $\omega = 2\pi n$ is the angular speed of the rotor, that can be expressed as a function of the slip by:

$$\omega = \omega_s(1 - s) \tag{10.28}$$

where $\omega_s = 2\pi n_s$ is the synchronous angular speed of the rotating magnetic field. From equations (10.27) and (10.28), the developed electromagnetic torque results in:

$$C_{dev} = \frac{P_{em}}{\omega_s(1-s)}$$

or,

$$C_{dev} = \frac{m\frac{r_2'(1-s)}{s}I_2'^2}{\omega_s(1-s)} = \frac{m\frac{r_2'}{s}I_2'^2}{\omega_s} \tag{10.29}$$

As a result:

$$C_{dev} = \frac{P_{12}}{\omega_s} \tag{10.30}$$

In conclusion, the developed electromagnetic torque is proportional to the power transferred to the rotor by electromagnetic action.

To acquire the developed electromagnetic torque as a function of the speed, it is necessary to express the rotor current as a function of the slip. To do that, we should represent the stator as its Thévenin equivalent, as figure 10.27 shows.

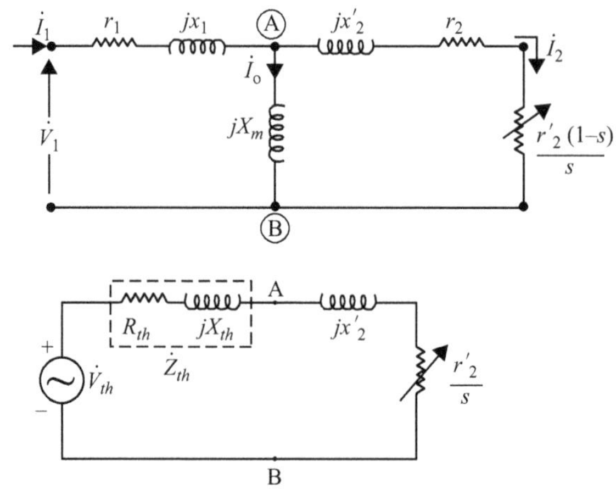

Figure 10.27. The Thévenin equivalent.

The Thévenin voltage is the open voltage at points A and B of the circuit shown in figure 10.26 that is given by:

$$V_{th} = \frac{X_m}{\sqrt{r_1^2 + (X_m + x_1)^2}} V_1 \tag{10.31}$$

The Thévenin impedance is the equivalent impedance at the points A and B, with the source short-circuited. That is:

$$\dot{Z}_{th} = R_{th} + jX_{th} = \frac{jX_m(r_1 + jx_1)}{r_1 + j(X_m + x_1)} \tag{10.32}$$

where:

$$R_{th} = Re[\dot{Z}_{th}]\text{: Thévenin resistance}$$

and

$$X_{th} = Im[\dot{Z}_{th}]\text{: Thévenin reactance}$$

Therefore, the magnitude of the rotor current referred to as the stator yields:

$$I_2' = \frac{V_{th}}{\sqrt{\left(R_{th} + \frac{r_2'}{s}\right)^2 + (X_{th} + x_2')^2}} \tag{10.33}$$

Substituting expression (10.33) into (10.29), results in:

$$C_{\text{dev}} = m\frac{r_2'}{s\omega_s} \times \frac{V_{th}^2}{\left(R_{th} + \frac{r_2'}{s}\right)^2 + (X_{th} + x_2')^2} \tag{10.34}$$

Equation (10.34) expresses the developed electromagnetic torque as a function of the slip. The general aspects of the representative curve of the torque will be discussed next.

To find the shape of this curve, we can start with $s \to 0$ and $s \to 1$. These two limits represent the motor operation.

For $s \to 0$ that means that $n \to n_s$, the term $\frac{r_2'}{s}$ is such that:

$$\frac{r_2'}{s} \gg R_{th} \gg X_{th} + x_2'$$

Imposing this condition in the expression (10.34) yields:

$$C_{\text{dev}} = \frac{m V_{th}{}^2}{r_2' \omega_s} s \text{ (for } s \to 0) \tag{10.35}$$

The plot of the expression (10.35) around $s = 0$ is shown in figure 10.28.

For $s \to 1$, that means $n \to 0$, the term $X' = X_{th} + x_2' \frac{r_2'}{s}$ is such that:

$$X' \gg \frac{r_2'}{s} + R_{th}$$

Imposing this condition in the expression (10.34) yields:

$$C_{\text{dev}} = \frac{m r_2' V_{th}{}^2}{s \omega_s X'^2} \text{ (for } s \to 1) \tag{10.36}$$

The plot of the expression (10.36) around $s = 1$ is shown in figure 10.29.

From equation (10.30), the developed torque can be expressed by:

$$C_{\text{dev}} = \frac{P_{12}}{\omega_s}$$

Therefore, the maximum value of developed torque, also called the *breakdown torque,* can be achieved by finding the slip that results in:

$$P_{12} = P_{12\text{max}}$$

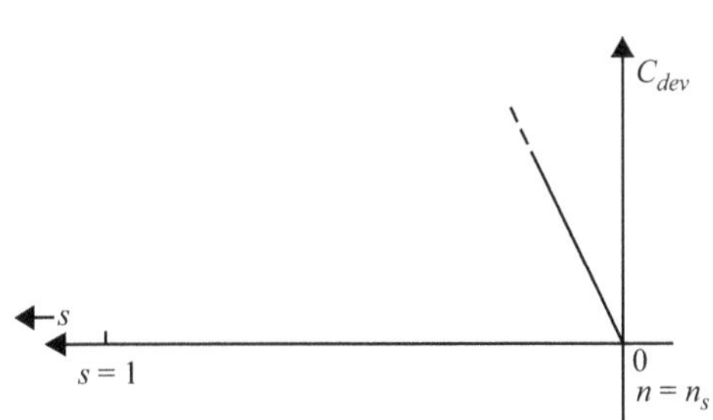

Figure 10.28. $C_{\text{dev}} = C_{\text{dev}}(s)$ is a straight line around $s = 0$. The angular coefficient $\frac{m V_{th}{}^2}{r_2' \omega_s}$. s is increasing at the left side.

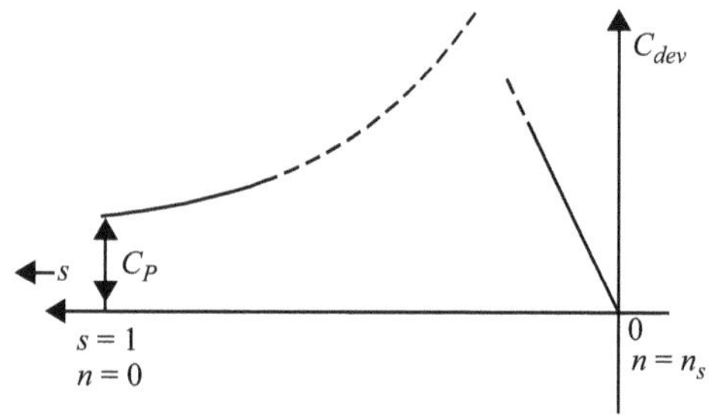

Figure 10.29. $C_{\text{dev}} = C_{\text{dev}}(s)$ is a hyperbole around $s = 1$. $C_p = C_{\text{dev}}(s = 1)$ is the starting torque or locked-rotor torque.

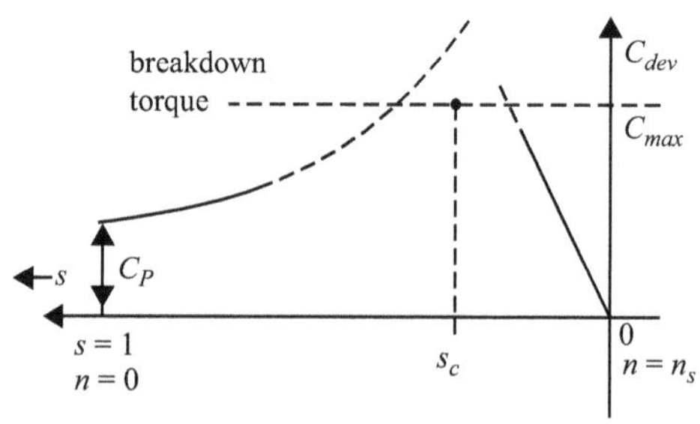

Figure 10.30. The breakdown torque. s_c is the slip when the breakdown torque occurs.

This condition is satisfied when the equivalent resistance of the Thévenin equivalent circuit from figure 10.27 matches the impedance $\sqrt{(R_{th})^2 + (X')^2}$, that is:

$$\frac{r_2'}{s_c} = \sqrt{(R_{th})^2 + (X')^2} \tag{10.37}$$

where s_c is the *breakdown slip*, that is the slip when the maximum torque occurs. Hence:

$$s_c = \frac{r_2'}{\sqrt{(R_{th})^2 + (X')^2}} \tag{10.38}$$

Figure 10.30 shows both curves on the same plot.

Substituting (10.38) in expression (10.34), the *breakdown torque* is given by:

$$C_{\max} = \frac{mV_{th}^2}{2\omega_s\left[\sqrt{(R_{th})^2 + (X')^2} + R_{th}\right]} \tag{10.39}$$

Figure 10.31 shows the complete plot of the developed torque characteristics of the asynchronous machine. Observe that the coordinate system represents the developed torque as a function of speed, highlighting not only the points of the starting C_{st} and

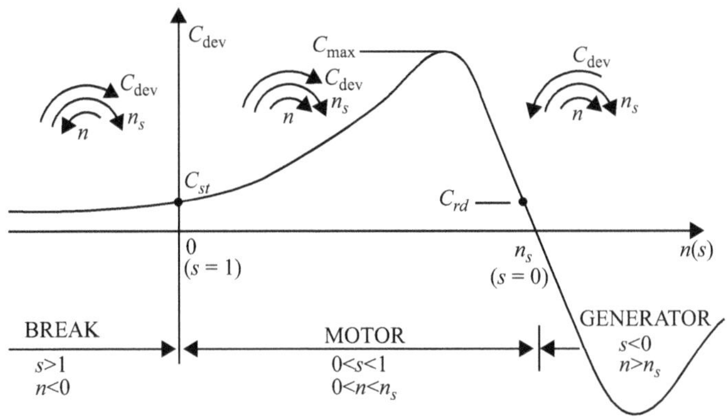

Figure 10.31. Characteristic torque × slip curve. Break—$s > 1$ and $n < 0$. Motor—$0 < s < 1$ and $0 < n < n_S$. Generator—$s > 1$ and $n > n_S$.

the breakdown torque C_{max}, but also the point associated with the rated torque C_{rd}, which is normally close to the synchronous speed.

Three slip ranges can be identified from the analysis of this characteristic torque versus speed.

First case ($0 < n < n_s$):

In this case the slip varies in the range $0 < s < 1$, that corresponds to $0 < n < n_S$.

The developed torque is positive, which means the torque is acting in the direction of the rotational speed. This is characteristic of the motor operation.

Second case ($n > n_s$):

In this situation of $s < 0$, the developed torque is negative and acts in the opposite direction to the rotational speed. This is characteristic of the electric generator, when the rotor is driven by a prime mover delivering mechanical power to the shaft. The mechanical power is converted to electric power that is delivered to the AC system connected to the stator.

Third case ($n < 0$):

When $n < 0$ (the opposite direction to the rotating magnetic field), it results in $s > 1$. In this case, the asynchronous machine is developing a negative electromagnetic torque, that is in the opposite direction to the rotational speed. So, the mechanical power is delivered to the rotor and at the same time the AC voltage source is supplying positive electric power to the stator. Both the electric power delivered to the stator and the mechanical power delivered to the shaft are converted into losses (Joule and rotational losses). This is characteristic of the break operation.

The motor operation is the most common operational condition for the asynchronous machine. Figure 10.32 highlights the characteristics of the induction motor.

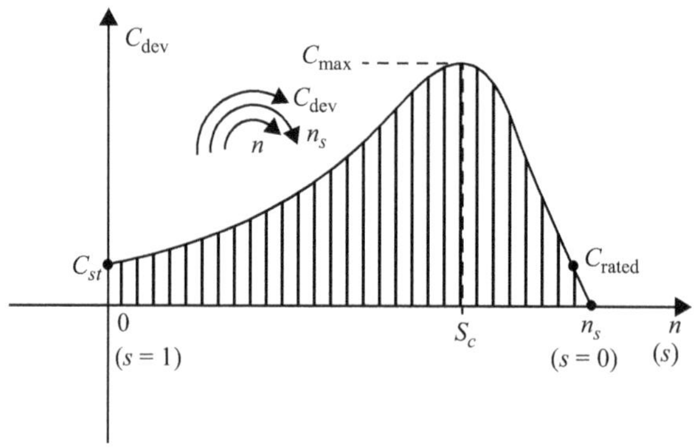

Figure 10.32. Torque versus speed characteristics—motor operation. C_{st}—starting torque. C_{max}—breakdown torque. C_{rated}—rated torque. s_c—breakdown slip.

Three special operation points should be discussed because they are very important in characterizing the performance of the machine.

Point A: This is the operational situation of the machine when the rotational speed is null. The electromagnetic torque developed by the motor is that delivered to the load immediately when the AC voltage system is connected to the stator. This is the starting torque (C_{st}).

Imposing $s = 1$(or $n = 0$) in the general equation of the developed torque (10.34), results in:

$$C_{st} = \frac{mr_2' V_{th}^2}{\omega_s[(R_{th} + r_2')^2 + (X_{th} + x_2')^2]} \tag{10.40}$$

Point B: This is the breakdown torque (C_{max}). This is the maximum developed torque that the motor can deliver to the load. If the load requires torque bigger than the breakdown torque, the motor will lock and can be burned if it is not disconnected from the AC system.

The expression for evaluating not only the breakdown torque but also its slip is given by:

$$C_{max} = \frac{mV_{th}^2}{2\omega_s\left[\sqrt{(R_{th})^2 + (X')^2} + R_{th}\right]} \tag{10.41}$$

$$s_c = \frac{r_2'}{\sqrt{(R_{th})^2 + (X')^2}} \tag{10.42}$$

Remarks: The rotor resistance does not affect the final value of the breakdown torque but does affect the slip that occurs. Therefore, the shape of the curve is also affected if the rotor resistance is changed.

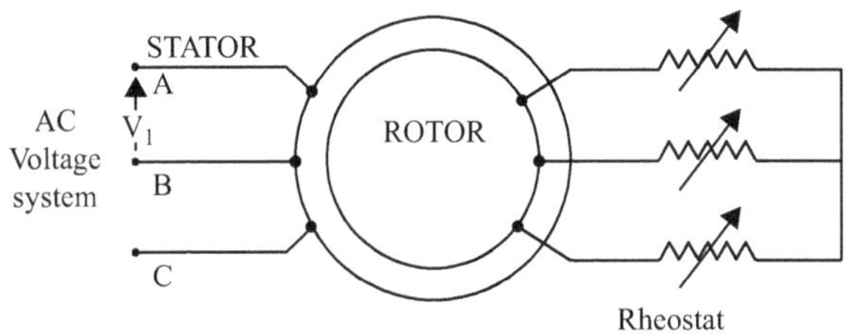

Figure 10.33. The slip ring motor with 3-phase rheostat.

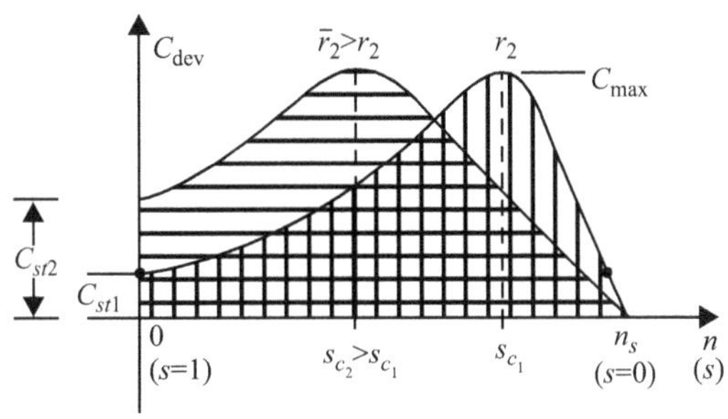

Figure 10.34. Effect of rotor resistance in the torque versus speed curve. $\bar{r}_2 = r_2 + R_{\text{rheostat}}$.

The slip ring motor is suitable for exploring this property, because it is possible to change the rotor resistance inserting an external rheostat in series with the rotor winding, as shown in figure 10.33.

Figure 10.34 shows the effect of the external resistance in the rotor windings. It is an important point that although the breakdown torque is unaltered with the changing rotor resistance, the starting torque grows when the rotor resistance is increased. This is an advantage of this kind of machine but it is not suitable for the squirrel-cage motor.

The possibility of introducing an external rheostat in the rotor winding enables us to control the starting torque of the machine. We can also have the breakdown torque occurring when the motor is stalled. For that, we just have to impose $s_c = 1$ and we get equation (10.43):

$$1 = \frac{r_2' + R_r'}{\sqrt{(R_{th})^2 + (X')^2}} \tag{10.43}$$

R_r' is the rheostat resistance referred to the stator. Therefore, the rheostat resistance (R_r) for this requirement is given by:

$$R_r = \left(\frac{N_{2\text{eff}}}{N_{1\text{eff}}}\right)^2 \left(\sqrt{(R_{th})^2 + (X')^2} - r_2'\right) \tag{10.44}$$

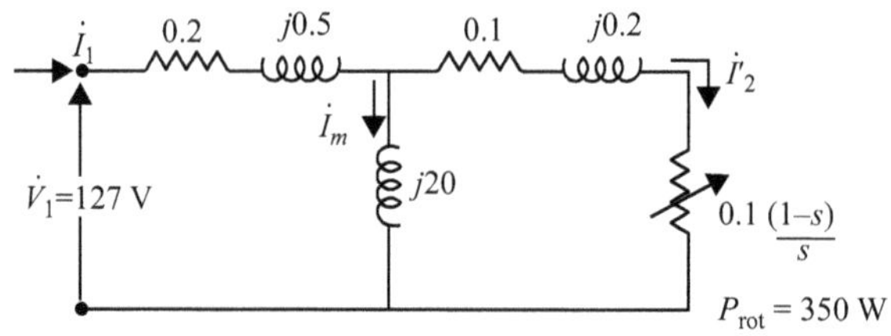

Figure 10.35. Phase-equivalent circuit.

Example 10.5

A 3-phase induction motor (60 Hz, wye connected, 4 poles) delivers 6.5 hp (1 hp = 746 W) to the load at rated speed. The parameters obtained from tests are:

$r_1 = 0.2\ \Omega$; $x_1 = 0.5\ \Omega$; $X_m = 20\ \Omega$; $r'_2 = 0.1\ \Omega$; $x'_2 = 0.2\ \Omega$.

The rotational losses can be considered constants and equal to 350 W. Determine:

(a) the starting torque;
(b) the breakdown torque and the slip that occurs;
(c) the developed torque at 1755 rot m^{-1};
(d) the efficiency for the condition of part (c).

Solution

The phase-equivalent circuit of the motor is the one given by figure 10.35.

The Thévenin equivalent of the stator circuit is evaluated by equations (10.31) and (10.32). The magnitude of the Thévenin voltage is:

$$V_{th} = \frac{X_m}{\sqrt{r_1^2 + (X_m + x_1)^2}} V_1 = \frac{20}{\sqrt{0.2^2 + (20 + 0.5)^2}} \times 127$$

$$V_{th} = 123.9\ (\mathrm{V})$$

and the Thévenin impedance is:

$$\dot{Z}_{th} = R_{th} + jX_{th} = \frac{jX_m(r_1 + jx_1)}{r_1 + j(X_m + x_1)} = \frac{j20(0.2 + j0.5)}{0.2 + j(20 + 0.5)}$$

$$\dot{Z}_{th} = R_{th} + jX_{th} = 0.19 + j0.49\ \Omega$$

Therefore, $R_{th} = 0.19\ \Omega$ and $X_{th} = 0.49\ \Omega$. So, the Thévenin equivalent circuit is shown in figure 10.36.

(a) The starting torque

As for $n = 0$ result in $s = 1$, the rotor current referred to the stator yields:

$$I'_2 = \frac{V_{th}}{\sqrt{\left(R_{th} + \frac{r'_2}{s}\right)^2 + (X_{th} + x'_2)^2}} = \frac{123.9}{\sqrt{(0.19 + 0.1)^2 + (0.49 + 0.2)^2}} = 165.5\ \mathrm{A}$$

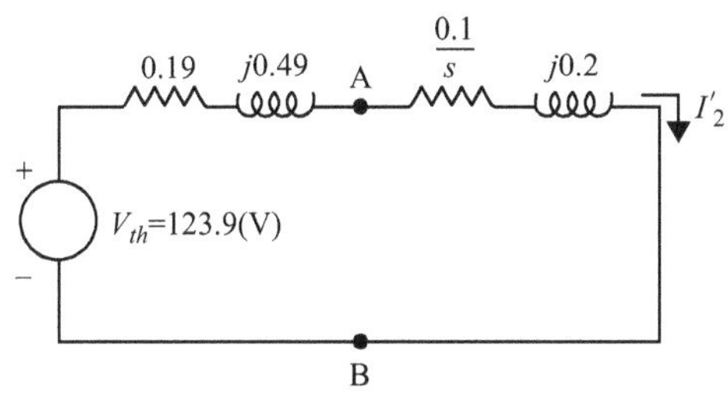

Figure 10.36. Thévenin equivalent circuit.

The power transferred to the rotor in this condition is:

$$P_{12} = m\frac{r_2'}{s}I_2'^2 = 3 \times 0.1 \times 165.5^2 = 8217\text{ W}$$

The starting torque is such that:

$$C_{st} = \frac{P_{12}(s = 1)}{\omega_s} = \frac{8217}{2\pi \times 30} = 43.6\text{ Nm}$$

(b) The breakdown torque

Based on equations (10.41) and (10.42), the result is:

$$C_{\max} = \frac{mV_{th}^2}{2\omega_s\left[\sqrt{(R_{th})^2 + (X')^2} + R_{th}\right]} = \frac{3 \times 123.9^2}{2 \times 60\pi[\sqrt{(0.19)^2 + (0.69)^2} + 0.19]}$$

that is:

$$C_{\max} = 135\text{ Nm}$$

The slip associated with the breakdown torque is evaluated by:

$$s_c = \frac{r_2'}{\sqrt{(R_{th})^2 + (X')^2}} = \frac{0.1}{\sqrt{(0.19)^2 + (0.69)^2}} = 0.14$$

(c) The rated torque

The power transferred to the rotor at rated speed is extracted from the circuit of figure 10.35, imposing $s = 0.025$. The rotor current is given by:

$$I_2' = \frac{123.9}{\sqrt{(0.19 + 4)^2 + (0.49 + 0.2)^2}} = 29.2\text{ A}$$

As a result: $P_{12} = m\frac{r_2'}{s}I_2'^2 = 3\frac{0.1}{0.025}29.2^2 = 10\,231\text{ W}$

The rated torque is given by:

$$C_{\mathrm{dev}} = \frac{P_{12}}{\omega_s} = \frac{10\,231}{2\pi \times 30} = 54.2 \text{ Nm}$$

Using the approach of low slip, the rated torque can also be evaluated by:

$$C_{\mathrm{dev}} = \frac{mV_{th}^2}{r_2'\omega_s}s = \frac{3 \times 123.9^2}{0.1 \times 60\pi} \times 0.025 \cong 61 \text{ Nm}$$

Comparing both results, we note that the difference is about 10%. That is enough for a fast evaluation of the quantity.

(d) The efficiency

As the power transferred to the rotor (P_{12}) has already been evaluated, it remains only to include the Joule losses in the stator winding to find the input power (P_{in}). The electric current in the stator winding is extracted for the circuit of figure 10.35. Using the current divider, we can establish that:

$$I_1 = \left\| \frac{(r_2' + jX_2') + (jX_m)}{jX_m} \right\| \times I_2' = \left\| \frac{0.1 + j(0.2 + 20)}{j20} \right\| \times 29.2$$

Since $I_2' = 29.2\ A$, resulting in:

$$I_1 = 29.5 \text{ A}$$

So that:

$$P_{js} = mr_1 I_1^2 = 3 \times 0.2 \times 29.5^2 = 522 \text{ W}$$

Note that $P_{\mathrm{in}} = P_{12} + P_{js}$, resulting in: $P_{\mathrm{in}} = 10753\ W$

As the useful power is such that $P_{\mathrm{u}} = P_{\mathrm{em}} - P_{\mathrm{rot}}$ and since $P_{\mathrm{em}} = (1 - s)P_{12} = 10484$ W, we have:

$$P_{\mathrm{u}} = P_{\mathrm{em}} - P_{\mathrm{rot}} = 10\,484 - 350 = 10\,134 \text{ W}$$

Therefore, the efficiency is:

$$\eta\% = \frac{P_u}{P_{\mathrm{in}}} \times 100 = \frac{10\,134}{10\,753} \times 100 = 94\%$$

The slip dependence of the stator current is important information not only for the performance of the machine, but also for decision-making concerning the electrical protection of the equipment.

Based on the standard equivalent electric circuit of figure 10.37 and on the current divider, we can write:

$$I_1 = \frac{\sqrt{(r_1)^2 + (X_m + x_2')^2}}{X_m} \times I_2' \qquad (10.45)$$

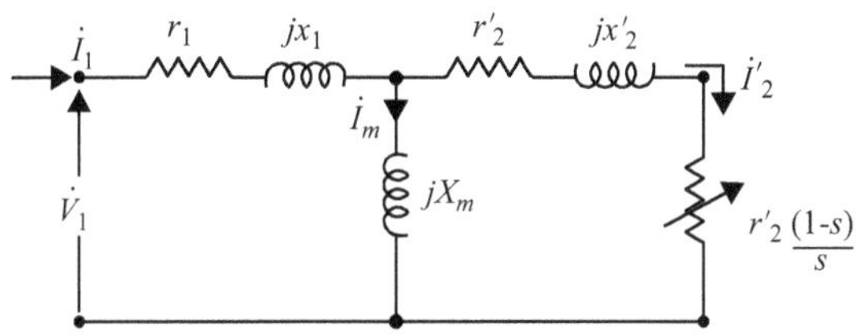

Figure 10.37. The standard equivalent electric circuit.

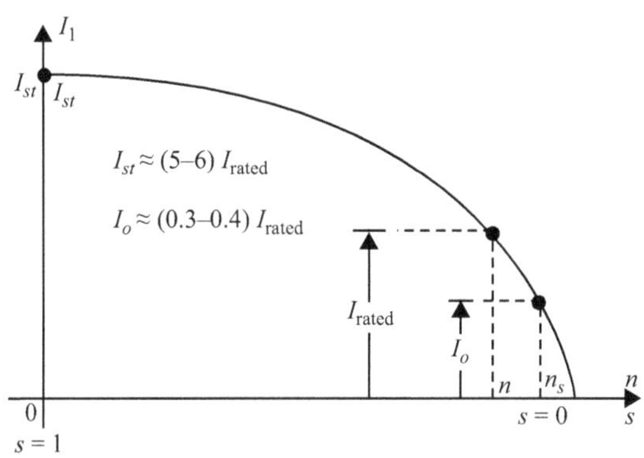

Figure 10.38. The current × slip characteristics.

as,

$$I_2' = \frac{V_{th}}{\sqrt{\left(R_{th} + \frac{r_2'}{s}\right)^2 + (X_{th} + x_2')^2}}$$

resulting in:

$$I_1 = \frac{\sqrt{(r_1)^2 + (X_m + x_2')^2}}{X_m\sqrt{\left(R_{th} + \frac{r_2'}{s}\right)^2 + (X_{th} + x_2')^2}} \times V_{th} \tag{10.46}$$

We plot expression (10.46) as a function of the slip, figure 10.38.

Three operation points should be highlighted:

Point A (I_{st}): The stator current at $s = 1$

This is the condition observed when the motor is connected to the AC power supply with the rotor stalled. The current at this point is also called the *starting current* that should be considered for sizing electrical protection devices. The magnitude of this current is normally 5 to 6 times the rated current. So, it is a high current that can promote stresses not only in the stator winding but also in the rotor winding. Several starting devices were designed to mitigate this problem.

Point B (I_{rated}): The stator current at rated speed

This is the ideal operation point for the induction motor. Normally, at this condition, the efficiency of the machine achieves its maximum value.

Point C (I_0): The stator current at no-load condition

In this case, the motor is running without driving any mechanical load. Different to the transformer, the no-load current of the induction motor is not negligible. Its magnitude can be achieved with 30% to 40% of the rated current.

Series rheostat

The feature of inserting a series rheostat into the rotor winding enables the slip ring induction motor to be more flexible and allows us to control its starting current. Figure 10.39 shows the evolution of the starting current when the slip ring induction motor is connected to the AC power supply.

The complete rheostat, composed of three series resistors, is inserted in the rotor winding ends. Normally, not only is an upper limit imposed for the stator current that is dependent on the acceleration time (I_{max}), but also a lower limit that is a little bigger than the rated current.

Such resistors are fitted based on the imposed limits of the starting current. They are short-circuited in sequence, when the stator current reaches its lower value.

The acceleration of the induction motor

Figure 10.40 shows the acceleration process of an induction motor that is driving a load at its shaft that offers a given resistant torque characteristics.

Curve (A) represents the developed torque characteristics (T_{dev}) of the induction motor and curve (B) the load resistance torque characteristics (T_{res}). For each speed, the difference between the curves defines the acceleration torque ($T_{acc} = T_{dev} - T_{res}$), that is responsible for the speed changing.

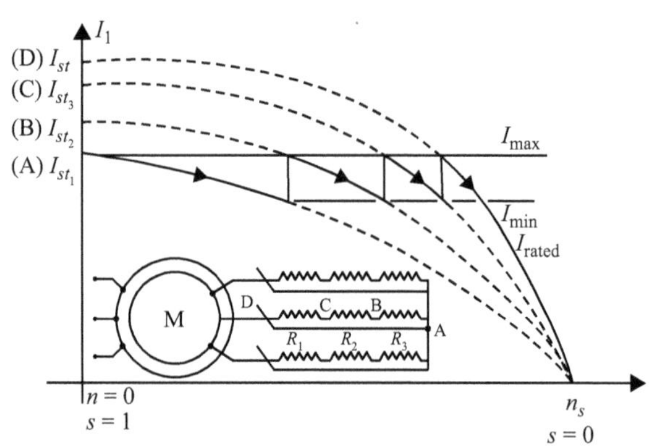

Figure 10.39. Starting current evolution with starting rheostat. $R_1 + R_2 + R_3$: point A. $R_1 + R_2$: point B. R_1: point C. Rheostat short-circuited: point D.

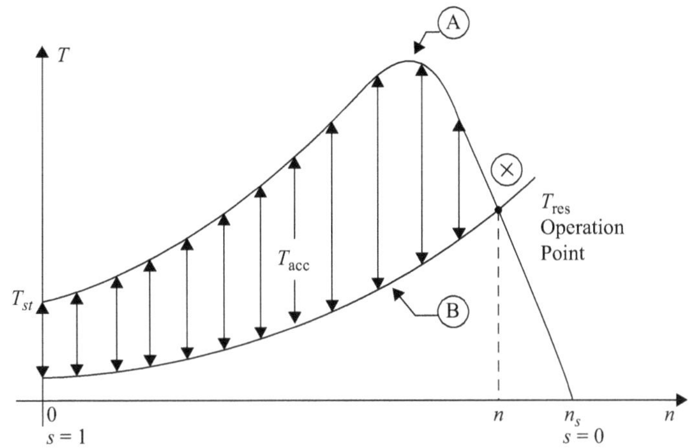

Figure 10.40. Acceleration process. $T_{acc} = T_{dev} - T_{res}$. For the operation point X, $T_{dev} = T_{res}$.

When the acceleration torque is non-null the speed changes and when the acceleration torque is null, the speed is constant at its operation point (X).

Speed control of the induction motor

Nowadays, the speed control of the induction motors via power electronics converters has became very popular, not only because of cost reductions in the last decades, but also because its features provide more versatility in the operation of the machine. The price of induction motors associated with converters is lower than a DC motor with lower maintenance costs. The speed control of the induction motor through the frequency converter, practically eliminated the use of controlled DC machine systems.

10.5 The single-phase induction motor

The 1-phase version of the induction motor is, perhaps, the most produced motor in the world. It is employed in several electrical appliances like washing-machines, food processors, fans etc.

The rotor of the single-phase induction motor is a squirrel-cage like the one used in the 3-phase induction motor. The stator hosts two independent windings.

The first is called the *main winding* where all energies are changed in steady state operation. Figure 10.41 shows a cross-section of the main winding that carries AC electric current.

As the winding is distributed, the associated magnetic flux density distribution can be considered as a sinusoidal function with a fundamental component that has a spatial period of $2\pi(m)$ and some harmonics.

We can consider in our analysis only the *fundamental*, since the magnitudes of harmonics are low and can be neglected. Different to the 3-phase windings, the sinusoidal distribution of magnetic flux density is neither rotating or static. As the

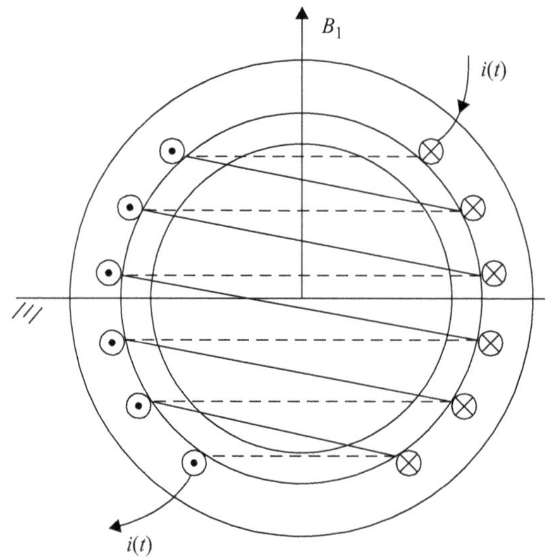

Figure 10.41. The main winding of a 2-poles single-phase induction motor.

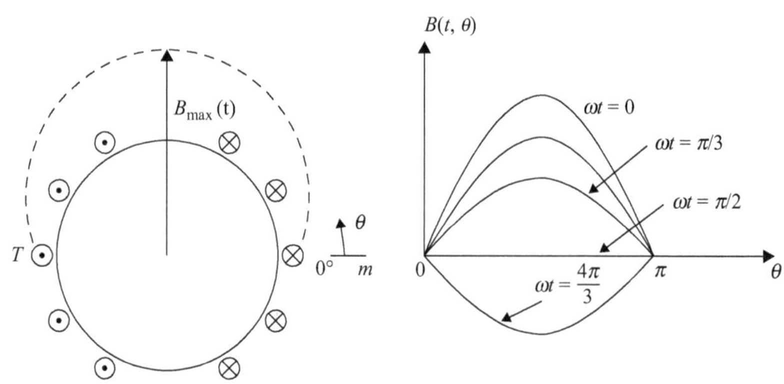

Figure 10.42. The sinusoidal pulsating magnetic flux density distribution.

current injected in the coil is a sinusoidal time-dependent function, the magnitude of the magnetic field distribution is also a time-dependent function.

This characteristic is called the sinusoidal pulsating function, as described in figure 10.42.

The mathematical expression of such a function is:

$$B(\theta, t) = B_{max}(t) \sin \theta \tag{10.47}$$

where:

$$B_{max}(t) = B_0 \cos \omega t \tag{10.48}$$

Therefore, it can be written as:

$$B(\theta, t) = B_0 \cos \omega t \times \sin \theta \tag{10.49}$$

Applying the identity:

$$2 \sin x \times \cos y = \sin(x - y) + \sin(x + y)$$

equation (10.49) becomes:

$$B(\theta, t) = \frac{1}{2} B_0 \sin(\theta - \omega t) + \frac{1}{2} B_0 \sin(\theta + \omega t) \tag{10.50}$$

Equation (10.50) is the sum of two rotating magnetic fields, such that:

$$B_f(\theta, t) = \frac{1}{2} B_0 \sin(\theta - \omega t) \tag{10.51}$$

It represents a magnetic flux density distribution that rotates in the direction $\theta > 0$ with the angular speed ω rad s^{-1} and magnitude $\frac{1}{2}B_0$, and:

$$B_r(\theta, t) = \frac{1}{2} B_0 \sin(\theta + \omega t) \tag{10.52}$$

It is the magnetic flux density distribution that rotates in the opposite direction $\theta < 0$ with the same angular speed ω rad s^{-1} and magnitude $\frac{1}{2}B_0$.

In conclusion, we can suppose that the single-phase induction motor is composed of two 3-phase induction motors that have magnetic field distributions rotating in opposite directions. These two motors are series associated and share its parameters fifty-fifty, as figure 10.43 shows.

The forward slip is given by:

$$s_f = \frac{n_s - n}{n_s} \tag{10.53}$$

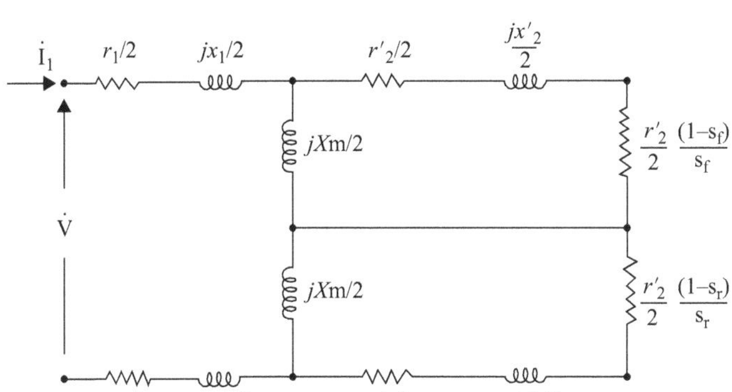

Figure 10.43. Preliminary equivalent circuit for a one-phase induction motor. s_f: forward slip. s_r: reverse slip.

And the reverse slip is:

$$s_r = \frac{-n_s - n}{-n_s} \tag{10.54}$$

The negative signal is due to the opposite sense of the rotating field. Extracting *n* from equations (10.53) and (10.54) and equalizing both, results in:

$$1 - s_f = -1 + s_r$$

or:

$$s_r = 2 - s_f \tag{10.55}$$

Renaming the forward slip such that ($s_f = s$), the equivalent electric circuit becomes the one presented in figure 10.44.

We can see that the single-phase induction machine is composed of two 3-phase induction motors, both with rotating magnetic fields running in the same stator but in opposite directions, the resultant characteristics of the developed torque is obtained by the superposition of them, as shown figure 10.45.

From the analysis of the developed torque characteristics, we observe that the single-phase induction motor has no *starting torque*. So, it is necessary to overcome this difficulty with another winding that carries an electrical current out-of-phase with the electrical current of the main winding.

The new winding, called the *auxiliary winding*, is spatially located 90° from the main winding, as shown figure 10.46.

To make both currents out-of-phase, the auxiliary winding is series associated with a capacitor—the *starting capacitor*—and parallel connected to the main winding, as shown figure 10.47.

The combination of the main and the auxiliary winding carried by its own current, produces a *fragile* rotating magnetic field able to develop a starting torque to move the mechanical load.

Once the motor starts moving, the auxiliary winding is no longer necessary. The switch in figure 10.47 is a centrifugal switch that is opened when the motor speed

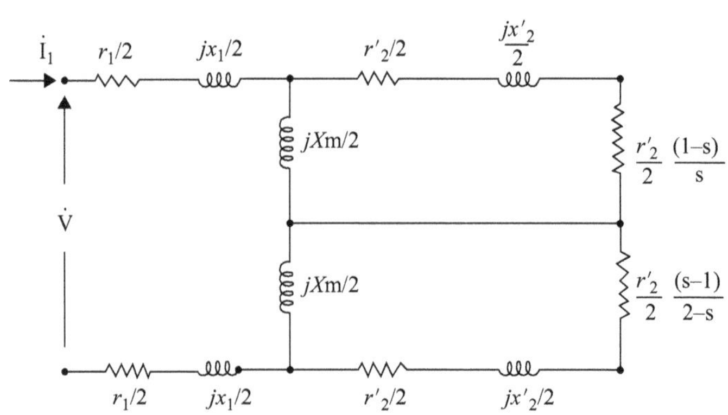

Figure 10.44. The final equivalent electric circuit of the single-phase induction motor.

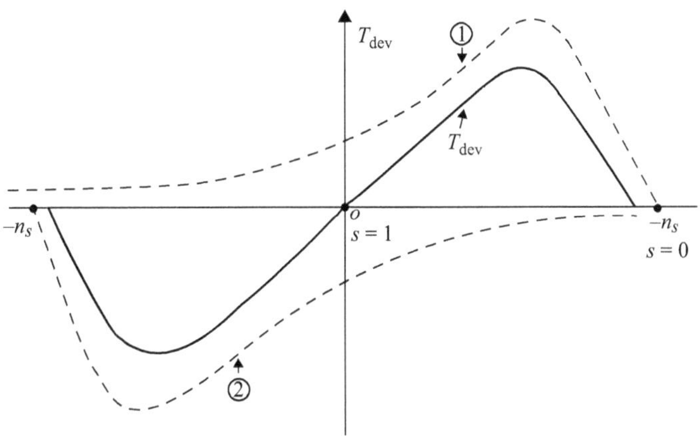

Figure 10.45. Developed torque characteristics of the single-phase induction motor. Dashed line 1: the developed characteristics due to the forward rotating magnetic field. Dashed line 2: the developed characteristics due to the reverse rotating magnetic field. Solid line: the resultant developed torque characteristics.

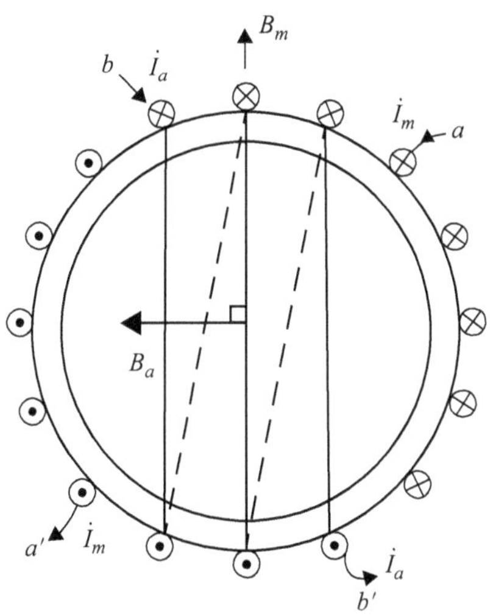

Figure 10.46. The auxiliary winding. I_m: main winding current; I_a: auxiliary winding current. B_m: main magnetic field; B_a: auxiliary magnetic field.

achieves about 80% of the final speed. As a result, the final developed torque characteristics is a composition of two curves, as figure 10.48 shows.

Remarks: Modern single-phase motors work with a permanent capacitor avoiding the centrifugal switch.

There is a kind of single-phase motor, called a *split-phase* motor, which has no starting capacitor. Its starting torque is lower than the one used for a *starting capacitor* but is sufficient for driving small loads.

Even with a permanent capacitor, the efficiency of a single-phase induction motor is not convenient for use in integral motors (more than 1 hp).

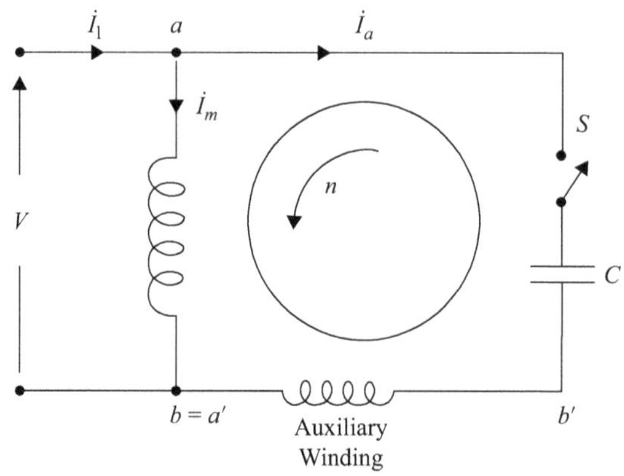

Figure 10.47. Winding connections of single-phase induction motor. C: starting capacitor; S: centrifugal switch. $\dot{I}_1 = \dot{I}_m + \dot{I}_a$.

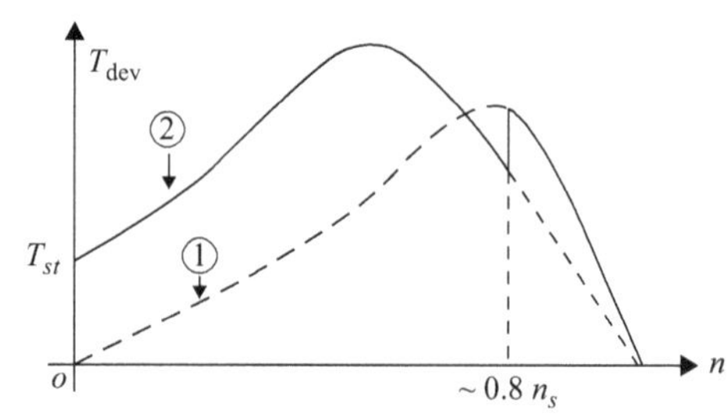

Figure 10.48. The developed torque characteristics of a single-phase induction motor. Curve 1: the torque characteristics of the main winding effect. Curve 2: the torque characteristics of both windings. Solid line: the final torque developed characteristics.

10.6 The shade-pole induction motor

Figure 10.49 shows a cross-section of a shade-pole induction motor. Three different windings can be identified. The first one is the main winding located on the poles. The main winding is fed by the AC voltage source that produces the main magnetic flux ϕ. The second one is the shade-coil, also located on the poles. This winding is a single coil short-circuited. Finally, the third is squirrel-cage winding located on the rotor.

The induced current of the shade-coil ($\dot{I}_s$) is not in phase with the one carried by the main winding ($\dot{I}$). This characteristic produces a polarity variation in the pole that is similar to the rotating magnetic field.

Figure 10.50 shows the polarities of the magnetic field at poles, dependent on the direction of the currents. At instant (1), the current in the main winding is positive (producing a south pole) and the one on the shade-coil is negative

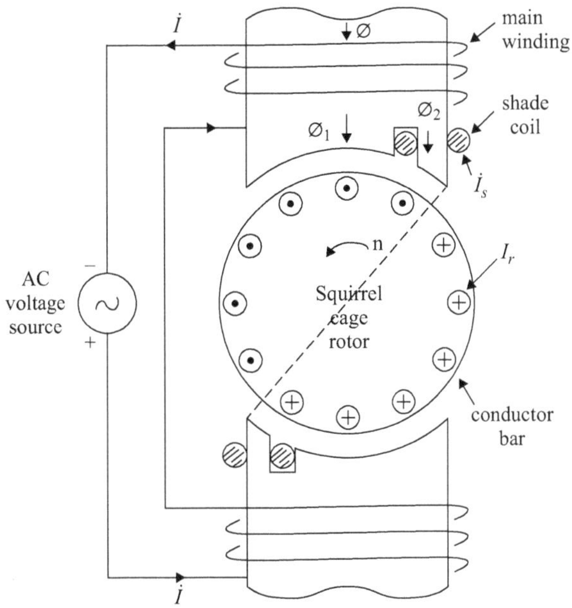

Figure 10.49. Cross-section of a shade-pole induction motor. I: main winding current; I_s: shade-coil induced current; I_r: rotor induced current.

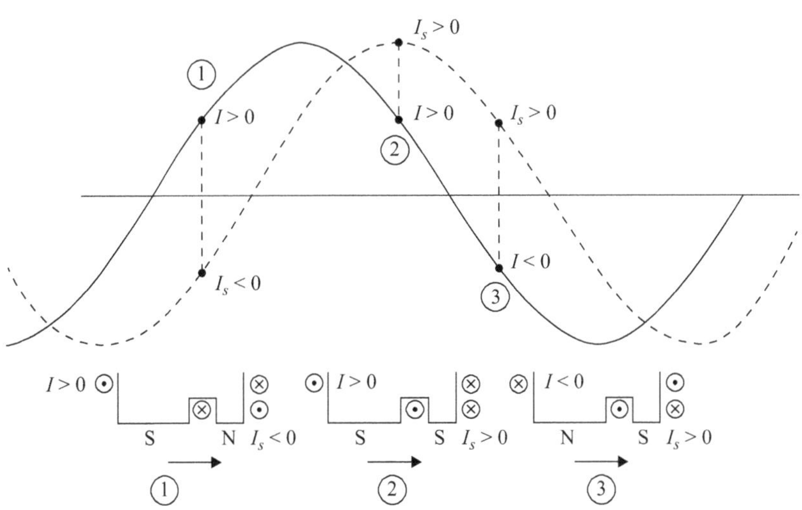

Figure 10.50. The shade-coil effect in the magnetic flux distribution. The sequence SN—SS—NS works like a rotating magnetic field.

(producing a north pole). At instant (2), both currents are positive therefore, both magnetic field polarities are the same. Finally, at instant (3), the main winding current is negative producing a north pole and the one at the shade-coil is positive, producing a south pole.

Comparing the polarities of both instants (1) and (3), we can see that the poles were moved. In a magnetic circuit of the motor, this movement acted as a rotating magnetic field sufficient to turn the rotor.

You will find this kind of motor in several electric appliances such as home-fans, blenders, mixers, etc.

10.7 Summary

This long chapter has introduced the operational characteristics of asynchronous machines. The first section started with a brief contextualization. The second section discussed the design aspects of this machine and the two types of the rotor were presented. The three modes of operation were presented in section 10.3: generator, breaking and motor. This section also discussed the representation of the asynchronous machine using equivalent electric circuits and the strategy to simplify the analysis that makes the rotor and the stator frequencies equal. The single-phase motor was also covered in section 10.5. We learned that no starting torque exists in this configuration, however, it is possible to apply methodologies to make it happen. The use of the torque curves is simple, but it can show the full range of possible operation of this kind of machine to help understand its principles.

Project

A 3-phase induction motor, 60 Hz, 4-pole, 220 V, should be designed to drive an elevator to lift a load of 4000 kg 120 m in 1 ($\pm 3\%$) minute (figure 10.51).

Due to the specification of the steel cable, the minimum radius of the pulley should be 25 cm.

The design should specify:

(a) the diameter of the steel cable;
(b) the speed ratio of the gearbox;
(c) the rated power of the induction motor.

Remarks: The designer should consider using information provided by manufactures to select the right components.

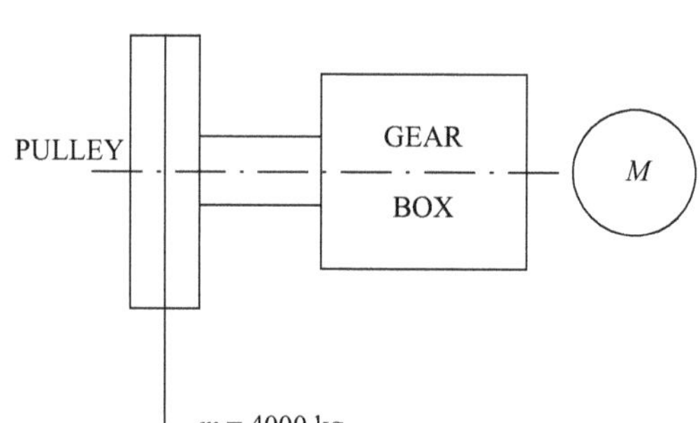

Figure 10.51. Design of an elevator.

Problems

10.1 Find a 3-phase squirrel-cage induction motor, and learn all you can about it by observation. They are used on machine tools, pumps, nearly any AC drive from 1-HP up. Write about two pages of description, including a diagram, giving the physical size and form, the nature of the load, power supply, speed, etc. Determine the number of poles, per cent slip on full load, method of starting. Count, if possible, the bars in the rotor.

10.2 The magnetic field of an induction motor is estimated to be 0.6 Wb m^{-2} (maximum). The motor, a 2-pole machine, has a synchronous speed of 3600 rpm. The rotor of the machine is 20 cm long and 30 cm in diameter. There are 52 rectangular rotor bars in the squirrel-cage, each one's dimensions are width 0.5 cm and depth 2.0 cm.

(a) What is the frequency of the line voltage?

(b) Find the induced emf (rms) in a rotor bar at standstill.

(c) Find the induced emf (rms) in a rotor bar with the machine running under full load at 3520 rpm. Make the approximations and assumptions that are necessary and reasonable.

Answers: (a) 60 Hz; (b) 4.8 V; (c) 0.1 V.

10.3 The magnetic field of an induction motor is estimated to be 0.8 Wb m^{-2} (maximum). The motor, a 4-pole machine, has synchronous speed of 1800 rpm; the speed under load is 1760 rpm. The rotor of the machine is 5.0 cm long and 15 cm in diameter. There are 45 rectangular rotor bars in the squirrel-cage, each one's dimensions are width 0.5 cm and depth 2.0 cm.

(a) What is the frequency of the line voltage?

(b) Find the induced emf (rms) in a rotor bar at standstill.

(c) Find the induced emf (rms) in a rotor bar at full load. Make the approximations and assumptions that are necessary and reasonable.

Answers: (a) 60 Hz; (b) 1.4 V; (c) 0.03 V.

10.4 Extend figure 10.52 so that the torque–speed curve is shown from speed equal to −1 to speed equal to +2 times rated speed. Explain how the

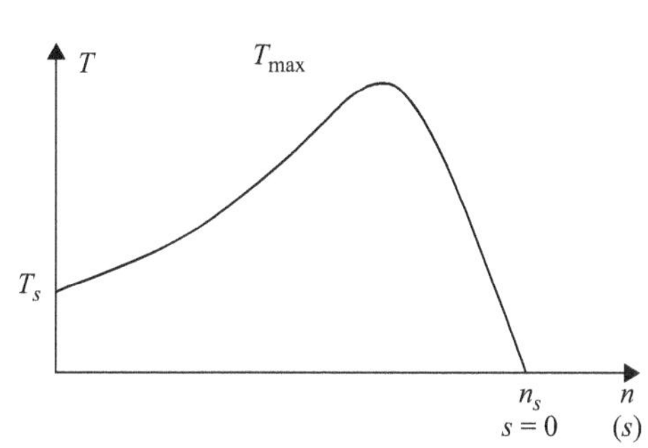

Figure 10.52. Problem 10.4.

machine can operate in the ranges of speed from −1 to 0, from 0 to +1, and from +1 to +2. Is power in or out on:

(a) the electrical side;
(b) the mechanical side.

10.5 The torque required to drive a centrifugal pump is $T = k_p n^2$, where k_p is a constant and n is the spinning speed in rev s^{-1}. An induction motor is used to drive the pump. In the practical operating range, the torque of the motor is $k_m S$, where ($S = n_s - n$) slip is in rev s^{-1}. At what speed will the motor drive the pump?

Answer: $\frac{kp}{km} \times \frac{1-\left(1+\frac{4kp}{km}\right)^{-0.5}}{2}$.

10.6 A 3-phase, 60 Hz, 6-pole induction motor, wye connected, is rated 220 V line-to-line. Stator resistance is 0.20 Ω, stator reactance is 0.50 Ω, and rotor quantities referred to the stator are 0.14 Ω (resistance) and 0.20 Ω (reactance). Magnetizing reactance referred to the stator is 20 Ω. When the slip is 0.020, what is:

(a) The motor speed?
(b) The mechanical power delivered to the shaft of the motor?
(c) The developed torque?
(d) The electrical power input?
(e) The maximum torque that can be obtained from the motor?
(f) How much must the rotor resistance be changed to make this the starting torque?

Answers: (a) 1176 rpm; (b) 926 W; (c) 1 Nm; (d) 955 W; (e) 160 Nm; (f) 0.472 Ω referred to the stator.

10.7 A 3-phase, 6-pole, 220 V, 60 Hz has $a = 0.8$. The effects of all resistances, leakage and magnetization inductances can be neglected. A balance 3-phase load, wye-connected, of 3 Ω in parallel with 2200 μF capacitance is connected to the rotor terminals. The motor is rotating at 350 rpm. Determine:

(a) The effective impedance as seen by the AC source.
(b) The total power delivered by the AC source.
(c) The power delivered to the rotor load.
(d) The mechanical power and the shaft torque.

Answers: (a) $2.71 + j1.53$ Ω; (b) 13.5 kW; (c) 13.5 kW, (d) 3.9 kW.

(h) 10.8. Draw the flow energy balance chart for the motor of problem 10.6 operating at the condition expressed by the following rotor speed:

(a) 1176 rpm.
(b) 1224 rpm.
(c) 48 rpm against the rotating magnetic field.

10.9 A 3-phase, 6-pole, wound rotor is to be employed as a variable-frequency source. For this purpose, the stator is excited from a 60 Hz, 440 V line-to-line, 3-phase supply; and the machine is driven at a variable speed so that the rotor terminals constitute the variable-frequency source. When the rotor is stationary, the line-to-line open-circuit rotor potential difference is 220 V.

(a) Determine the speed range required to give a frequency range of $20 < f < 150$ Hz.

(b) Determine the corresponding range of open-circuit rotor potential difference.

(c) Regard the machine as ideal, determine the relative magnitudes of the power supplied or absorbed at both limits of frequency ranges by

(i) the stator source;

(ii) the driving machine;

(iii) the rotor circuit.

Answers: (a) 800–1800 rpm; (b) 73–550 V; (i) $\frac{p_{20\ \mathrm{HZ}}}{p_{150\ \mathrm{HZ}}} = 7.5$; (ii) $\frac{p_{20\ \mathrm{HZ}}}{p_{150\ \mathrm{HZ}}} = -3.33$; (iii) $\frac{p_{20\ \mathrm{HZ}}}{p_{150\ \mathrm{HZ}}} = 1$.

10.10 A slip ring induction 4-pole motor, has the following equivalent circuit parameters: $r_1 = 0.260\ \Omega$; $r_2 = 0.182\ \Omega$; $l_1 = 1.06$ mH; $l_2 = 0.803$ mH; $L_m = 19.9$ mH and; $a = 1.15$.

The machine is to be employed as a frequency converter and for this purpose it is mechanically coupled to an 8 pole, 60 Hz, synchronous motor, and the stator terminals of the induction motor are connected to a 3-phase, 60 Hz source of 220 V line-to-line.

A balance load, wye connected, with resistance of $5\ \Omega$ and an inductance of 7 mH, is then connected to the rotor terminals. Assume that the set is running in the direction that gives the higher of the two possible rotor frequencies, and determine:

(a) The current in the load circuit.

(b) The flow energy balance chart

Answers: (a) 14.2 A; (b) figure 10.53.

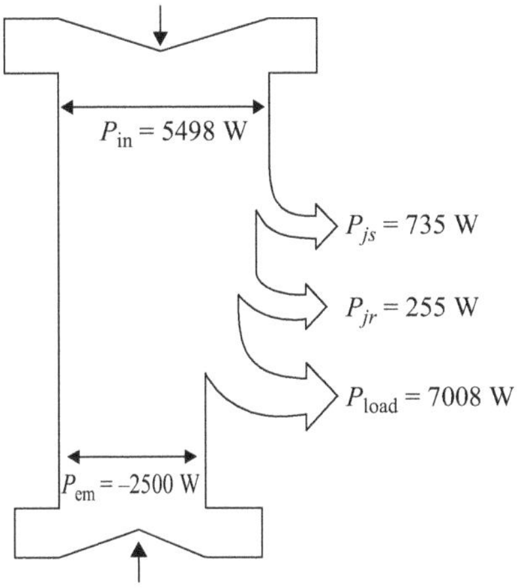

Figure 10.53. Flow energy balance chart. Answer for 10.10(b).

Further reading

[1] Langsdorf A S 1955 *Theory of Alternating Current Machinery* 2nd edn (New York: McGraw-Hill)

[2] Fitzgerald A E, Kingsley C, Jr and Umans S D 1992 *Electric Machinery* 5th edn (New York: McGraw-Hill)

[3] Say M G 1968 *The Performance and Design of Alternating Current Machines* 3rd edn (London: Pitman)

[4] Kostenko M and Piotrovski L 1979 *Maquinas Eletricas* (Lopes Silva Editora) (in Portuguese)

[5] Falcone A G 1979 *Eletromecanica* (Sao Paulo: Edgard Blucher Ltda) (in Portuguese)

[6] Miller T J E 1989 *Brushless Permanent-Magnet and Reluctance Motor Drives* (Oxford: Oxford University Press)

[7] Nasar S A and Unnewehr L E 1983 *Electromechanics and Electrical Machines* 2nd edn (New York: Wiley)

[8] Slemon G R and Straughen A 1980 *Electric Machines* (Reading, MA: Addison-Wesley)

[9] Skilling H H 1962 *Electromechanics: A First Course in Electromechanical Energy Conversion* (New York: Wiley)

[10] Krause P, Wasynczuk O and Pekarek S 2012 *Electromechanical Motion Devices* 2nd edn (Hoboken, NJ: Wiley)

[11] Hughes A and Drury B 2013 *Electric Motors and Drives: Fundamentals, Types, and Applications* 4th edn (Amsterdam: Elsevier)

[12] Lipo T A 2017 *Introduction to AC Machine Design* 1st edn (New York: Wiley)

[13] Boldea I and Syed A N 2010 *The Induction Machines Design Handbook* 2nd edn (Boca Raton, FL: CRC Press)

[14] Gieras J F 2016 *Electrical Machines: Fundamentals of Electromechanical Energy Conversion* 1st edn (Boca Raton, FL: CRC Press)

[15] Boldea I 2013 *Linear Electric Machines, Drives, and MAGLEVs Handbook* 1st edn (Boca Raton, FL: CRC Press)

[16] Gonen T 2011 *Electrical Machines with MATLAB®* 2nd edn (Boca Raton, FL: CRC Press)

[17] Emadi A 2014 *Advanced Electric Drive Vehicles* 1st edn (Boca Raton, FL: CRC Press)

[18] Toliyat H A *et al* 2012 *Electric Machines: Modeling, Condition Monitoring, and Fault Diagnosis* 1st edn (Boca Raton, FL: CRC Press)

[19] Mohan N 2012 *Electric Machines and Drives: A First Course* 1st edn (New York: Wiley)

[20] Pyrhonen J, Jokinen T and Hrabovcova V 2013 *Design of Rotating Electrical Machines* 2nd edn (New York: Wiley)

[21] Boldea I and Tutelea L N 2010 *Electric Machines: Steady State, Transients, and Design with MATLAB* 1st edn (Boca Raton, FL: CRC Press)

IOP Publishing

Electromechanical Energy Conversion Through Active Learning

J R Cardoso, M B C Salles and M C Costa

Chapter 11

Special electrical machines

11.1 DC motor

11.1.1 Design aspects of a DC machine

In chapter 4 we have presented the *elementary DC machine* derived from a single coil moving under a spatially variable magnetic flux density distribution. At both ends of this coil a special switch device is installed, called a *commutator*, that changes the coil's contact connection during the translational movement. The function of such a device is to maintain a voltage polarity that is invariable with the displacement of the coil. Although the commutator presented in chapter 4 is mounted in a planar magnetic system, its performance is more useful in a rotating magnetic system.

Figure 11.1 shows a cross section of such a device, called a DC machine. In each pole core the coils of the field winding are located that are series associated obeying polarity concordance.

The rotor hosts the armature winding that is composed of conductors inserted in uniformly distributed slots connected to the commutator device, as we will see in the next figure.

11.1.2 How a DC machines works

The cross section of an elementary rotating 2-poles DC motor is shown in figure 11.2. The armature winding is composed of four full-pitch coils numbered 1–1', 2–2', 3–3' and 4–4'.

The ends of each coil are connected to the *commutator* that is composed of a segmented metallic cylinder. Each segment has no electrical contact with its neighbor and the thickness of the electrical insulation is small. Figure 11.2 shows a four segmented commutator numbered from 1 to 4 at the center.

doi:10.1088/978-0-7503-2084-9ch11

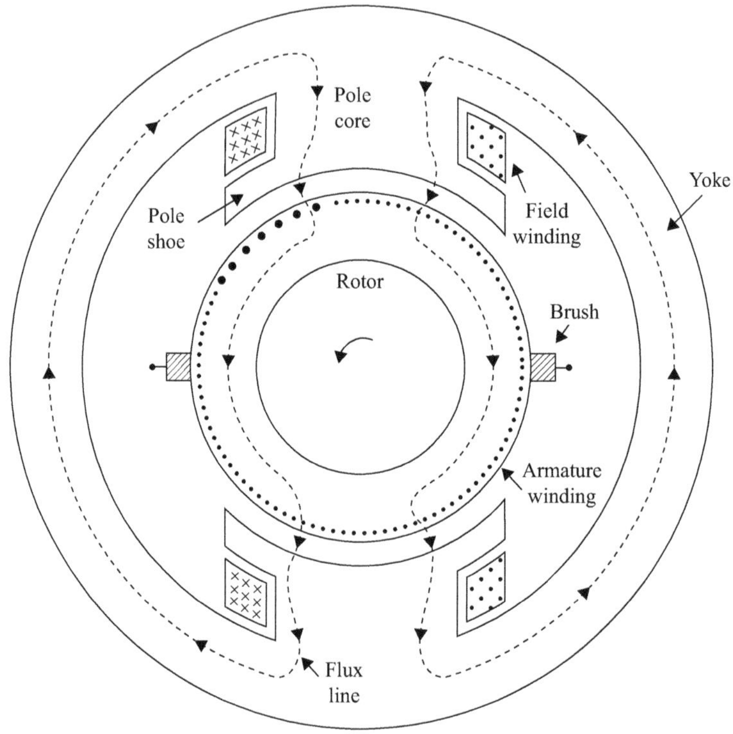

Figure 11.1. Magnetic system of a 2-pole DC machine.

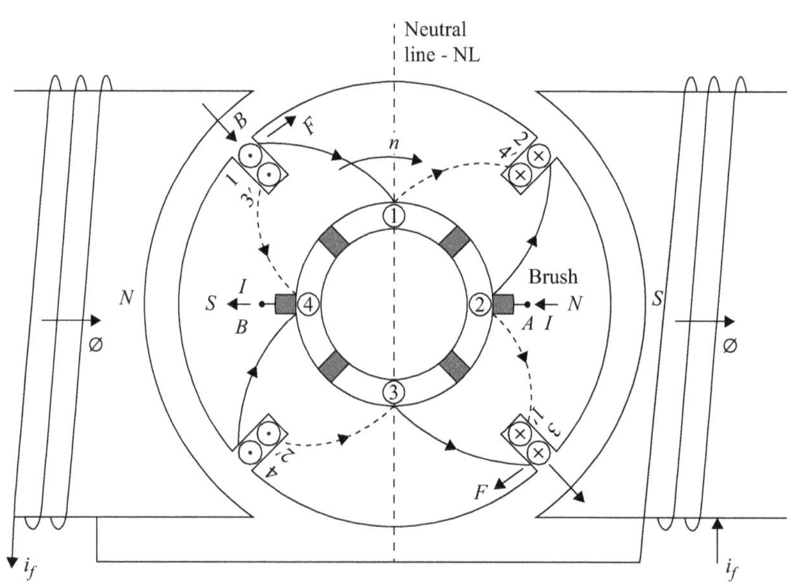

Figure 11.2. The cross section of an elementary 2-pole DC motor. ⊗: the current direction down into the plane of the figure. ⊙: the current direction up out of the plane of the figure.

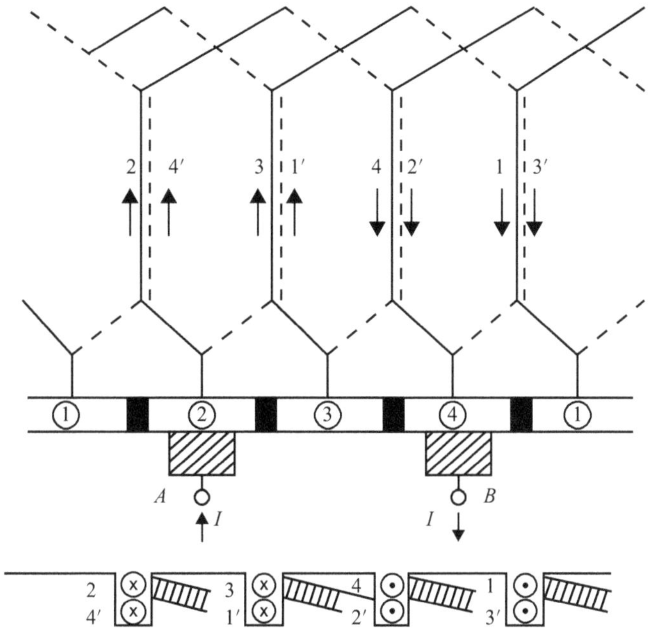

Figure 11.3. The armature winding developed.

Let us follow the circuit using the flat visualization of the machine in figure 11.3:

- Brush A is electrically connected with both ends 2 and 1′. If a DC current is injected in brush A, the electrical current (I) is equally shared in coils 2–2′ and 1′–1.
- Note the current direction because this is important for understanding the sense of the developed torque. As the current in conductor 2 is down into the plane of the figure, the current in the conductor 2′ is up out of the plane. They are the conductors of the same coil.
- The same is seen for the direction of the current in the conductors 1′ and 1.
- The currents that are up out in both the conductors 2′ and 1 are injected in the conductor 3 and 4′, respectively, through the segments (3) and (1) of the commutator.
- Both currents will arise in the conductors 3′ and 4 that are connected to the brush B closing the circuit through the source.

Remarks: The current direction of the conductors located in the same slot is concordant.

Figure 11.4 shows the electric circuit in this condition. The interaction of the armature current distribution with the magnetic flux density produced by the field winding develops a torque that moves the mechanical load connected to the rotor shaft.

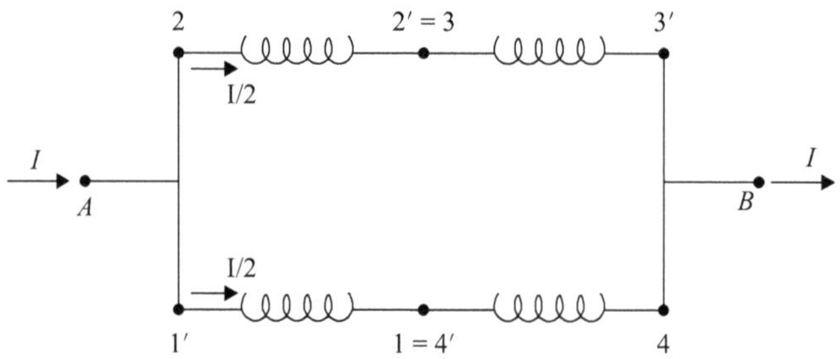

Figure 11.4. Electric circuit of a DC motor.

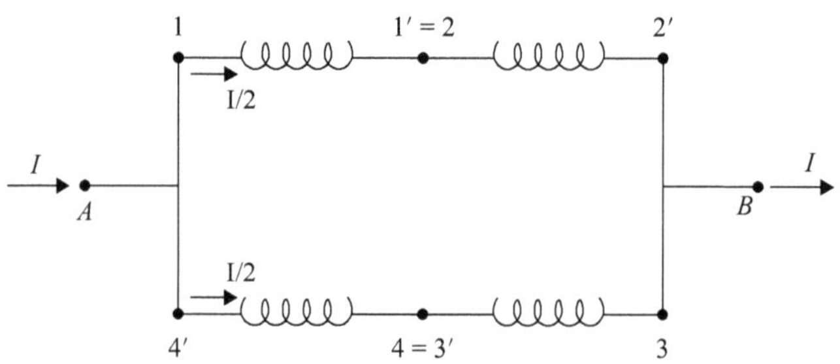

Figure 11.5. Electric circuit of DC motor after a 90° angular displacement.

The direction of the torque is determined observing the action of the electromagnetic force in each conductor. Regarding conductor 1 of figure 11.2, the direction of the electromagnetic force, determined by the right-hand rule, is such that it develops an electromagnetic torque in the clockwise direction.

This torque does not change direction with the movement of the conductors because the commutator acts commuting the connections of the coils to maintain both the current and the force direction under the pole shoes, producing the torque in the same rotational direction.

Figure 11.5 shows the conductor connections after a 90° angular displacement.

In figure 11.5, we can verify that the current sense in coil 1′−1, and in coil 3–3′, are inverted. As the direction of the magnetic flux density distribution changes (from north to the south pole, seen by the conductor), the sense of the torque remains the same.

11.1.3 The induced electromotive force

As the conductors are moving under a magnetic flux density distribution, each conductor has an induction of an emf that can be calculated by $e = blv$. The association of these emfs gives a resultant emf (E_0) such that:

$$E_0 = K_e \phi n \tag{11.1}$$

where:

ϕ: the magnetic flux produced by the field winding (W_b);
n: rotor speed in rev s^{-1};

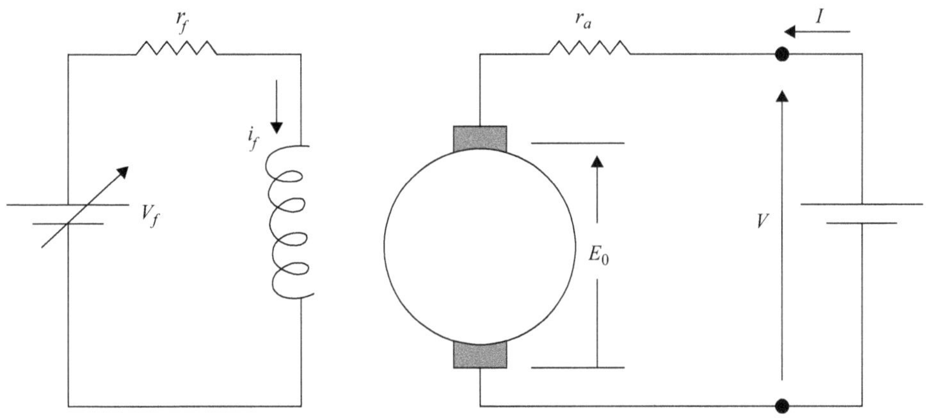

Figure 11.6. The representative electric circuit for the DC motor; i_f: field winding current (A); r_a: armature resistance (Ω); r_f: field winding resistance (Ω); $V_f = r_f i_f$: excitation voltage source (V); $E_0 = K_e\phi n$: The armature electromotive force (V); V: the armature voltage (V); I_a: the armature current (A).

K_e: electromotive force constant that depends on the constructive characteristics of the motor.

As the magnetic flux is directly proportional to the field current, this electromotive force is also directly proportional to the field current, since the rotor speed could be considered constant.

The representative electric circuit for the DC motor is shown in figure 11.6.

11.1.4 The rotor speed equation

Based on the representative electric circuit of figure 11.6, we can write for the armature:

$$V = E_0 + r_a I_a \tag{11.2}$$

Substituting E_0 for its value of (11.1), the rotor speed is given by:

$$n = \frac{V - r_a I_a}{K_e \phi} \tag{11.3}$$

11.1.5 Operation under constant field current

Operating under constant field current, the magnetic flux produced by the field winding is also constant. Once the denominator of equation (11.3) is constant, the behaviour of the rotor speed is dependent on the armature current, that is the load. Figure 11.7 shows the rotor speed as a function of armature current (I_a).

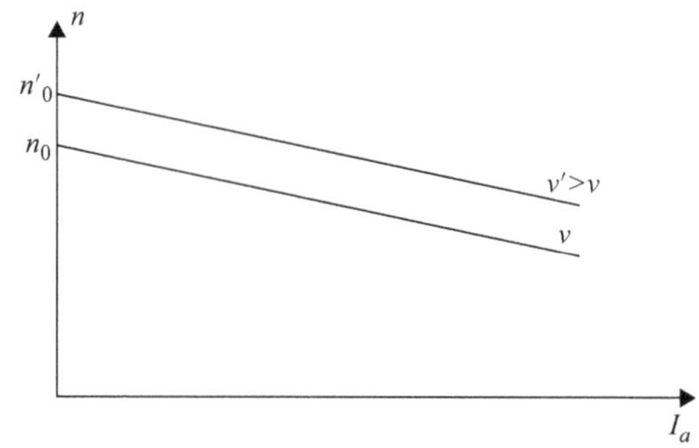

Figure 11.7. The rotor speed as a function of the armature current; $n_0 = \frac{V}{Ke\phi}$: the rotor speed at no-load ($I_a \to 0$).

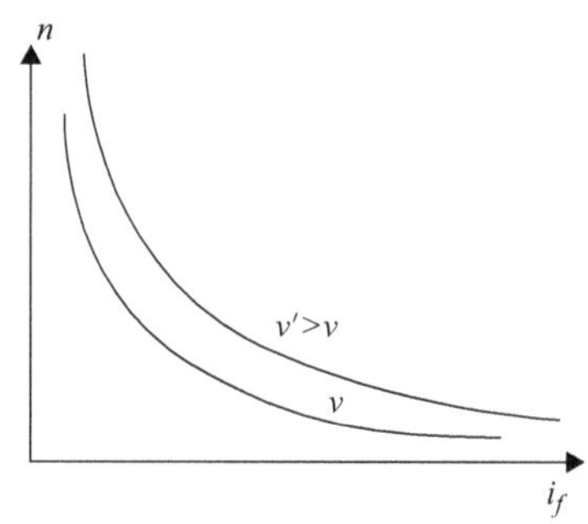

Figure 11.8. The rotor speed as a function of the field current.

11.1.6 Operating under constant voltage

As the DC voltage source is considered constant, the speed variation is obtained by the field current control. Considering that the DC motor has a small armature resistance (except for very small motors), the drop voltage in the armature resistance is such that $r_a I_a \ll V$.

As the field current is directly proportional to the magnetic flux, from equation (11.3) the speed variation as a function of the field winding current will be the one shown in figure 11.8.

For low values of field current, the rotor speed can be very high. To avoid any mechanical damage, all DC motors have a protection device for detecting low values of field current.

11.1.7 The developed torque

Each conductor of the armature that carries an electric current is also travelling under a magnetic flux density distribution. The interaction of this current and the magnetic field distribution, issued by the field winding, produces a unidirectional electromagnetic torque that can be used to move a rotating load. To evaluate it, the start point is the voltage equation (11.2). Multiplying it by I_a in both terms, results in:

$$VI_a = E_0 I_a + r_a I_a^2 \tag{11.4}$$

Interpreting each term of (11.4), we conclude:

VI_a: the power input in the DC motor by the DC electric source (W);

$r_a I_a^2$: the Joule losses in the armature (W);

$E_0 I_a$: the electromechanical power, that is the power converted from electrical to mechanical.

Neglecting any kind of additional losses, the electromechanical power $E_0 I_a$ is the one offered to the load, normally called the *useful power* (P_u).

The useful power is related to the developed torque by the equation:

$$P_u = T_{dev}\omega \tag{11.5}$$

where:

T_{dev}: the developed torque applied to the load (Nm);

$\omega = 2\pi n$: the angular speed of the armature (rad s^{-1});

n: the rotor speed in rev s^{-1}.

As a result, we get:

$$T_{dev} = \frac{E_0 I_a}{2\pi n} = \frac{K_e \phi I_a}{2\pi} \tag{11.6}$$

So, we can write:

$$T_{dev} = K_t \phi I_a \tag{11.7}$$

where:

$K_t = \frac{Ke}{2\pi}$: is the torque constant that depends on the constructive characteristics of the DC motor.

Remarks: If the DC motor works with constant field current, the torque is directly proportional to the armature current. The permanent magnet DC motor has this kind of characteristic.

11.1.8 Configurations of the DC motor

The DC motor we have studied in the previous section is composed of two windings fed by two independent electrical sources. The main source is the one connected to the armature, because it is the source of the main power for moving the load. The source that feeds the field winding supplies enough power to establish the required field current. As the field winding is made by good conductor material, the power involved in the excitation process is low.

Even though the characteristics of the sources are completely different, it is not convenient to have two different electrical sources to drive the DC motor.

11.1.8.1 The shunt DC motor

The *shunt* connections concern the use of only one source for feeding both the armature and the field windings. To do such an operation, the field winding is parallel connected to the armature. But, as the current of the field winding should be controlled to provide a suitable speed control, a rheostat (called a *field rheostat*) is connected in series with the field winding. Consequently, the entire set of components—field winding and field rheostat—are parallel connected with the armature, as figure 11.9 shows.

The shunt motor is suitable for industrial applications where the speed control should be accurate. When working with constant field current, the magnetic flux is also constant and the torque characteristics become linear.

Figure 11.10 shows the torque characteristics of a shunt DC motor operating with two different values of field current.

An efficient control of both the field current and the armature voltage makes the shunt DC motor one of the most flexible electrical motors for industrial applications.

11.1.8.2 The series DC motor

The field winding of the shunt DC motor is parallel connected to the armature winding. On the other hand, in the *series* DC motor the field winding is series connected to the armature winding, as shown in figure 11.11.

As the field winding is series connected to the armature, the field current is the armature current, that is:

$$i_f = I_a \tag{11.8}$$

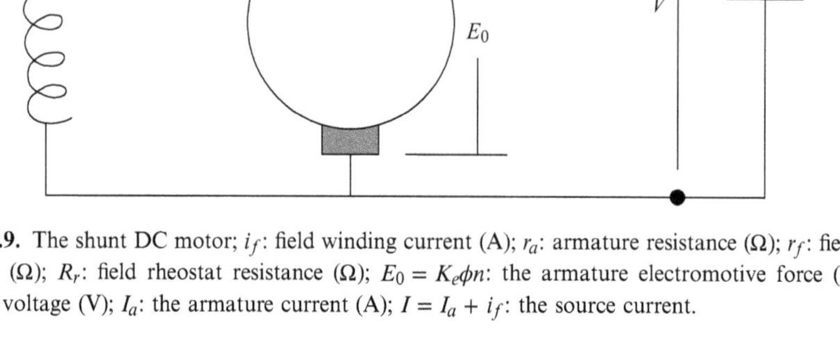

Figure 11.9. The shunt DC motor; i_f: field winding current (A); r_a: armature resistance (Ω); r_f: field winding resistance (Ω); R_r: field rheostat resistance (Ω); $E_0 = K_e\phi n$: the armature electromotive force (V); V: the armature voltage (V); I_a: the armature current (A); $I = I_a + i_f$: the source current.

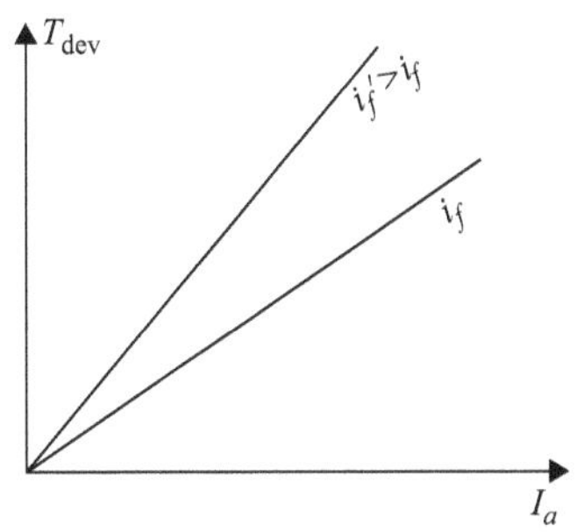

Figure 11.10. Torque characteristics of the shunt DC motor $i_f' > i_f \longrightarrow (\phi' > \phi)$.

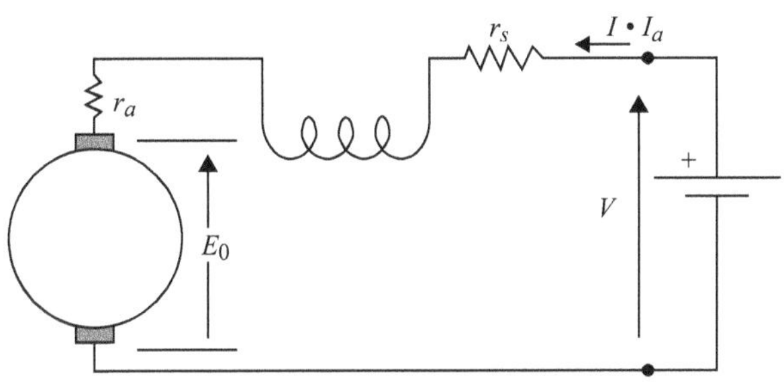

Figure 11.11. The series DC motor.

Due to this property, the field winding of a series motor is composed of a few turns with conductors of large cross section, because the armature current is the load current of the motor. This is completely different to the field winding of the shunt motor that has a large number of turns and carries a small value of current.

As the field current in a series motor is the armature current, the magnetic flux produced by the field winding is in direct proportion to the armature current.

Taking the speed equation of the DC motor:

$$n = \frac{V - r_a I_a}{K_e \phi} \tag{11.9}$$

we observe that the rotor speed varies in inverse proportion to the armature current, because the magnetic flux is directly proportional to the armature current, that is:

$$\phi = K_\phi I_a \tag{11.10}$$

At no load, when the armature current is very small, the rotor speed could achieve prohibitive limits as shown in figure 11.12. Therefore, no series DC motor must work at this condition.

The developed torque of the series DC motor is very convenient for several applications that require a high starting torque, for example, vehicular and crane applications (figure 11.13).

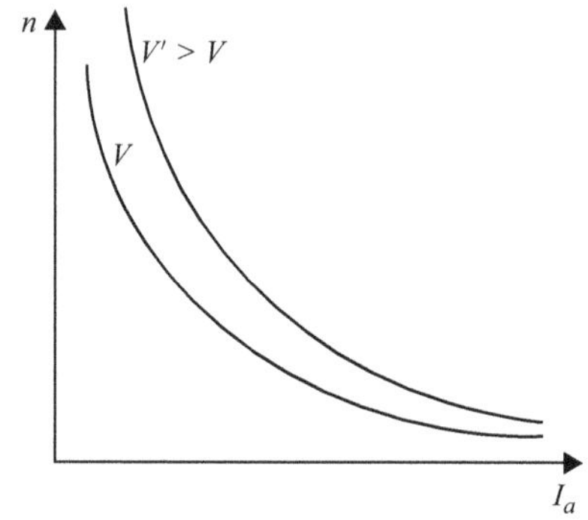

Figure 11.12. The rotor speed of series motor as a function of the armature current. In a series motor $I_a = i_f$.

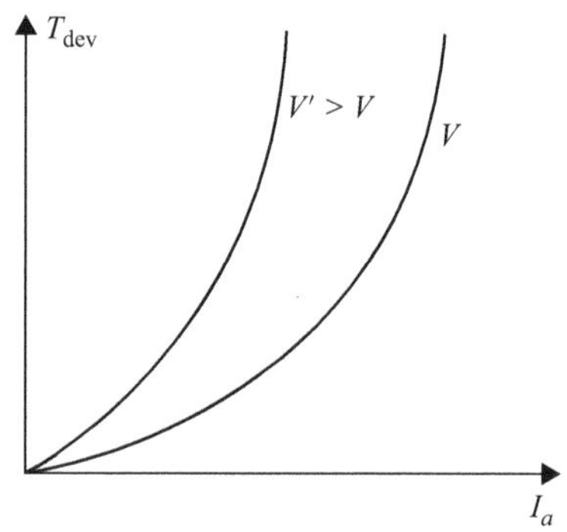

Figure 11.13. Developed torque as a function of the armature current. These curves are voltage source parametrized.

As the torque equation is:

$$T_{dev} = K_t \phi I_a \tag{11.11}$$

and considering that the magnetic flux is directly proportional to the excitation current, we can write that $\phi = K_\phi I_a$, consequently:

$$T_{dev} = K_\phi K_t I_a^2 \tag{11.12}$$

As a result, the developed torque is dependent on the armature current squared. As the starting current is very high in a series DC motor, the starting torque is very high, sufficient to drive a heavy load from inertia.

The developed torque as a function of the rotor speed is extracted from the voltage equation:

$$V = K_e \phi n + r_a I_a \tag{11.13}$$

Multiplying both terms by $\frac{I_a}{2\pi}$, equation (11.13) becomes:

$$\frac{VI_a}{2\pi} = T_{dev} n + \frac{r_a I_a^2}{2\pi}$$

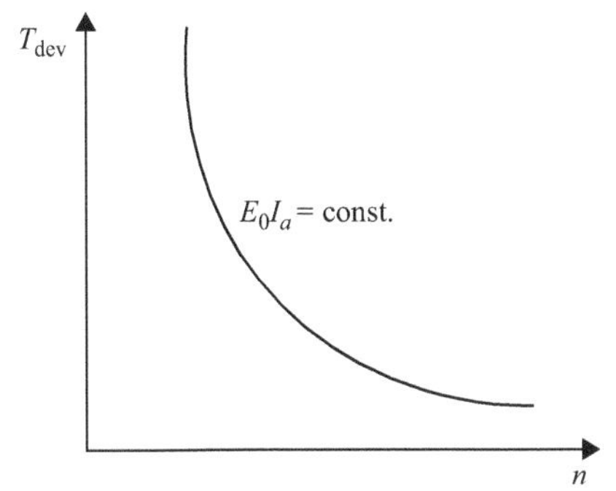

Figure 11.14. Developed torque as a function of rotor speed at useful constant power.

then:

$$T_{dev} = \frac{VI_a - r_aI_a^2}{2\pi n} = \frac{E_0I_a}{2\pi n} \tag{11.14}$$

Operating at useful constant power, that is, E_0I_a = constant, the developed torque is in inverse proportion to the rotor speed, as figure 11.14 shows. This feature is very convenient for vehicular applications like electric busses, electric traction, subways, tramways, etc. The power control is made by suitable electronic hardware.

Example 11.1

A *shunt* DC motor has its field current adjusted for 2 A at no-load operation. The rotor speed at this condition is 2000 rpm at 250 VDC voltage source. Analyze the effect of the variation of these quantities in the speed of the DC motor.

(a) What is the new rotor speed if the field current changes from 2 A to 1.5 A, maintaining the voltage source in 250 V DC?
(b) What is the new rotor speed if the voltage source changes from 250 V DC to 200 V DC, maintain constant the field current?
(c) What is the new rotor speed if both the field current and the voltage source change simultaneously for the values indicate in (a) and (b)?
(d) Supposing that the armature resistance is 0.1 Ω, what will the new speed of the DC motor be if the armature current is 100 A, the field current 2.5 A and the DC voltage source 250 V?
(e) What is the developed torque in the condition of (d)?
(f) What is the new field current, when the DC motor develops 250 Nm at the same armature current?

Solution

(a) The variation of the field current

If the shunt DC motor is operating at no-load, it results in $I_a \approx 0$. Consequently, we can establish that:

$$V \approx E_0 = K_e\phi n$$

as the magnetic flux is in direct proportion to the field current, the previous equation can be written as:

$$V \approx E_0 = K i_f n$$

The first statement establishes that for $V = 250$ V and $i_f = 2$ A, we get $n = 2000$ rpm. Therefore:

$$250 = K \times 2 \times 2000$$

As the only variation is the field current from 2 A to 1.5 A, we can also write:

$$250 = K \times 1.5 \times n$$

Dividing this expression by the previous one, results in:

$$n = 2666 \text{ rpm}$$

(b) The variation of the voltage source

As the only variation is the DC voltage source, we can write:

$$200 = K \times 2 \times n$$

Dividing this expression by the last one in (a), results in:

$$n = 1600 \text{ rpm}$$

(c) The variation of both the DC voltage source and the field current

In this situation, we can write:

$200 = K \times 1.5 \times n$

Dividing this expression by the last one in (a), results in:

$$n = 2133 \text{ rpm}$$

(d) In the specified load condition, we know that:

$V = 250$ V, $I_a = 100$ A, $r_a = 0.1\ \Omega$ that implies:

$$E_0 = 250 - 0.1 \times 100 = 240 \text{ V}$$

as the field current is 2.5 A, we can write:

$$240 = K \times 2.5 \times n$$

Dividing this expression by the last one in (a), results in:

$$n = 1536 \text{ rpm}$$

(e) The developed torque in such a condition is:

$$T_{\text{dev}} = \frac{E_0 I_a}{2\pi n} = \frac{240 \times 100}{2\pi \frac{1536}{60}} = 149 \text{ Nm}$$

Remarks: In the International Systems, the speed should be expressed in rotation by second (rev s^{-1}).

(f) To extract a developed torque of 250 Nm with the same armature current, the field current must change.

As $\phi = k_f i_f$, we can write for the condition of (d):

$$149 = K_t K_f \times 2.5 \times 100$$

and for the condition of the (f):

$$250 = K_t K_f \times i_f \times 100$$

Dividing both equations results in:

$$i_f = 4.2 \text{ A}$$

11.1.8.3 The universal motor

The *series* DC motor is a flexible electric motor since it can be driven by both a DC or an AC voltage source. As both, the excitation current and the armature current are the same, the direction of the developed torque does not change due to the alternating currents.

Some additional parameters should be considered, when the *series* motor is fed by an AC voltage source. These parameters are the reactance of the armature and the field windings (figure 11.15). The introduction of these reactances implies a lag between the applied voltage and the armature current. It decreases not only its efficiency, but also introduces some malfunction in the operation of the commutation process, characterized by the contact flashover between the commutator and the brushes.

The developed torque of the universal motor should be evaluated considering the lag between the internal emf E_0 and the armature current, such that:

$$T_{\text{dev}} = \frac{E_0 I_a}{2\pi n}\cos\delta \tag{11.15}$$

where δ is the lag between E_0 and I_a.

The voltage equation in AC operation becomes:

$$\dot{V} = \dot{E}_0 + (r_a + r_s + jx_a + jx_s)\dot{I}_a \tag{11.16}$$

Example 11.2

A *series* DC motor presents the following parameters:

- armature resistance $r_a = 0.06\ \Omega$;

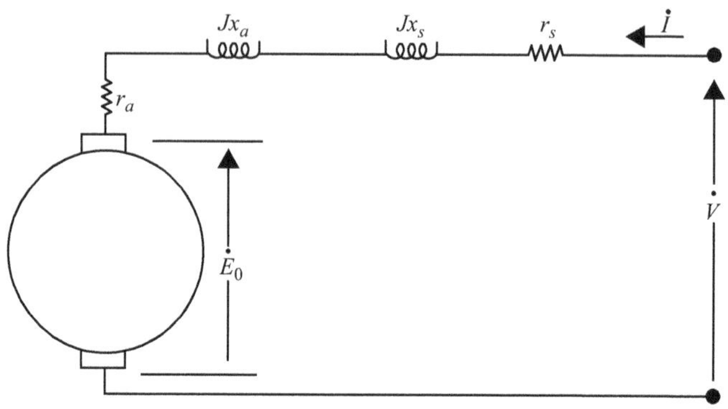

Figure 11.15. Equivalent electric circuit for a universal motor; r_a: armature resistance (Ω); r_s: series field resistance (Ω); $x_a = 2\pi l_a$: armature reactance (Ω); $x_s = 2\pi l_s$: series field reactance (Ω).

- series field resistance $r_s = 0.04\ \Omega$.

When fed by a 200 V DC, the armature current is 100 A for a 1800 rpm rotor speed.

(a) What is the developed torque in this operational condition?
(b) What is the armature current for a 150 Nm developed torque?
(c) What is the rotor speed for the condition of part (b)?

Solution

(a) The developed torque

From the voltage equation, we get:

$$E_0 = V - (r_a + r_s)I_a = 200 - (0.06 + 0.04) \times 100$$

that is:

$$E_0 = 190\ \text{V}$$

therefore:

$$T_{\text{dev}} = \frac{E_0 I_a}{2\pi n} = \frac{190 \times 100}{2\pi \times \dfrac{1800}{60}} = 100\ \text{Nm}$$

(b) The armature current for 150 Nm.

As the developed torque in a *series* DC motor is given by:

$$T_{\text{dev}} = K_\phi K_t I_a^2$$

using the data from part (a) results in:

$$K_\phi K_t = \frac{T_{\text{dev}}}{I_a^2} = \frac{100}{100^2} = 0.01$$

a developed torque of 150 Nm yields:

$$I_a = \sqrt{\frac{T_{\text{dev}}}{K_\phi K_t}} = \sqrt{\frac{150}{0.01}} = 122 \text{ A}$$

The new induced emf is given by:

$$E_0 = V - (r_a + r_s)I_a = 200 - (0.06 + 0.04) \times 122 = 187.8 \text{ V}$$

as:

$$E_0 = K_e\phi n$$

and

$$\phi = K_\phi I_a$$

yields:

$$E_0 = K_e K_\phi I_a n$$

Using the data from part (a) the result is:

$$K_e K_\phi = \frac{E_0}{I_a n} = \frac{190}{100 \times 30} = 0.633$$

Consequently, the new rotor speed will be:

$$n = \frac{E_0}{K_e K_\phi I_a} = \frac{187.8}{0.633 \times 122} = 24.3 \text{ rps (1458 rpm)}$$

11.2 Switched reluctance motor

11.2.1 Design aspects of the switched reluctance motor

The switched reluctance motor (SRM) is a doubly-salient pole machine, where only the stator poles host concentrated windings (see figure 11.16).

Although each pole has a single coil, the number of phases is half of the pole number because one phase comprises coils on opposite poles. The 8:6 pole SRM has four-phase windings, there is one phase per each two poles (north and south).

The rotor is not only free of windings, but is neither permanent magnet nor squirrel-cage. For good performance and efficiency, the air-gap should be as small as possible. Therefore, precision mechanics should be applied to ensure no-eccentricity in the rotor of the SRM.

The SRM has some characteristics that give advantages:

(1) The rotor is simple and has low inertia.
(2) The stator winding is very simple.
(3) As the SRM has no PM, the machine is able to operate at high temperature.
(4) As the rotor has no mounting parts, extremely high speeds are possible.

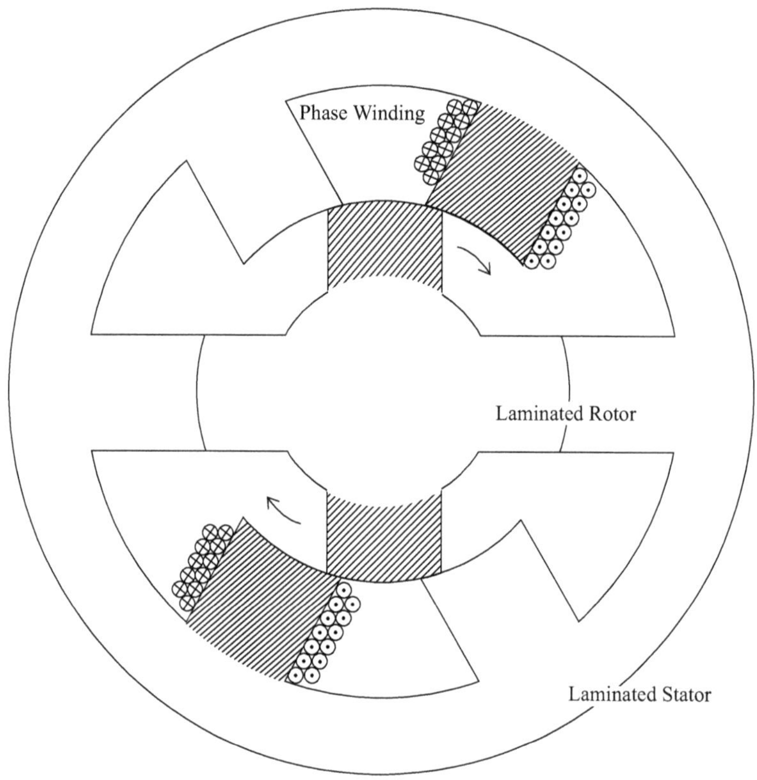

Figure 11.16. The switched reluctance motor; 6:4 pole—three-phase.

11.2.2 How switched reluctance motor works

Although the SRM has several phases, they are not fed simultaneously. Each phase is fed by a square wave voltage source dependent on the rotor position.

If the number of poles in the rotor is p_r, the fundamental switching frequency of the voltage source is:

$$f = np_r \tag{11.17}$$

where n is the rotor speed in rev s^{-1}.

If there are m phases, there are mp_r steps per revolution and supposing that the stator poles usually exceeds the number of rotor poles, the angular displacement of two consecutive square wave voltage, called 'step angle' or 'stroke angle', is given by:

$$\varepsilon = \frac{2\pi}{mp_r} \text{ (rad)} \tag{11.18}$$

For each stroke, we can evaluate the developed torque computing the electromagnetic energy variation (figure 11.17) by the difference of areas, as we discussed in chapter 5.

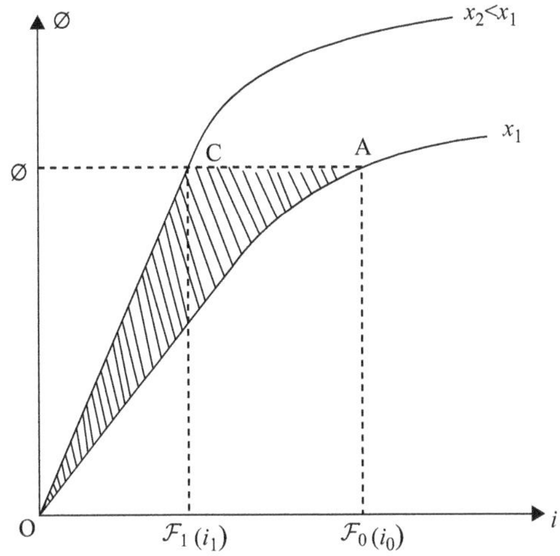

Figure 11.17. Electromechanical energy variation ($\triangle E_{em}$); $\triangle E_{em} = -(\text{area OC } \phi - \text{area OA } \phi)$; x_1: unaligned position; x_2: aligned position.

The average *torque per stroke* is such that:

$$T_{dev} = \frac{\triangle E_{em}}{2\pi n}\ (\text{Nm}) \tag{11.19}$$

Remarks: In a linear magnetic circuit, the *torque per stroke* is evaluated directly by:

$$T_{dev} = \left[\frac{1}{2}i^2\frac{dL(\theta)}{d\theta}\right]_{av} \tag{11.20}$$

In one revolution, each phase conducts as many strokes (n_s), as there are rotor poles, that yields:

$$n_s = mp_r \tag{11.21}$$

where m is the number of phases.

As a result, the average torque developed by the SRM is given by:

$$T_{av} = \text{torque per stroke} \times \text{number of strokes/revolution} \tag{11.22}$$

that yields:

$$T_{av} = \frac{\triangle E_{em}}{2\pi n} \times mp_r$$

or in a linear magnetic circuit:

$$T_{av} = \left[\frac{1}{2}i^2\frac{dL(\theta)}{d\theta}\right]_{av} \times mp_r \tag{11.23}$$

11.2.3 The torque production

Neglecting both the fringing effect and the saturation, the phase inductance represented as a function of rotor position can be considered as a linear function, such as the one shown in figure 11.18.

To have an effective motoring torque, we must turn on the phase at the position θ_1, corresponding to the minimum inductance point (unaligned position). The turn off should be at position θ_2, corresponding to the maximum inductance point (aligned position).

As we need to know the rotor position for switching the voltage source, the driver should identify it continuously. Even though, there are some sensorless techniques for identifying the rotor position, the most common practice is to use a *high definition encoder*, because the accuracy is very important in this kind of converter. The time interval that the voltage should be applied to the coil is given by:

$$\Delta t = \frac{\theta_2 - \theta_1}{2\pi n} \tag{11.24}$$

where n is the rotor speed in rev s^{-1}.

When the voltage of the coil is turned off at position θ_2, the nearby winding located in the opposite direction of the movement should be commuted to continue the process in the next pair of poles (figure 11.19).

Example 11.3

The 8:6 four-phases SRM of figure 11.20 is built to run at 2000 rpm. The main dimensions are:

- inner diameter: $D = 68$ mm;

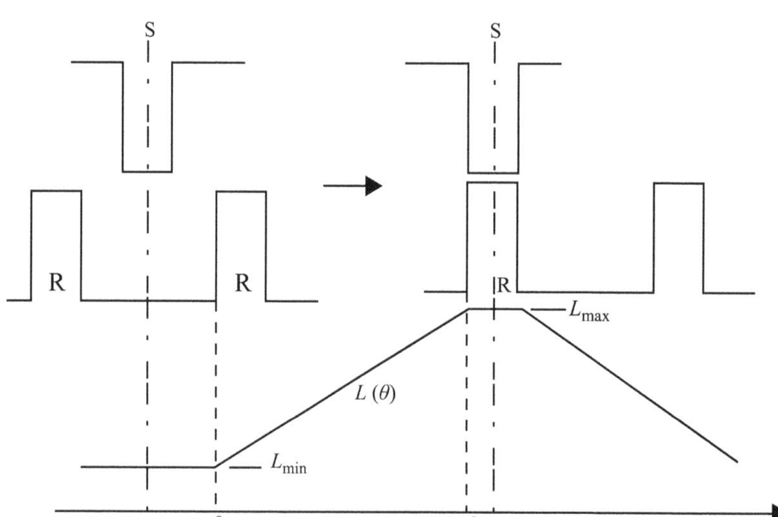

Figure 11.18. Phase inductance as a function of rotor position. Right: misaligned poles. Left: aligned poles. S: stator. R: rotor.

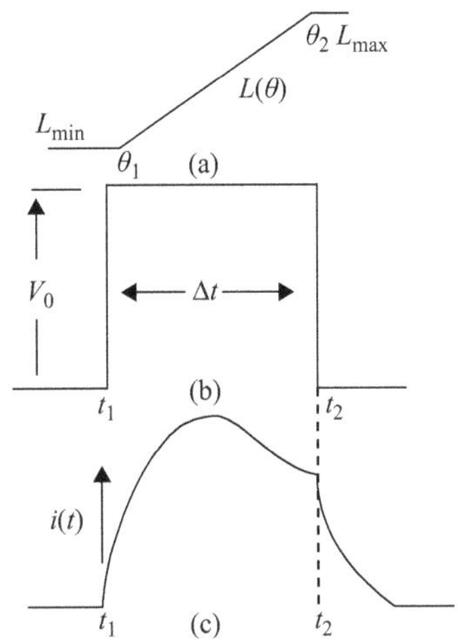

Figure 11.19. Commutation process of each coil of the SRM; (a) idealized inductance variation, neglecting fringing and saturation; (b) idealized voltage waveform; (c) idealized current waveform for motoring (low and medium speed).

Figure 11.20. Cross section of a four-phase 8:6 SRM.

- stator pole width: $2w = 12.5$ mm;
- air-gap: 0.25 mm.

Let us determine:
(a) the step angle;
(b) the commutation frequency;
(c) the time interval of the square wave applied by the voltage source.

Solution
(a) The step angle
From equation (11.18), the step angle is such that:

$$\varepsilon = \frac{2\pi}{mp_r} = \frac{2\pi}{4 \times 6} = 0.26 \text{ rad } (15° \text{ degrees})$$

(b) The commutation frequency
From equation (11.17):

$$f = np_r = \frac{2000}{60} \times 6 = 200 \text{ Hz}$$

(c) The time interval of the square wave

From equation (11.24), we have:

$$\Delta t = \frac{\theta_2 - \theta_1}{2\pi n}$$

As $\theta_1 = \text{arc sin}\,\frac{2w}{D} = 10.6°$ and $\theta_2 = 45° - \theta_1 = 34.4°$, resulting in:

$$\Delta t = \frac{\theta_2 - \theta_1}{2\pi n} = \frac{34.4 - 10.6}{2\pi \times \frac{2000}{60}} \times \frac{\pi}{180} = 4 \text{ ms}$$

11.3 Stepper motor

The operation of several mechatronics processes need to convert digital pulses into accurate mechanical shaft rotation. The stepper motor is very suitable for this kind of application because it is possible to divide every revolution into a discrete number of steps. Each step is incremented by a voltage pulse. A series (or a train) of 10 pulses will increment 10 positions in the shaft, it is common to have 200 or more steps in a full rotation.

The SRM has this feature because for each pulse applied to the coil a movement is introduced until there is a new alignment of the poles. However, if we need a high number of steps, improvements must be made.

11.3.1 The elemental stepper motor

Figure 11.21 shows an elemental stepper motor with four poles in the stator and four poles in the rotor. In this figure, all rotor poles are aligned with the stator poles and no movement is admissible because this is a stable position.

We can now imagine another identical stack inserted behind it with its stator displaced 45° to the first one, remaining constant not only with the rotor geometry but also with its position relative to the first stack, as shown in figure 11.22.

With this approach, the device is able to move continuously if a suitable sequence of pulses in its coils is applied. Note that if we excite the coils following the sequence A–E–B–F–C–G–D–H, eight steps are accomplished for each turn.

If instead of using two stacks we use three, displaced 30° from each other, four more steps are integrated in the process, completing twelve steps in each turn (30° the angular interval between two consecutive steps). We can improve this as we want to not only introduce more stacks, but also introduce more teeth in both stator and rotor poles.

Figure 11.21. 4:4 poles single stack stepper motor.

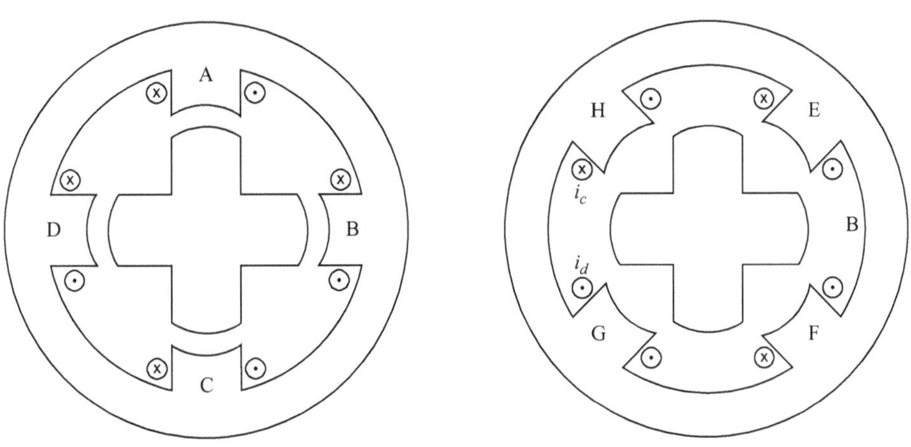

Figure 11.22. 4:4 poles two-stack stepper motor; both stacks are displaced 45° from each other.

It is time now to introduce some parameters of the stepper motor. Let us consider that:

p_{ts}: number of stator teeth per stack. If there are no teeth on the stator pole, p_{ts} is the number of stator poles per stack.

p_{tr}: number of rotor teeth per stack. If there are no teeth on the rotor pole, p_{tr} is the number of rotor poles per stack.

N_s: number of stacks.

τ_p: tooth pitch.

S_p: step length.

As a result, we can establish that:

$$\text{Tooth pitch} = \frac{360}{\text{Number of rotor teeth per stack}}$$

That is:

$$\tau_p = \frac{360}{p_{tr}} \tag{11.25}$$

The tooth pitch is also related to the step length by:

$$\text{Tooth pitch} = \text{Number of stacks} \times \text{Step length}$$

That is:

$$\tau_p = N_s S_p \tag{11.26}$$

We can substitute (11.25) into (11.26) and obtain:

$$S_p = \frac{360}{N_s p_{tr}} \tag{11.27}$$

The steps lengths of multistack stepping motors typically range from 2° to 15°.

Example 11.4

For the stepper motor in figure 11.23, identify its parameters: p_{ts}, p_{tr}, N_s, τ_p and S_p.

Solution

$p_{ts} = 8$: number of stator teeth per stack (four magnetic pole)

$p_{tr} = 8$: number of rotor teeth per stack.

$N_s = 3$: number of stacks

$\tau_p = \frac{360}{p_{tr}} = \frac{360}{8} = 45°$: tooth pitch

$S_p = \frac{360}{N_s p_{tr}} = \frac{360}{3 \times 8} = 15°$: step length.

11.3.2 The hybrid stepper motor—digital motor

Figure 11.24 shows a view of a typical hybrid stepper motor (HY motor) that has an odd number of poles. The slotted poles establish several teeth on its surface. The

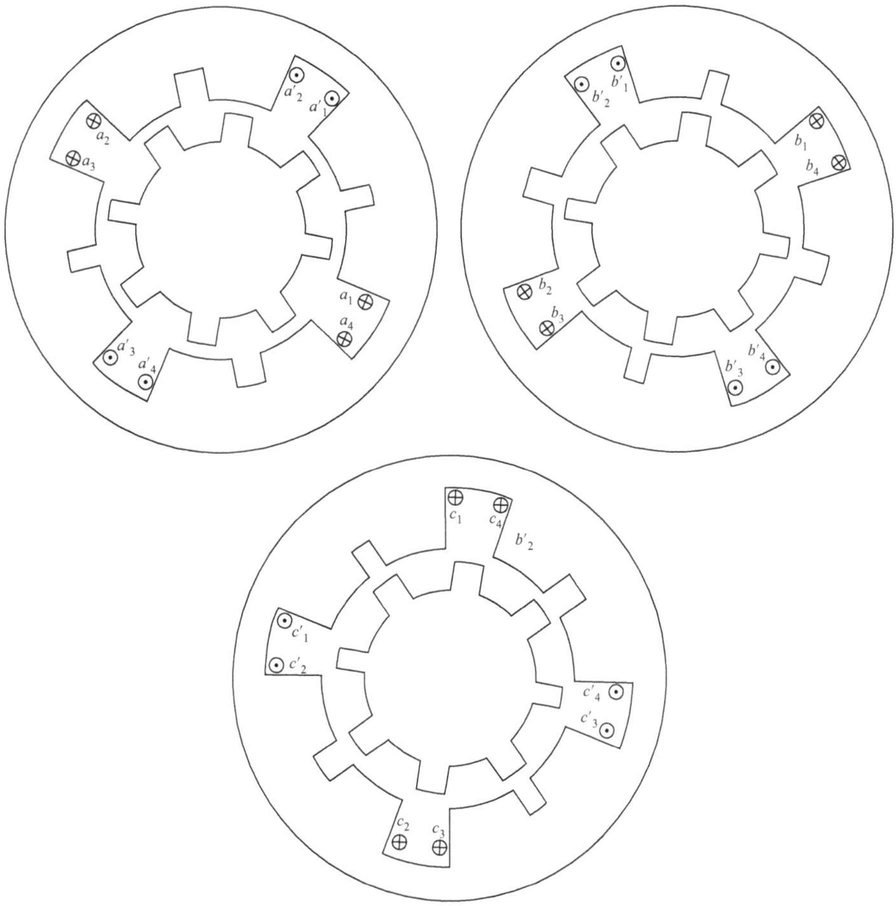

Figure 11.23. Four-pole, three-stack stepper motor.

cross-section of the poles is the same for its entire length. The simplest rotor is composed of three poles: two unaligned slotted cups with half teeth pitch displaced. Inside them, there is an axially magnetized permanent magnet. The number of both stator and rotor teeth are different.

The number of teeth impacts on the number of motor steps. Two or four hundred steps by turn is common in this kind of machine that implies steps of 1.8° or 0.9°.

The hybrid stepper motor is known for both its accurate position without control and high electromagnetic torque.

The rotor can also have more sections to improve both the developed torque and step number.

11.3.3 How the hybrid stepper motor works

The very simple HY motor, shown in figure 11.25, is given just to clarify its operation. All figures refer to the same HY motor. Figure 11.25(a) shows the front view and, figure 11.25(c) the rear view of the same machine. Both views are attached

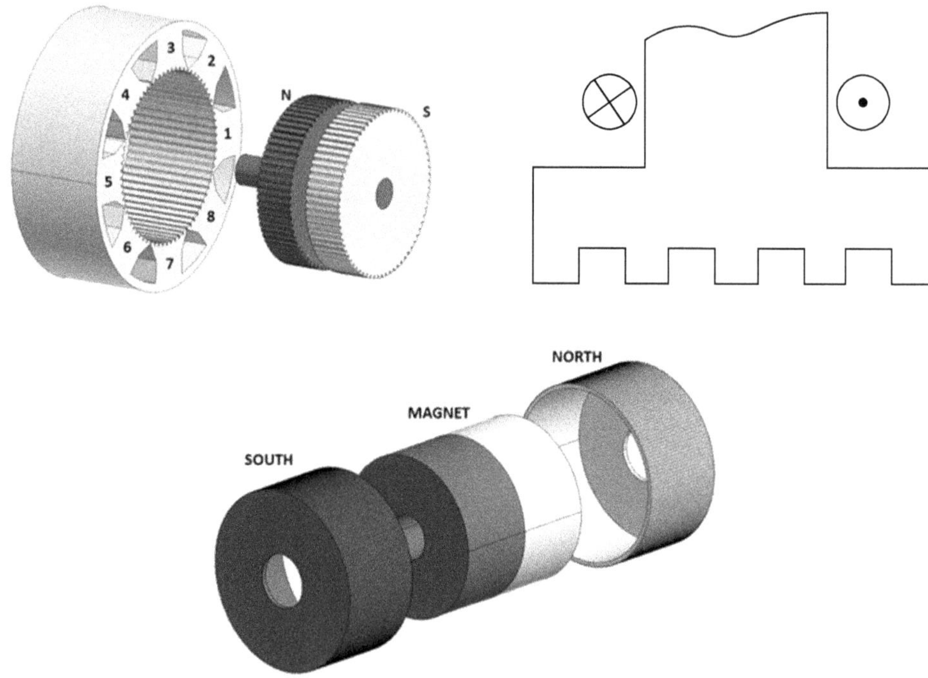

Figure 11.24. The HY stepper motor; left: components of 8:98 HY stepper motor; right: toothed pole of an HY stepper; bottom: the rotor cups with PM.

through an axial PM configuration. Due to PM, all poles of the front view are north magnetized and all poles of the rear view are south magnetized. The magnetic flux produced by the PMs flows axially and then goes to the stator through the poles. The main flux crosses the air gap under the pole in opposite magnetization.

Remarks: The rotor teeth of the front view are displaced one half a tooth pitch from the teeth in the rear view. No displacement is observed in the rotor poles.

This HY motor has two independent coils, one for each pair of poles. Each coil receives positive current, negative current, and zero current. As a convention, the positive current produces a south pole in the pole. When one coil is energized, its winding attracts the teeth of one rotor pole. When the next coil is energized, it attracts the teeth of the other rotor pole. If you apply a sequence of currents such that i_1, i_2, $-i_1$, $-i_2$, i_1, … the rotor moves in counterclockwise stepping rotation. Clockwise rotation is achieved by i_1, $-i_2$, $-i_1$, i_2, …

As the cups of the rotor structure are displaced half a tooth pitch of each other, the front teeth of the hybrid motor are displaced half a tooth from the rear teeth, hence the reason the tooth pitch is half of the stepper motor tooth pitch.

Likewise, we can establish that:

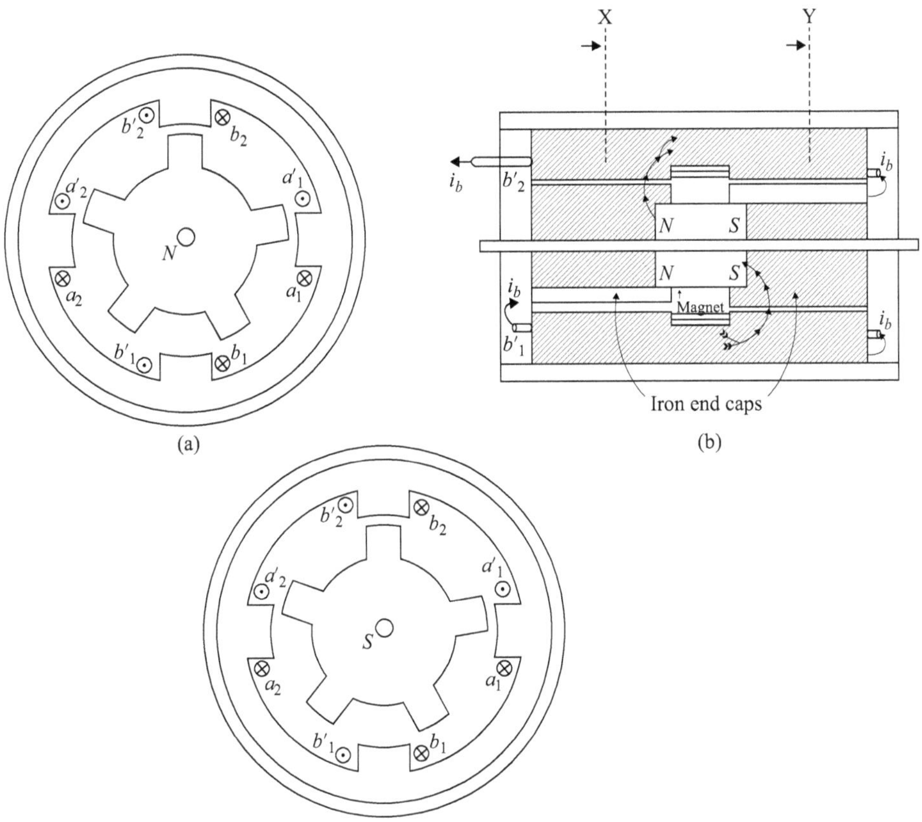

Figure 11.25. HY motor operation—axial view; (a) front view north magnetized; (b) axial view; (c) rear view south magnetized.

$$\text{Tooth pitch} = \frac{180}{\text{Number of rotor teeth per stack}}$$

that is:

$$\tau_p = \frac{180}{p_{tr}} \tag{11.28}$$

As the tooth pitch is related to the step length by:

$$\tau_p = N_s S_p \tag{11.29}$$

it results in:

$$S_p = \frac{180}{N_s p_{tr}} \tag{11.30}$$

The HY motor in figure 11.25, yields:

$p_{ts} = 4$: number of stator teeth per stack (four magnetic pole);

$p_{tr} = 5$: number of rotor teeth per stack;
$N_s = 2$: number of stacks;
$\tau_p = \frac{180}{p_{tr}} = \frac{180}{5} = 36°$: tooth pitch;
$S_p = \frac{180}{N_s p_{tr}} = \frac{180}{2 \times 5} = 18°$: step length.

We identify that HY motors have two disadvantages due to their complexity:
(1) the HY motor is significantly more expensive than the regular stepper motor;
(2) the structure is not only larger but also heavier than the regular stepper motor.

11.4 Brushless DC motor

The beginning of the 1960s was remarkable for the development of power electronics that saw the appearance of a new feature for the permanent magnet synchronous machine. Brushless DC (BLDC) motors first appeared on the scene in 1962, when T G Wilson and P H Trickey unveiled what they called 'a DC motor with solid state commutation'. Indeed, the possibility of having a non-commutator electrical machine controlled by a DC voltage source was a revolutionary milestone.

Although they use the name *DC motor*, the brushless DC motor is supplied by a multiphase pulsed voltage source that is obtained from a DC/AC pulsed converter.

It is convenient to classify the BLDC motor into two categories: the small machines, like those used by manufacturers and in small electronic appliances, and integral machines, like those required for industrial applications or for large equipment like submarines, ships, electric vehicles, and so on.

The small brushless DC motor is suitable for several uses such as computer disk drives, drones, robotics and in aircraft. Depending on the application, the small BLDC can be mounted as both an *inrunner* or *outrunner* assemblage. In the *inrunner* BLDC motor, the rotor turns inside the stator, while in the *outrunner* BLDC the rotor turns outside the stator. It is common to see many small *inrunner* iron-less BLDC motors. The lack of iron cores implies no iron losses improving the efficiency, but at the same time it decreases the torque capacity production. This kind of machine is suitable for high speed applications.

The *outrunner* BLDC motor is suitable for applications that require more torque than the *inrunner* motor at low speed. The *outrunners* motors are very popular for CD/DVD drives. Their high torque has also made them popular in the remote-controlled aircraft community. As we will see later, those kinds of machine do not need a sensor to control the pulsed voltage source.

The integral BLDC motor has, commonly, more that tens of HP and has become very popular in high power applications due to the introduction of new power electronics that are suitable for high current control.

This kind of machine is always *inrunner* assembled with the same number of magnetic poles in both stator and rotor, different to the small machine that often has different numbers of poles, like the switched reluctance motor.

The integral BLDC motor requires that emf and mmf should be synchronized to develop the required electromagnetic torque. The synchronization of both emf and mmf, as we will see later, requires a position sensor to trigger the pulsed voltage.

11.4.1 Small BLDC motor

The small BLDC motor is very similar to the switched reluctance motor because it works based on the same principles. The difference between them is that in an *inrunner* motor the rotor is a surface mounted permanent magnet cylinder that has a different pole number to the stator.

Figure 11.26 shows a 6:4 pole BLDC motor. Energizing its coils with a suitable pulse sequence, the rotor spins to follow the change of the current.

The sequence of the coil current variation determines the direction of the rotor spin.

No sensor is required for detecting the rotor position. Establishing the sequence, the rotor will follow the coil energized like a greyhound following a rabbit in a race.

11.4.2 Integral BLDC motor

Figure 11.27 shows a cross section of 2-pole magnetic structure of BLDC motor. The rotor is a cylinder of high magnetic permeability that has magnets mounted on its surface facing an air-gap. The stator structure is similar to that of an induction motor.

In this ideal motor, the winding is composed of just one full pitch N-turns coil. Although this machine is not feasible, it is suitable to use it to discuss the magnetic phenomenon involved in its operation.

Figure 11.28 shows a developed representation of the cross-section and the magnetic flux density distribution derived from the action of the PM.

Supposing that the rotor is spinning at a rotational speed n (rev s^{-1}), a magnetic field distribution will induce an emf that can be obtained from ($e = Blv$) the equation such that its maximum value is given by:

$$E_0 = 4\pi N B_0 L R n (\mathrm{V}) \tag{11.31}$$

Figure 11.26. Small 3-phase BLDC motor—6:4 poles.

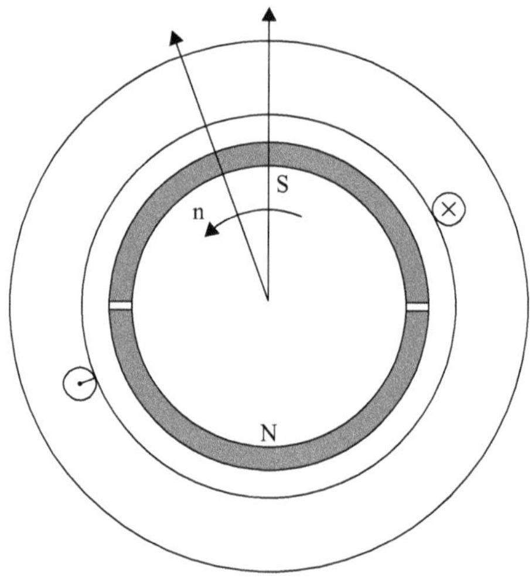

Figure 11.27. Cross section of a two pole—1-phase—BLDC motor.

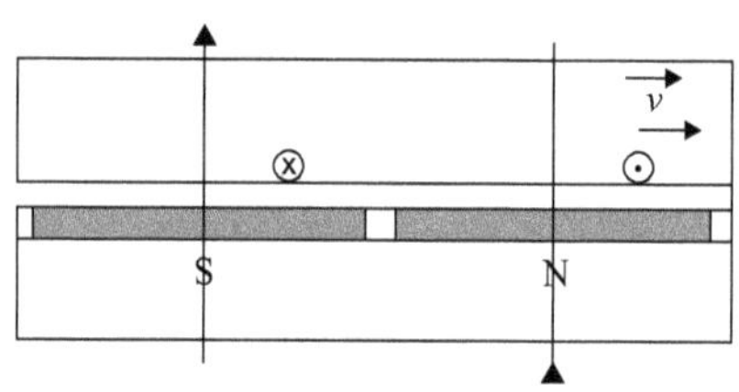

Figure 11.28. Developed representation of BLDC motor.

where:

L: length of the stator;
R: stator radius at the air-gap;
$v = 2\pi n$: peripheral speed.

If a suitable pulsed current is injected in the stator winding, a pulsed developed torque can be extracted. The maximum value of the developed torque is such that:

$$T_{\max} = \frac{E_0 I}{2\pi n} = 2NB_0 LRI \quad (11.32)$$

Figure 11.29 shows not only the emf square wave, but also a suitable square wave for the current and the resultant developed torque.

The borderline that separates the north pole from the south pole establishes a neutral zone ($B = 0$), such that when the coil sides are located near $\theta = 0$ and $\theta = \pi$, not only the emf, but also the developed torque becomes null and the motor stops.

11.4.3 The multiphase BLDC motor

The multiphase BLDC motor is the solution for avoiding the time variation shape curve of the developed torque becoming null in a fixed position. The principle of a

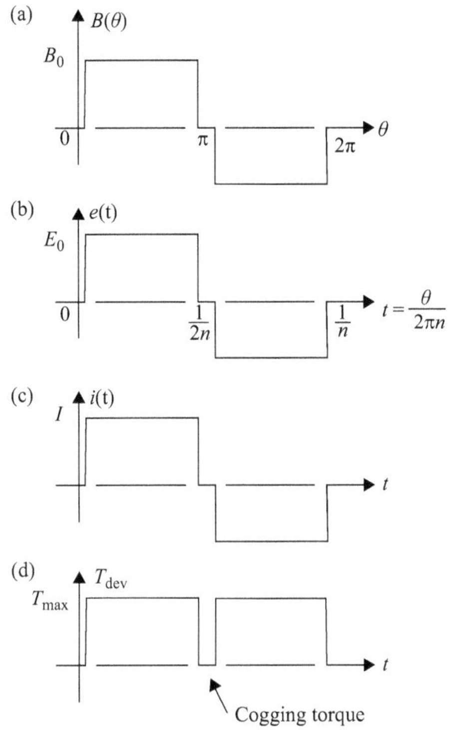

Figure 11.29. The main quantities involved in BLDC motor operation; (a) magnetic flux density distribution produced by a PM; (b) EMF time variation when the rotor is spinning at n (rev s^{-1}); (c) electrical current that should supplied by the voltage source; (d) developed torque time variation evidenced by the cogging torque.

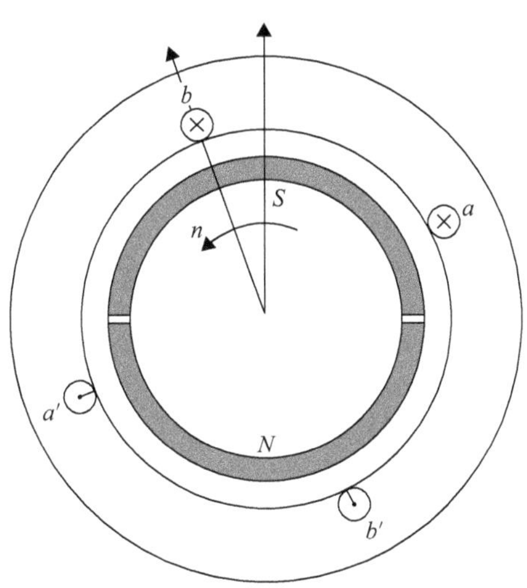

Figure 11.30. Two-phase BLDC motor; phase A: coils sides a-a′; phase B: coils side b-b′.

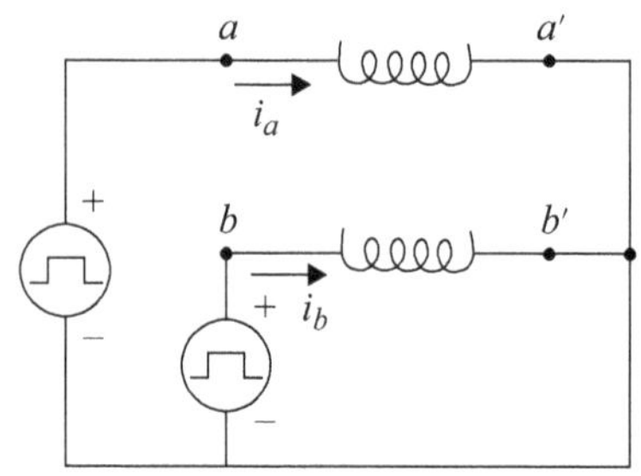

Figure 11.31. The connection of winding of a two-phase BLDC motor.

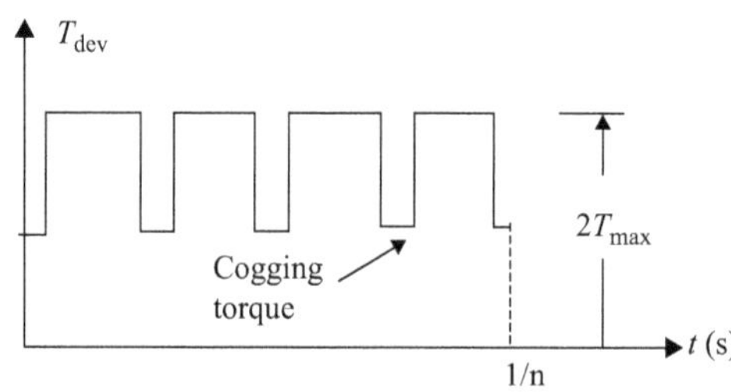

Figure 11.32. Developed torque for a two-phase BLDC motor.

multiphase winding of a BLDC motor is very similar to the multiphase winding of an AC machine.

For a two-phase machine, the stator should be composed of two identical windings displaced $\frac{\pi}{2}$rad from each other, as shown in figure 11.30. Consequently, the square waves of the injected currents should also be displaced $\frac{\pi}{2\omega}$ or $\frac{1}{4n}$ (s) from each other.

The connection of the windings is shown in figure 11.31. Each voltage source supplies an electrical current with both the same shape and displaced $\frac{1}{4n}$ (s) from each other. The effect of each current is the same as that shown in figure 11.30.

Concerning the developed torque, if we add the two curves of the developed torque displaced $\frac{1}{4n}$ (s) from each other, the resultant developed torque becomes that presented in figure 11.32.

In a multiphase BLDC motor, the cogging torque is such that no null developed torque appears, but it introduces undesirable vibrations and noises in the operation. The triphasic winding is common in large BLDC motors and additional care should be undertaken to mitigate the cogging torque.

The maximum value of the developed torque is such that:

$$T_M = mT_{max} = \frac{mE_0I}{2\pi n} = 2mNB_0LRI \tag{11.33}$$

where m is the number of phases and T_{max} is the maximum developed torque of each phase.

Example 11.5

A two-phase BLDC motor presents the following datasheet:

Inside diameter: 40 cm—length: 40 cm—rotational speed: 1500 rev m^{-1}—number of turns per phase: 32—PM coverage: 165°—magnetic flux density in the air-gap: 0.5 T.

The voltage source applies a 450 V/phase square wave voltage with a suitable frequency to synchronize it with the induced emf. The resistance of the winding is 0.5 Ω/phase and its inductance can be neglected. Determine:

(a) the induced emf per phase;
(b) the electrical current injected per phase by the voltage source;
(c) the maximum value of the developed torque;
(d) sketch the time variation of the developed torque;
(e) the average developed torque;
(f) the frequency of the voltage source.

Solutions

(a) The induced emf

Based on equation (11.31), the induced emf is given by:

$$E_0 = 4\pi N B_0 L R n = 4\pi \times 32 \times 0.5 \times 0.4 \times 0.2 \times \frac{1500}{60}$$

resulting in: $E_0 = 402$ V

(b) The electric current

Neglecting the phase inductance, the voltage equation is such that:

$$V = E_0 + r_a I$$

therefore:

$$I = \frac{V - E_0}{r_a} = \frac{450 - 402}{0.5} = 96 \text{ A}$$

(c) The maximum torque

Based on equation (11.33), the maximum torque is given by;

$$T_M = mT_{\max} = \frac{mE_0 I}{2\pi n} = \frac{2 \times 402 \times 96}{2\pi \times \frac{1500}{60}} = 491 \text{ Nm}$$

(d) The time-variation developed torque

Figure 11.33 shows the time variation of developed torque for both phases and their composition. The time lapse of the cogging torque is because the PM does not cover the full pole pitch. The corresponding time is such that:

$$\frac{\pi}{6 \times 2\pi n} = \frac{1}{12n} = 3.33 \text{ ms}$$

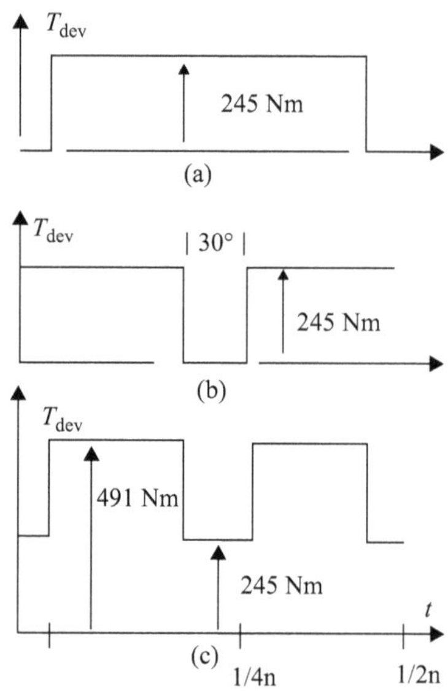

Figure 11.33. Time variation of developed torque—half period; (a) developed torque of phase 'a'; (b) developed torque of phase 'b'; (c) developed torque of phases 'a + b'.

(e) The average torque

The average torque is the average value of the time variation developed torque represented in figure 11.33(c), that yields:

$$T_{av} = \frac{165}{180} \times T_M = \frac{165}{180} \times 491 = 450 \text{ N}$$

(f) The frequency of the source

As the period of the induced emf is $\frac{1}{n}$s, the resultant frequency is such that:

$$f = n = 25 \text{ Hz}$$

11.5 Synchronous reluctance motor

11.5.1 Design aspects of synchronous reluctance motor

The synchronous reluctance motor (SyRM) is a kind of synchronous machine that has a stator very similar to the one used in a regular synchronous machine. Its rotor does not have neither winding nor permanent magnet and is built using ferromagnetic material. The rotor poles are designed like a salient pole machine.

Figure 11.34 shows a cross section of a 3-phase 2-pole SyRM.

11.5.2 How the SyRM works

As the stator winding is excited by an *m-phase* (normally 3-phase) AC voltage source, a rotating magnetic flux distribution that rotates at synchronous speed

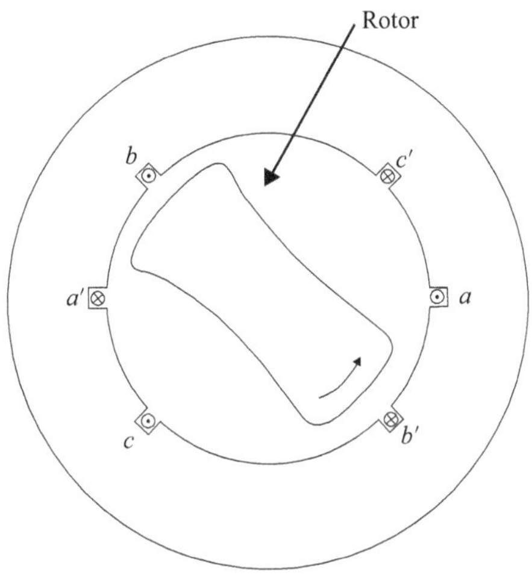

Figure 11.34. Synchronous reluctance motor; voltage source: *3-phase*; frequency: *f*; number of poles: 2.

is established. The rotational speed of the magnetic flux density distribution is given by:

$$n_s = \frac{f}{p} \text{rev s}^{-1} \tag{11.34}$$

Under a magnetic flux density distribution, any ferromagnetic part looks for the position where the system reluctance is maximum. Hence, the rotor poles will also synchronize with the rotating magnetic flux density distribution.

Normally, the SyRM operates at low power and it is not expensive due to the simplicity of the rotor construction. The SyRM is widely used in textile industries, where hundreds of small SyRM with synchronized spin in big looms are fed by only one inverter.

Although, these features are very interesting, SyRMs have a big problem: they have no starting torque! To overcome this difficulty, a squirrel-cage is installed in the rotor to provide enough starting torque to enable it to synchronize with the rotating magnetic field. The lack of field winding or permanent magnet in the rotor makes the power factor of the machine low.

Figure 11.35 shows two different cross sections of an SyRM rotor. Figure 11.35(a) shows a salient pole rotor provided by a squirrel-cage located in its pole-shoes. Figure 11.35(b) shows a cylindrical rotor, whose magnetic asymmetry is provided by air holes called a *flux guide.*

The flux guide gives a preferential route for the magnetic flux. In this structure the squirrel-cage is located at the periphery of the rotor.

11.5.3 PM synchronous reluctance motor

The easiest way to improve the performance of a SyRM is to provide it with a permanent magnet (PM) installed in the rotor magnetic structure. The PM not only improves the efficiency but also increases the torque.

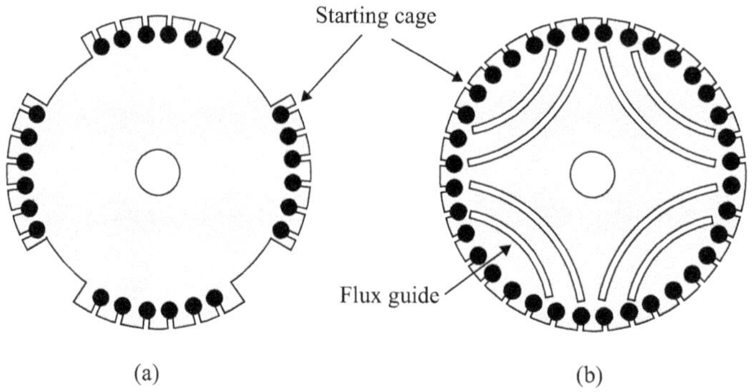

Figure 11.35. SyRM with squirrel-cage; (a) salient pole with squirrel-cage at the pole-shoes; (b) non-salient pole with flux guide and squirrel-cage at the periphery; Flux guide: magnetic barriers composed of convenient air holes; the easy route for the magnetic flux is the space between barriers. Reprinted from [27], copyright (2013), with permission from Elsevier.

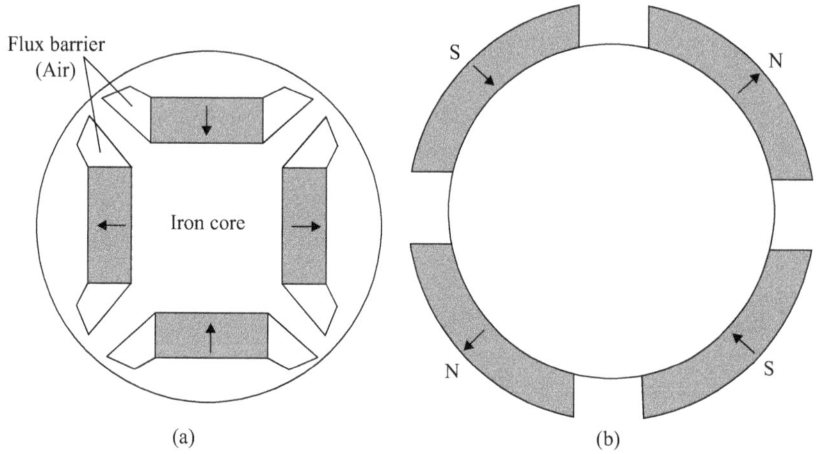

Figure 11.36. PM SyRM rotor; (a) embedded; (b) surface mounted (reproduced from [26]).

Several topologies of a PM SyMR are available and the choice depends on the rotational speed. For high speed, an embedded structure is suitable to protect the PM against the high values of centrifugal force (figure 11.36(a)). Otherwise, surface mounted is suitable for low speed due to its reduced centrifugal force (figure 11.36(b)).

The air-gap magnetic flux density of both geometries is lower than the one presented by PM.

We can achieve an air-gap magnetic flux density higher than the PM magnetic flux density by embedding it as shown in figure 11.37(b).

11.6 Linear induction motor

The linear version of the induction motor has became very popular recently due to its application in electric traction railways. Although there is research applying it in heavy compositions, its use is limited to small vehicles like the ones used in airports.

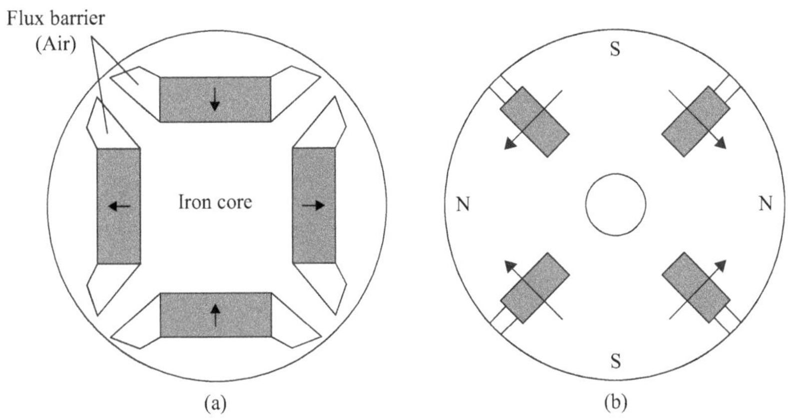

Figure 11.37. Embedded PMSyRM; (a) the air-gap magnetic field is lower than for the PM; (b) the air-gap magnetic field is higher than for the PM due to the concentration of flux (reproduced from [26]).

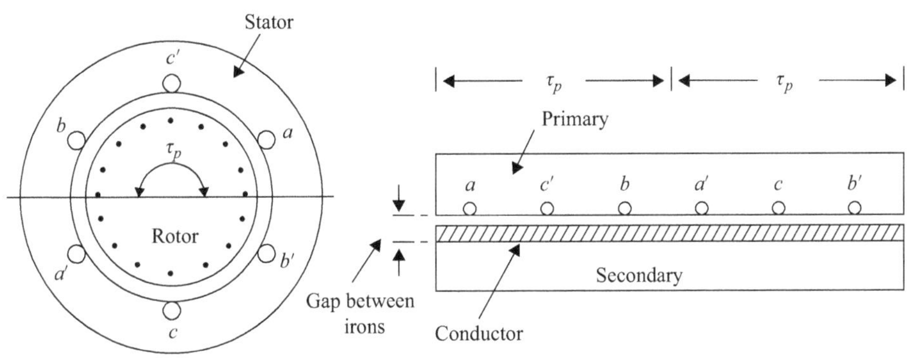

Figure 11.38. Single-side LIM; τ_p: pole pitch (m); the gap between irons includes the air gap and thickness of the metallic slab.

The mechanical difficulty of assuring a small air-gap is the limitation that limits the wide range application of this kind of machine in electrical traction railways.

Nonetheless, the linear induction motor (LIM) is very useful for driving moving parts in machine tools, operating sliding doors, and so on.

The concept of LIM is derived from the round motor. Figure 11.38 shows both a classical two-pole induction motor with a 3-phase concentrated winding and its linear version.

Comparing these two concepts, we identify some differences that impose loss of performance in the LIM. The LIM's *secondary* is a metal slab, normally made of aluminum, and the conductors of the rotor are the metal bars of the squirrel cage.

As the bars are mounted perpendicular to the magnetic flux density distribution, its efficiency is high. In the LIM the metal slab does not guide the current line perpendicular to the magnetic flux density, as figure 11.39 shows.

As significant amounts of the current lines are not perpendicular to the magnetic field, they do not develop electromagnetic forces but generate Joule losses, increasing the *secondary* resistance of LIM.

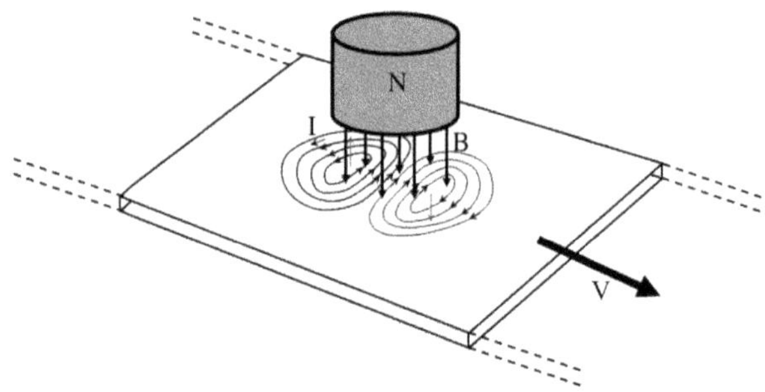

Figure 11.39. Secondary current paths.

Another constructive aspect that contributes to the loss of efficiency of the LIM is its extremities that do not exist in a round motor.

Both, the forward and the backward discontinuity induces current in the secondary that acts as a break, imposing additional losses that are not present in the round motor. Finally, the presence of the LIM's metal slab imposes a large *gap between irons*. While in an ordinary classical induction motor used by electric traction the air-gap is about 1 mm, the *gap between irons* of LIM for the same application is about 25 mm. The high air-gap decreases the magnetization reactance, as a result the magnetization current increases affecting the Joule losses.

The synchronous speed is evaluated by the equation:

$$v_s = \frac{2\tau_p}{T} = 2\tau_p f \frac{\text{m}}{\text{s}} \tag{11.35}$$

where:

τ_p: pole pitch (m)

$T = \frac{1}{f}$: period of the AC voltage source (s).

The thrust curve of the LIM is very similar to the classical induction motor with high rotor resistance. Figure 11.40 shows two thrust curves of the same LIM drawn by two different frequencies. An inverter that improves the performance of the speed control normally feeds the LIM.

One of the most important features of both the classical induction motor and the LIM when fed by inverters is the possibility to have a regeneration of energy. The point P_1 of figure 11.41 is the normal motor operation point of the LIM. If we want to break it by electromagnetic effect, we only need to reduce the frequency of the inverter. As the speed does not change immediately, the reduction of the frequency from f_1 to $f_2 < f_1$ makes the new operation point P_2. Observe that point P_2 is located at the generator part of the thrust curve. Therefore, the kinetic energy is converted to electrical energy delivered to the source imposing a reaction force in the opposite direction to the movement.

There are several structures of LIM dependent on the application. For small vehicles, such as are used in airports, the configuration of a long secondary is common. Figure 11.42 shows the configuration of this kind of motor.

The single-sided LIM for this kind of application is not suitable for high speed, because it is difficult to control the air-gap resulting from the cabin instability.

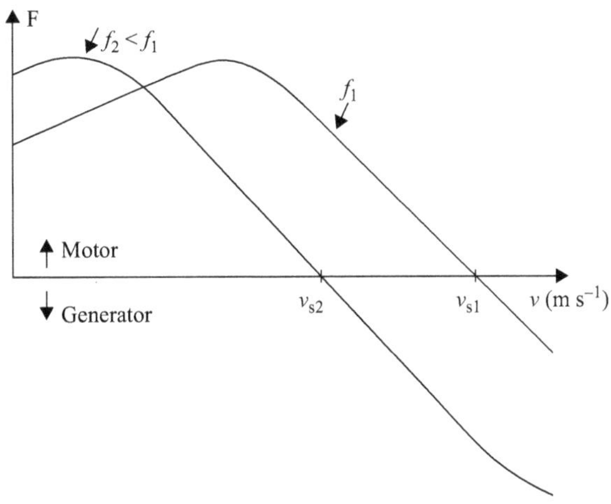

Figure 11.40. Thrust curves of the LIM for two different frequencies.

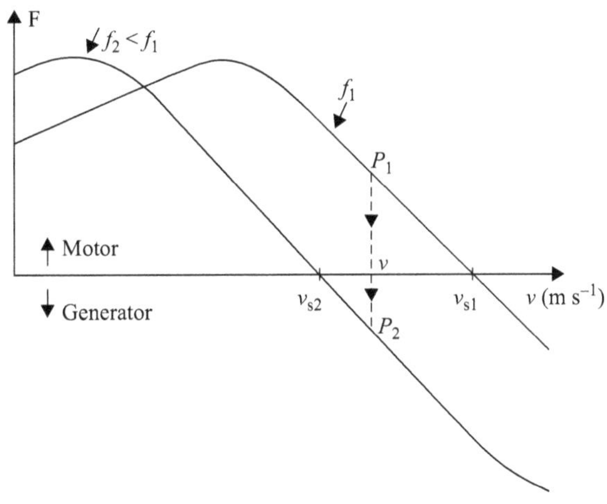

Figure 11.41. Electromagnetic breaking with regeneration energy; point P_1: motor operation at frequency f_1; point P_2; generator operation at frequency $f_2 < f_1$.

The double-sided LIM is very common for machine tool applications, sliding doors and for pumping liquid metals. The two-primary sides of the LIM are series connected, as shown in figure 11.43.

The double-sided motor is more efficient that the single-sided. However, its mechanical structure is more complex.

A case study presented by Matthew Scarpino in his book *Motors for Makers* is the LINIMO Train Line that was built in Japan, during the 2005 World Expo. It was the first time that the viability of this kind of technology for public transportation was demonstrated. This line links Fujigaoka to Yagusa and uses a short-stator single-sided LIM, whose primary is part of the train.

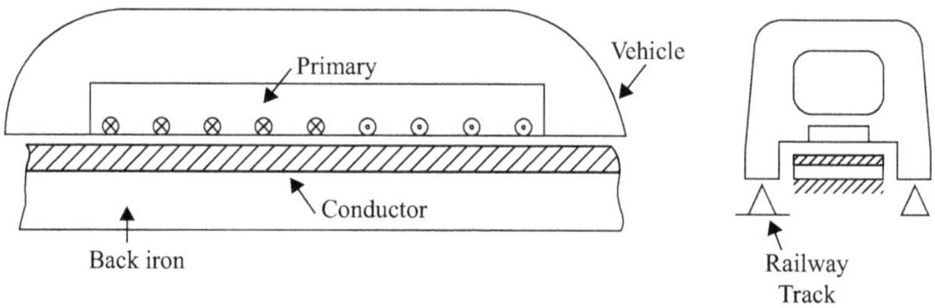

Figure 11.42. The long secondary single-sided LIM; primary is on the vehicle; secondary is on the track.

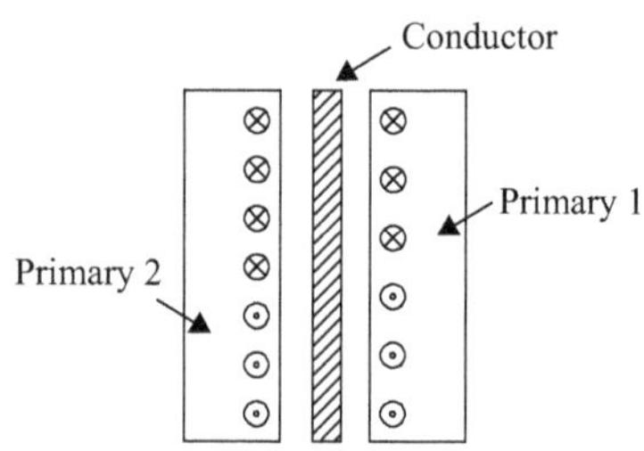

Figure 11.43. Double-sided LIM; two-primaries are series connected.

11.7 Summary

Electrical machine technology has developed an extensive list of different types of electrical motors that is practically impossible to cover in only one book.

This chapter starts by describing the electric DC motor. Although this kind of machine is no longer used in modern industrial drives, it is still working in several plants and electric traction. Both *shunt* and *series* are the most common connections still used in DC electric motor drives. We have not discussed the special connections derived from the combination of the previous ones.

The universal motor is a *series* DC motor operating under AC voltage excitation, it can be found in electric drills and old electric appliances.

The switched reluctance motor has become very useful due to the great progress in power electronics technology. As the SRM has no winding in the rotor, its efficiency and reliability are very high.

The family of stepper motors is suitable for drives that require a step movement for each voltage pulse. One can be built using only one or several stacks dependent on the step angle required.

The hybrid stepper motor is a kind of stepper motor that provides both high accuracy and torque due to the PM being inserted in the rotor magnetic structure. The HySM is not only more expensive but also heavier that the elemental one.

Today, the brushless DC motor has wide uses from drones to nuclear submarine drives. Its high efficiency, flexibility and reliability make the BLDC motor the main option when accurate speed control is required. The small BLDC motor does not require a sensor to identify the rotor position; otherwise, the integral BLDC motor requires an accurate sensor position to trigger the current synchronized with the back emf to develop the electromagnetic torque.

The synchronous reluctance motor is also a free rotor winding machine that has a starting torque provided by the squirrel-cage installed in the rotor periphery. The SyRM is widely used in textile industries for synchronizing hundreds of motors at the same speed imposed by only one AC inverter. As the SyRM exhibits a low power factor, the introduction of a PM in the rotor mitigate this undesirable characteristic. The choice of a suitable rotor topology is strongly attached to the rotor speed due to the centrifugal forces developed in the rotor periphery.

The linear induction motor is an induction machine that develops a linear movement. It has some disadvantages when we compare it with a rotate induction motor, mainly due to the high dimensions of its air-gap. Today, the main application of LIMs is the electric traction railway for small vehicles.

Finally, the authors think that the reader should have acquired sufficient competence to analyze any electromagnetic phenomenon involving the operation of electrical devices. We suggest you spend more time on this exciting technology, extending your knowledge with the suggested reading and with some practical skills on your undergraduate or graduate course or in industry.

11.7.1 Project

A typical underground convoy is composed of six motorized cars. Each car has eight series DC motors that are series connected during starting operation and parallel connected during cruising speed.

Figure 11.44 shows not only the tractive force supplied by the DC electric motors, but also the motion resistance force when the convoy is running on a flat track.

The first stage of the tractive force curve is the starting operation where the DC motors are series connected and are fed by a constant current. The second stage is the cruising operation where all motors are parallel connected and fed by the rated voltage.

The rated data of each DC motor is 100 kW – 1500 V DC – 1800 rpm that is accomplished when the convoy is running at 80 km h^{-1}.

The motion resistance force is composed of several components like the friction force, pullout resistance capacity, curve resistance, etc.

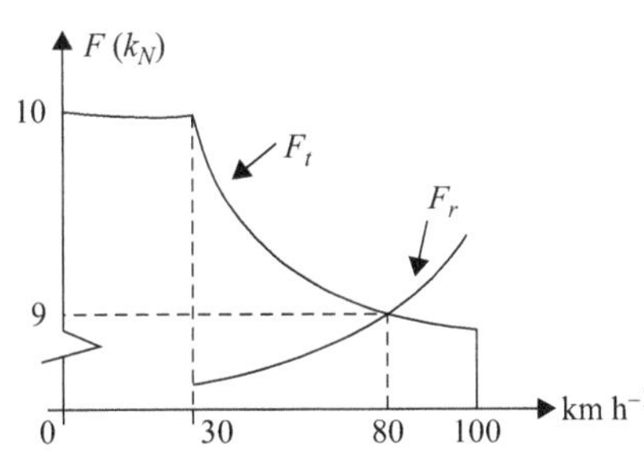

Figure 11.44. The tractive and resistive force for a flat track.

The typical wheel for this kind of convoy has a diameter of 860 mm. The shaft of the wheel is connected to a gearbox that transforms the rotative motion of the DC motor to a linear motion.

As an engineer you should specify:

(a) the transmission relation (k_G) of the gearbox given by;

$$k_G = \frac{n_{\text{motor}}}{n_{\text{shaft}}}$$

(b) the power delivered to each motor of the car by the power system, neglect all types of losses;
(c) the torque delivered to the car by each motor;
(d) the developed torque by each DC motor;
(e) The electric current of each DC motor.

Further reading

[1] Meisel J 1984 *Principles of Electromechanical-Energy Conversion* (Malabar, FL: R.E. Krieger)
[2] Miller T J E 1989 *Brushless Permanent-Magnet and Reluctance Motor Drives* (Oxford: Clarendon Press)
[3] Sacarpino M 2016 *Motors for Makers: A Guide to Steppers, and Other Electrical Machines* (Indianapolis, IN: Que Publishing)
[4] Nasar S A and Unnewehr L E 1983 *Electromechanics and Electrical Machines* 2nd edn (New York: Wiley)
[5] Slemon G R and Straughen A 1980 *Electric Machines* (Reading, MA: Addison-Wesley)
[6] Skilling H H 1962 *Electromechanics: A First Course in Electromechanical Energy Conversion* (New York: Wiley)
[7] Allenbach J M, Chapas P, Comte M and Kaller R 2008 *Traction Electrique* 2nd edn (Presses Polytechniques et Univeristaires Romandes – EPFL) (in French)
[8] Hanselman D 2006 *Brushless Permanent Magnet Motor Design* 2nd edn (Lebanon, OH: Magna Physics Publishing)
[9] Krause P, Wasynczuk O and Pekarek S 2012 *Electromechanical Motion Devices* 2nd edn (Hoboken, NJ: Wiley)
[10] Gourishankar V 1966 *Electromechanical Energy Conversion* (Scranton: International Textbook Company)
[11] Hendershot J R and Miller T J E 1994 *Design of Brushless Permanent-Magnet Motors* (Hillsboro, OH: Magna Physics Pub.)
[12] Hughes A and Drury B 2013 *Electric Motors and Drives: Fundamentals, Types, and Applications* 4th edn (Amsterdam: Elsevier)
[13] Boldea I and Tutelea L 2018 *Reluctance Electric Machines: Design and Control* 1st edn (Boca Raton, FL: CRC Press)
[14] Lipo T A 2017 *Introduction to AC Machine Design* 1st edn (Hoboken, NJ: Wiley)
[15] Boldea I and Syed A N 2010 *The Induction Machines Design Handbook* 2nd edn (Boca Raton, FL: CRC Press)
[16] Hamdi E S 1994 *Design of Small Electrical Machines* 1st edn (New York: Wiley)

[17] Gieras J F 2016 *Electrical Machines: Fundamentals of Electromechanical Energy Conversion* 1st edn (Boca Raton, FL: CRC Press)
[18] Boldea I 2013 *Linear Electric Machines, Drives, and MAGLEVs Handbook* 1st edn (Boca Raton, FL: CRC Press)
[19] Gonen T 2012 *Electrical Machines with MATLAB®* 2nd edn (Boca Raton, FL: CRC Press)
[20] Emadi A 2014 *Advanced Electric Drive Vehicles* 1st edn (Boca Raton, FL: CRC Press)
[21] Toliyat H A *et al* 2012 *Electric Machines: Modeling, Condition Monitoring, and Fault Diagnosis* 1st edn (Boca Raton, FL: CRC Press)
[22] Mohan N 2012 *Electric Machines and Drives: A First Course* 1st edn (New York: Wiley)
[23] Pyrhonen J, Jokinen T and Hrabovcova V 2014 *Design of Rotating Electrical Machines* 2nd edn (New York: Wiley)
[24] Boldea I and Tutelea L N 2010 *Electric Machines: Steady State, Transients, and Design with MATLAB* 1st edn (Boca Raton, FL: CRC Press)
[25] Langsdorf A S 1959 *Principles of Direct-Current Machines* (New York: McGraw-Hill)
[26] Liu X, Lin Q and Fu W 2017 Optimal design of permanent magnet arrangement in synchronous motors *Energies* **10** 1700
[27] Hughes A and Drury B 2013 *Electric Motors and Drives: Fundamentals, Types and Applications* 4th edn (Amsterdam: Elsevier)

www.ingramcontent.com/pod-product-compliance
Ingram Content Group UK Ltd.
Pitfield, Milton Keynes, MK11 3LW, UK
UKHW051906100926
8556IPUK00005B/781

9 780750 320825